T0335681

Noradrenergic Signaling and Astroglia

To our families

Noradrenergic Signaling and Astroglia

Edited by

Nina Vardjan and Robert Zorec

Laboratory of Neuroendocrinology-Molecular Cell Physiology, Institute of Pathophysiology, Faculty of Medicine, University of Ljubljana, Ljubljana, Slovenia; Laboratory for Cell Engineering, Celica Biomedical, Ljubljana, Slovenia

Academic Press is an imprint of Elsevier
125 London Wall, London EC2Y 5AS, United Kingdom
525 B Street, Suite 1800, San Diego, CA 92101-4495, United States
50 Hampshire Street, 5th Floor, Cambridge, MA 02139, United States
The Boulevard, Langford Lane, Kidlington, Oxford OX5 1GB, United Kingdom

Copyright © 2017 Elsevier Inc. All rights reserved.

No part of this publication may be reproduced or transmitted in any form or by any means, electronic or mechanical, including photocopying, recording, or any information storage and retrieval system, without permission in writing from the publisher. Details on how to seek permission, further information about the Publisher's permissions policies and our arrangements with organizations such as the Copyright Clearance Center and the Copyright Licensing Agency, can be found at our website: www.elsevier.com/permissions.

This book and the individual contributions contained in it are protected under copyright by the Publisher (other than as may be noted herein).

Notices
Knowledge and best practice in this field are constantly changing. As new research and experience broaden our understanding, changes in research methods, professional practices, or medical treatment may become necessary.

Practitioners and researchers must always rely on their own experience and knowledge in evaluating and using any information, methods, compounds, or experiments described herein. In using such information or methods they should be mindful of their own safety and the safety of others, including parties for whom they have a professional responsibility.

To the fullest extent of the law, neither the Publisher nor the authors, contributors, or editors, assume any liability for any injury and/or damage to persons or property as a matter of products liability, negligence or otherwise, or from any use or operation of any methods, products, instructions, or ideas contained in the material herein.

British Library Cataloguing-in-Publication Data
A catalogue record for this book is available from the British Library

Library of Congress Cataloging-in-Publication Data
A catalog record for this book is available from the Library of Congress

ISBN: 978-0-12-805088-0

For Information on all Academic Press publications
visit our website at https://www.elsevier.com/books-and-journals

Working together
to grow libraries in
developing countries

www.elsevier.com • www.bookaid.org

Publisher: Mara Conner
Acquisition Editor: Natalie Farra
Editorial Project Manager: Kristi Anderson
Production Project Manager: Lucía Pérez
Designer: Alan Studholme

Typeset by MPS Limited, Chennai, India

Contents

List of Contributors xiii
Preface xvii
Acknowledgments xix

1. **Locus Coeruleus Noradrenergic Neurons and Astroglia in Health and Disease**
Robert Zorec, Nina Vardjan and Alexei Verkhratsky

Locus Coeruleus: Anatomy and Pathophysiology 2
Astroglia and Neurodegeneration 6
Presymptomatic Stage of Neurodegeneration Involves Astrogliosis 10
Dysregulation of Astrocytic Vesicle Dynamics in Neurodegeneration 12
Conclusions 15
Abbreviations 16
Acknowledgements 16
References 16

2. **Astroglial Adrenergic Receptor Signaling in Brain Cortex**
Leif Hertz and Ye Chen

Astrocytic and Neuronal Adrenergic Receptor Expression 26
β-Adrenergic Signaling Pathways 28
Signaling Pathways for Astrocytic α-Adrenergic Receptor Subtypes 31
Metabolic Effects of α- and β-Adrenergic Stimulation of Astrocytes 34
Glucose Uptake 34
Glucose Metabolism 34
Glycogen Turnover 38
Importance of Glycogen Turnover for Brain Function 41
β_1-Adrenergic Stimulation of the Astrocytic Na^+,K^+-ATPase 44
Importance of Adrenergic Stimulation During Culturing of Astrocytes 44
Conclusions 47
Abbreviations 47
References 48

3. White Matter Astrocytes: Adrenergic Mechanisms

Maria Papanikolaou and Arthur Morgan Butt

Introduction 64
White Matter Glia Ensure Rapid Neuronal Signaling Over Long Distances 65
Neuroglial Communication in White Matter: The "Nodal Synapse" 66
Adrenergic Mechanisms in White Matter 68
Adrenergic Signaling in White Matter Physiology 70
Adrenergic Signaling in White Matter Pathology 71
Adrenergic Signaling Regulates Blood Flow 72
Conclusions 73
Abbreviations 73
References 74

4. Role for Astroglial α_1-Adrenergic receptors in Glia-Neuron Communications and Aging-Related Metaplasticity in the Neocortex

Ulyana Lalo and Yuriy Pankratov

Introduction 82
Role for Astroglia in Brain Signaling and Metaplasticity 83
Astrocytic Ca^{2+} Signaling: Specific Role for Adrenergic Receptors 84
Adrenergic Receptors Induce the Release of Gliotransmitters From Neocortical Astrocytes 87
Age- and Environment-Related Alterations in Astroglial Adrenergic Signaling 89
Astroglial α_1-Adrenergic Receptors Modulate Synaptic Transmission and Plasticity in the Neocortex 91
Summary and Perspectives 99
Abbreviations 99
References 100

5. Adrenergic Ca^{2+} and cAMP Excitability: Effects on Glucose Availability and Cell Morphology in Astrocytes

Robert Zorec, Marko Kreft and Nina Vardjan

Introduction 104
Adrenergic Modulation of Cytosolic Ca^{2+} and cAMP Excitability in Cultured Astrocytes 105
Adrenergic Activation Triggers Phasic Ca^{2+} and Tonic cAMP/PKA Responses in Cultured Astrocytes 106
Simultaneous Activation of α- and β-ARs Potentiates Ca^{2+} and cAMP/PKA Responses in Astrocytes 108
Characteristics of Adrenergic Ca^{2+} Signaling in Astrocytes In Situ and In Vivo 109

Adrenergic Excitability and Availability of Glycogen-Derived Cytosolic Glucose in Astrocytes 110
Adrenergic Excitability and Astrocyte Morphologic Plasticity 113
β-Adrenergic Activation and Stellation of Astrocytes 114
Adrenergic Activation and Astrocyte Morphology In Vivo: Prevention of CNS Cellular Edema 117
Conclusions 118
Abbreviations 119
References 120

6. Adrenergic Receptors on Astrocytes Modulate Gap Junctions

Eliana Scemes, Randy F. Stout, Jr, and David C. Spray

Gap Junction Subtypes in Glia and Their Consensus Sites of Phosphorylation by Adrenergic Receptor—Mediated Processes 128
Direct Effects of Adrenergic Receptors on Gap Junctions 130
Gap Junction Formation and Degradation 133
Indirect Effects of Adrenergic Signaling on Coupling Within the Astrocyte Network 136
Calcium Signaling 136
Diffusion of cAMP 137
Diffusion of Metabolites 138
Conclusions 140
Abbreviations 140
References 141

7. Fluxes of Lactate Into, From, and Among Gap Junction-Coupled Astroglia and Their Interaction With Noradrenaline

Gerald A. Dienel

Introduction 146
Aerobic Glycolysis 146
Lactate Release vs Lactate Shuttling-Oxidation 148
Thematic Sequence 149
Lactate Fluxes During Brain Activation 149
Parallel Glucose Utilization Assays Reveal Increased Glycolysis During Brain Activation 149
Lactate is the Predominant Labeled Metabolite of Glucose Released From Brain 151
Impact of Lactate Spreading and Release on Functional Imaging of Brain Activation 151
Astrocytic Lactate Trafficking Via Gap Junctions 152
Dye Coupling 152
Selectivity of Gap Junctional Trafficking of Molecules Involved in Glycolysis 154

Lactate Uptake and Shuttling 155
Glucose Shuttling 157
Summary 157
Perivascular Routes for Metabolite Discharge From Activated Brain Structures 158
Influence of Noradrenaline on Astrocytic Lactate Fluxes 158
Adrenergic Signaling and Aerobic Glycolysis 158
β_2-Adrenergic Vagus Nerve Signaling by Adrenaline and Noradrenaline in Blood 159
Excitatory and Inhibitory Effects of Lactate and Influence on Brain Noradrenaline Release 159
Influence of Noradrenaline on Astrocytic Metabolism 161
Conclusions 162
Abbreviations 163
References 163

8. Dialogue Between Astrocytes and Noradrenergic Neurons Via L-Lactate

Anja G. Teschemacher and Sergey Kasparov

The Noradrenaline-to-Astrocyte Signaling Axis 168
L-Lactate Release by Astrocytes 169
L-Lactate as a Gliotransmitter Feed Forward Signal to Noradrenergic Neurons? 172
Further Potential Signaling Roles of L-Lactate in the Brain 174
Conclusions 178
Abbreviations 178
Acknowledgments 178
References 178

9. Noradrenergic System and Memory: The Role of Astrocytes

Manuel Zenger, Sophie Burlet-Godinot, Jean-Marie Petit and Pierre J. Magistretti

Introduction 184
Brain Noradrenergic System and its Weight on Cerebral Energy Metabolism 184
Noradrenergic Pathways and Receptors 184
The Specific Action of Noradrenaline on Glycogen Metabolism 187
Noradrenaline and Memory 188
Noradrenaline Action on Synaptic Plasticity: Neurons as Targets 188
Noradrenaline Action on Synaptic Plasticity: Astrocytes as Targets 189
Role of Noradrenaline in Memory Paradigms 189
Modulation of Astrocytic Energy Metabolism by Noradrenaline: Impact on Memory 190
Brain Energy Metabolism and Memory 190
The Central Role of Glycogen 191

Influence of the Sleep–Wake Cycle 193
Conclusions 194
Abbreviations 194
References 195

10. Hippocampal Noradrenaline Regulates Spatial Working Memory in the Rat

Rosario Gulino, Anna Kostenko, Gioacchino de Leo, Serena Alexa Emmi, Domenico Nunziata and Giampiero Leanza

Introduction 202
Methods 203
Subjects and Experimental Design 203
Lesion and Transplantation Surgery 204
Behavioral Tests 204
Morris Water Maze 205
Postmortem Analyses 206
Microscopic Analysis and Quantitative Evaluation 207
Results 208
General Observations 208
Behavioral Analyses 209
Morphological Analyses 212
Effects of the Lesion and of Transplants 212
Discussion 214
Effects of the Anti-DBH-Saporin Lesion 214
Effects of Transplants 215
Conclusions 216
Abbreviations 216
Acknowledgments 217
References 217

11. Enteric Astroglia and Noradrenergic/Purinergic Signaling

Vladimir Grubišić and Vladimir Parpura

Introduction 222
Innervation of the Gut Wall 222
Enteric Glia–Essentials 224
Enteric Glia Cells Respond to the Direct Sympathetic Input: Ca^{2+} Excitability 226
Enteric Glial Ca^{2+} Responses Regulate Gut Motility 227
Other Selected Roles of Sympathetic Innervation and Enteric Glia in the Gut 228
Sympathetic Nervous System and Enteric Glia in GI Disorders/Diseases 231
Conclusions 233
Abbreviations 234
Acknowledgment 234
References 234

12. Noradrenaline Drives Structural Changes in Astrocytes and Brain Extracellular Space

Ang D. Sherpa, Chiye Aoki and Sabina Hrabetova

The Noradrenergic System—General Remarks 242
Diversity of Noradrenergic Receptor Expression Underlies Diversity of Astrocytic Responses 242
Noradrenergic System Relates to Function of Astrocytes 243
Noradrenergic System's Effects on Astrocytes In Vitro 245
Noradrenergic System's Effects on Astrocytes In Situ 245
Brain Extracellular Space 247
Noradrenergic System's Effects on Extracellular Space Structure 249
Conclusions 251
Abbreviations 252
Acknowledgments 252
References 252

13. Signaling Pathway of β-Adrenergic Receptor in Astrocytes and its Relevance to Brain Edema

Baoman Li, Dan Song, Ting Du, Alexei Verkhratsky and Liang Peng

Introduction 258
β_1-Adrenergic Receptor 259
Extracellular Ions During Ischemia and/or Reperfusion 262
MAPK/$ERK_{1/2}$ Signaling Pathway During Ischemia and/or Reperfusion 263
Effect of β_1-Adrenergic Receptor Antagonist on Brain Edema During Ischemia/Reperfusion 264
Conclusions 266
List of Abbreviations 267
Acknowledgments 268
References 268

14. Noradrenaline, Astroglia, and Neuroinflammation

José L.M. Madrigal

Introduction 274
Astrocytes and Neuroinflammation 274
Noradrenaline Depletion in Neurodegenerative Diseases 275
Astrocyte Activation in Neurodegenerative Diseases 276
Noradrenaline Regulation of Astrogliosis 277
Noradrenaline Regulation of Astroglial Chemokines 280
Conclusions 282
Abbreviations 282
References 283

15. **Astrocytic β_2-Adrenergic Receptors and Multiple Sclerosis**

Jacques De Keyser

Introduction 290
Downregulation of Astrocytic β_2-Adrenergic Receptors in Multiple Sclerosis 291
β_2-Adrenergic Receptors in Multiple Sclerosis and Progressive Multifocal Leukoencephalopathy 292
Underlying Mechanism of Astrocytic β_2-Adrenergic Receptor Downregulation 293
Pathophysiological Role in Focal Inflammatory Lesions 294
Pathophysiological Role in Axonal Degeneration 294
Pathophysiological Role in Both Axonal Degeneration and Oligodendrogliopathy 295
Abbreviations 296
Acknowledgments 296
References 296

16. **Potentiation of β-Amyloid-Induced Cortical Inflammation by Noradrenaline and Noradrenergic Depletion: Implications for Alzheimer's Disease**

Douglas L. Feinstein and Michael T. Heneka

Introduction: The Locus Coeruleus and Noradrenaline Function in Alzheimer's Disease 302
Neuroinflammation in Alzheimer's Disease 302
Antiinflammatory Actions of Noradrenaline 303
Locus Coeruleus Damage in Alzheimer's Disease 305
Locus Coeruleus Damage in Mouse Models of Alzheimer's Disease 305
Clinical Trials to Modulate Noradrenaline Levels in Alzheimer's Disease Patients 306
Conclusions 307
Abbreviations 307
Acknowledgments 307
References 308

Index 313

List of Contributors

Chiye Aoki ✉
Center for Neural Science, New York University, New York, NY, United States
ca3@nyu.edu

Sophie Burlet-Godinot
LNDC, Brain Mind Institute, École Polytechnique Fédérale de Lausanne, Lausanne, Switzerland; Center of Psychiatric Neuroscience, Department of Psychiatry, Centre Hospitalier Universitaire Vaudois, Prilly, Switzerland
sophie.burletgodinot@epfl.ch

Arthur Morgan Butt ✉
Institute of Biomedical and Biomolecular Sciences, School of Pharmacy and Biomedical Sciences, University of Portsmouth, Portsmouth, United Kingdom
arthur.butt@port.ac.uk

Ye Chen
Henry M. Jackson Foundation, Bethesda, MD, United States
chenye503@yahoo.com

Jacques De Keyser ✉
Department of Neurology, University Hospital Brussel, Brussels, Belgium; Center for Neurosciences, Vrije Universiteit Brussel (VUB), Brussels, Belgium; University Medical Center Groningen, University of Groningen, Groningen, The Netherlands
jacques.dekeyser@uzbrussel.be

Gerald A. Dienel ✉
Department of Neurology, University of Arkansas for Medical Sciences, Little Rock, AR, United States; Department of Cell Biology and Physiology, University of New Mexico, Albuquerque, NM, United States
gadienel@uams.edu

Ting Du
Laboratory of Metabolic Brain Diseases, Institute of Metabolic Disease Research and Drug Development, China Medical University, Shenyang, P.R. China

Serena Alexa Emmi
B.R.A.I.N. Laboratory for Neurogenesis and Repair, Department of Life Sciences, University of Trieste, Trieste, Italy
emmiserena90@gmail.com

Douglas L. Feinstein ✉
Department of Anesthesiology, University Illinois at Chicago, Chicago, IL, United States; Jesse Brown VA Medical Center, Chicago, IL, United States
dlfeins@uic.edu

Vladimir Grubišić ✉
Department of Physiology, Neuroscience Program, Michigan State University, East Lansing, MI, United States
grubisic@msu.edu

Rosario Gulino
Department of Biomedical and Biotechnological Sciences, Physiology Section, University of Catania, Catania, Italy
rogulino@unict.it

Michael T. Heneka
Department of Neurodegenerative Disease and Gerontopsychiatry, University of Bonn, Bonn, Germany; German Center for Neurodegenerative Disease (DZNE), Bonn, Germany
michael.heneka@ukb.uni-bonn.de

Leif Hertz ✉
Laboratory of Metabolic Brain Diseases, Institute of Metabolic Disease Research and Drug Development, China Medical University, Shenyang, P.R. China
lhertz538@gmail.com

Sabina Hrabetova ✉
Department of Cell Biology and The Robert Furchgott Center for Neural and Behavioral Science, State University of New York Downstate Medical Center, Brooklyn, NY, United States
sabina.hrabetova@downstate.edu

Sergey Kasparov ✉
School of Physiology, Pharmacology and Neuroscience, University of Bristol, Bristol, United Kingdom
sergey.kasparov@bristol.ac.uk

Anna Kostenko
B.R.A.I.N. Laboratory for Neurogenesis and Repair, Department of Life Sciences, University of Trieste, Trieste, Italy
akostenka@gmail.com

Marko Kreft
Laboratory of Neuroendocrinology-Molecular Cell Physiology, Institute of Pathophysiology, Faculty of Medicine, University of Ljubljana, Ljubljana, Slovenia; Laboratory for Cell Engineering, Celica Biomedical, Ljubljana, Slovenia; Department of Biology, Biotechnical Faculty, University of Ljubljana, Ljubljana, Slovenia
marko.kreft@mf.uni-lj.si

Ulyana Lalo ✉
School of Life Sciences, University of Warwick, Gibbet Hill Campus, Coventry, United Kingdom
u.lalo@warwick.ac.uk

Giampiero Leanza ✉
B.R.A.I.N. Laboratory for Neurogenesis and Repair, Department of Life Sciences, University of Trieste, Trieste, Italy
gleanza@units.it

Gioacchino de Leo
B.R.A.I.N. Laboratory for Neurogenesis and Repair, Department of Life Sciences, University of Trieste, Trieste, Italy
gioacchino.deleo@libero.it

Baoman Li
Laboratory of Metabolic Brain Diseases, Institute of Metabolic Disease Research and Drug Development, China Medical University, Shenyang, P.R. China

José L.M. Madrigal ✉
Department of Pharmacology, School of Medicine, Universidad Complutense de Madrid, Madrid, Spain; Centro de Investigación Biomédica en Red de Salud Mental (CIBERSAM), Instituto de Investigación Neuroquímica and Instituto de Investigación Sanitaria, Hospital 12 de Octubre, Madrid, Spain
jlmmadrigal@med.ucm.es

Pierre J. Magistretti ✉
LNDC, Brain Mind Institute, École Polytechnique Fédérale de Lausanne, Lausanne, Switzerland; King Abdullah University of Science and Technology, Thuwal, Kingdom of Saudi Arabia
pierre.magistretti@kaust.edu.sa

Domenico Nunziata
B.R.A.I.N. Laboratory for Neurogenesis and Repair, Department of Life Sciences, University of Trieste, Trieste, Italy
dome.nunziata@gmail.com

Yuriy Pankratov ✉
School of Life Sciences, University of Warwick, Gibbet Hill Campus, Coventry, United Kingdom; Institute for Chemistry and Biology, Immanuel Kant Baltic Federal University, Kaliningrad, Russia
y.pankratov@warwick.ac.uk

Maria Papanikolaou
Institute of Biomedical and Biomolecular Sciences, School of Pharmacy and Biomedical Sciences, University of Portsmouth, Portsmouth, United Kingdom
maria.papanikolaou@port.ac.uk

Vladimir Parpura ✉
Department of Neurobiology, University of Alabama School of Medicine, Birmingham, AL, United States
vlad@uab.edu

Liang Peng ✉
Laboratory of Metabolic Brain Diseases, Institute of Metabolic Disease Research and Drug Development, China Medical University, Shenyang, P.R. China
hkkid08@yahoo.com

Jean-Marie Petit ✉
LNDC, Brain Mind Institute, École Polytechnique Fédérale de Lausanne, Lausanne, Switzerland; Center of Psychiatric Neuroscience, Department of Psychiatry, Centre Hospitalier Universitaire Vaudois, Prilly, Switzerland
jean-marie.petit@epfl.ch

Eliana Scemes ✉
Dominick P Purpura Department of Neuroscience, Albert Einstein College of Medicine, Bronx, NY, United States
eliana.scemes@einstein.yu.edu

Ang D. Sherpa ✉
Department of Cell Biology, State University of New York Downstate Medical Center, Brooklyn, NY, United States; Center for Neural Science, New York University, New York, NY, United States
ads420@nyu.edu

Dan Song
Laboratory of Metabolic Brain Diseases, Institute of Metabolic Disease Research and Drug Development, China Medical University, Shenyang, P.R. China

David C. Spray ✉
Dominick P Purpura Department of Neuroscience, Albert Einstein College of Medicine, Bronx, NY, United States
david.spray@einstein.yu.edu

Randy F. Stout Jr. ✉
Dominick P Purpura Department of Neuroscience, Albert Einstein College of Medicine, Bronx, NY, United States
randy.stout@einstein.yu.edu

Anja G. Teschemacher ✉
School of Physiology, Pharmacology and Neuroscience, University of Bristol, Bristol, United Kingdom
anja.teschemacher@bristol.ac.uk

Nina Vardjan ✉
Laboratory of Neuroendocrinology-Molecular Cell Physiology, Institute of Pathophysiology, Faculty of Medicine, University of Ljubljana, Ljubljana, Slovenia; Laboratory for Cell Engineering, Celica Biomedical, Ljubljana, Slovenia
nina.vardjan@mf.uni-lj.si

Alexei Verkhratsky ✉
Faculty of Life Sciences, University of Manchester, Manchester, United Kingdom; Achucarro Center for Neuroscience, IKERBASQUE, Basque Foundation for Science, Bilbao, Spain; Department of Neurosciences, University of the Basque Country UPV/EHU and CIBERNED, Leioa, Spain; University of Nizhny Novgorod, Nizhny Novgorod, Russia; Laboratory of Neuroendocrinology-Molecular Cell Physiology, Institute of Pathophysiology, Faculty of Medicine, University of Ljubljana, Ljubljana, Slovenia; Laboratory for Cell Engineering, Celica Biomedical, Ljubljana, Slovenia
alexej.verkhratsky@manchester.ac.uk

Manuel Zenger ✉
LNDC, Brain Mind Institute, École Polytechnique Fédérale de Lausanne, Lausanne, Switzerland
manuel.zenger@epfl.ch

Robert Zorec ✉
Laboratory of Neuroendocrinology-Molecular Cell Physiology, Institute of Pathophysiology, Faculty of Medicine, University of Ljubljana, Ljubljana, Slovenia; Laboratory for Cell Engineering, Celica Biomedical, Ljubljana, Slovenia
robert.zorec@mf.uni-lj.si

Preface

Noradrenaline is one of the most important signaling molecules in the brain. During wakefulness, attention, and situations of stress it activates the brain. Although neuronal noradrenergic responses have been extensively addressed in the past, the role of glial cells in the noradrenergic system has only recently been emphasized. The present book has been compiled to highlight the newest findings on noradrenergic signaling and astroglia, an abundant type of glial cells in the central nervous system.

The book begins with the background on anatomy and (patho)physiology of the *locus coeruleus*, the principal site for noradrenaline synthesis in the central nervous system. The *locus coeruleus* noradrenergic neurons project to almost all areas of the brain, where they release noradrenaline diffusely from numerous varicosities activating G-protein coupled α- and β-adrenergic receptors located on the surface of cells. Already in 1992 electron microscopy studies demonstrated that there is a much higher density of β-adrenergic receptors on astrocytic compared to neuronal processes in the adult neocortex. However, recent studies on neocortical brain slices (2015) revealed that astroglia and not neurons predominantly respond to noradrenaline with calcium transients, indicating that in the neocortex the primary target of noradrenergic activation are astroglial adrenergic receptors. In the following chapters the up-to-date findings on the adrenergic activation of astroglia are discussed, in particular the molecular nature of intracellular signaling pathways and downstream cellular processes, and how the latter affect the global brain function. The chapters deal with the adrenergic regulation of vesicular transmitter release from astroglia that leads to enhancement of the long-term synaptic plasticity. Adrenergic activation also regulates astroglial energy metabolism including the glycogen turnover and aerobic lactate production, the gap junction permeability and intercellular lactate fluxes, which affect metabolic coupling between astrocytes and neurons at global brain scale. Moreover, astroglial morphological plasticity is controlled by noradrenergic system, influencing the extracellular volume, diffusion of signaling molecules, metabolites, and waste products in the interstitial space. Toward the end chapters address the role of noradrenergic system and astroglia in synaptic plasticity and memory formation.

The final chapters deal with astroglial adrenergic receptor activation and *locus coeruleus* impairment in neurological disorders including brain edema, neuroinflammation, multiple sclerosis, neurodegeneration, with a view on emerging noradrenergic therapeutic paradigms.

With the contributions of leading scientists in the field, the book *Noradrenergic Signaling and Astroglia* positions astrocytes as the main target of the brain noradrenergic system. It will be a valuable source of new knowledge in the emerging field of glial pathophysiology, with astroglia no longer viewed entirely as neuron-subservient entities, but central players in the brain and a new target for the development of therapies.

Nina Vardjan and Robert Zorec
Ljubljana, January 2017

Acknowledgments

We would like to thank the authors for their contributions. We also thank Jan Jagodič (Kabinet01) for the cover design.

Chapter 1

Locus Coeruleus Noradrenergic Neurons and Astroglia in Health and Disease

Robert Zorec[1,2,✉], **Nina Vardjan**[1,2,✉] **and Alexei Verkhratsky**[1,2,3,4,5,6,✉]
[1]*Celica Biomedical, Ljubljana, Slovenia,* [2]*University of Ljubljana, Ljubljana, Slovenia,* [3]*University of Manchester, Manchester, United Kingdom,* [4]*IKERBASQUE, Basque Foundation for Science, Bilbao, Spain,* [5]*University of the Basque Country UPV/EHU and CIBERNED, Leioa, Spain,* [6]*University of Nizhny Novgorod, Nizhny Novgorod, Russia*

Chapter Outline

Locus Coeruleus: Anatomy and Pathophysiology 2
Astroglia and Neurodegeneration 6
Presymptomatic Stage of Neurodegeneration Involves Astrogliosis 10
Dysregulation of Astrocytic Vesicle Dynamics in Neurodegeneration 12
Conclusions 15
Abbreviations 16
Acknowledgements 16
References 16

ABSTRACT
The diffuse canvas of neuronal projections, arising from the brainstem locus coeruleus (LC) nucleus, is the primary source of noradrenaline (NA) in the central nervous system (CNS). During development the network of LC neurons develop in parallel with the neocortex, while projections of these neurons innervate virtually all areas in the CNS. The dense vascularization of the LC, and its proximity to the ventricles indicates high metabolic activity and vulnerability of these cells, which, in pathologic conditions can lead to the LC cell death, an early event in Parkinson's and Alzheimer's diseases. Reduced availability of NA affects astrocytes, the key homeostasis-providing cells in the CNS. Thus maintenance in the LC nucleus and/or enhancement of mechanisms that mimic the action of NA, delays the onset of clinical signs in neurodegeneration, a new strategy to mitigate neurodegeneration.

✉Correspondence address
E-mail: robert.zorec@mf.uni-lj.si; nina.vardjan@mf.uni-lj.si; alexej.verkhratsky@manchester.ac.uk

Noradrenergic Signaling and Astroglia. DOI: http://dx.doi.org/10.1016/B978-0-12-805088-0.00001-3
© 2017 Elsevier Inc. All rights reserved.

Keywords: Locus coeruleus; noradrenaline; astrocytes; neurodegeneration; gliocrine system; regulated exocytosis; gliosignaling molecules; vesicle dynamics; cAMP; cytosolic Ca^{2+}; signaling; glucose metabolism; aerobic glycolysis

LOCUS COERULEUS: ANATOMY AND PATHOPHYSIOLOGY

The history of locus coeruleus (LC) goes back to the times of French revolution: the LC was discovered in 1784 by Félix Vicq-d'Azyr (1748–94), a French physician and neuroanatomist[1,2] (Fig. 1.1). Later the LC was redescribed by Johann Christian Reil in 1809[3] and named "locus coeruleus" by the Wenzel brothers in 1812.[4,5] This small nucleus is located in the posterior area of the rostral pons in the lateral floor of the fourth ventricle. It is composed of neurons containing neuromelanin granules. Because of its color, LC is also known as the nucleus pigmentosus pontis[6] meaning "heavily pigmented nucleus of the pons." The neuromelanin, that colors the structure, is formed by the polymerization of noradrenaline (NA) and is analoguous to the black dopamine-based neuromelanin in the substantia nigra. In adult humans (ageing 19–78 years) the LC has around 50,000 pigmented neurons with mean cell volume of 35,000–49,000 μm^3.[3,7] High monoamine oxidase activity in the rodent LC was found in 1959, monoamines were identified in 1964 and noradrenergic ubiquitous projections to the central nervous system (CNS) in the 1970s.[4] It is generally acknowledged that LC is the prime source of NA in the CNS,[8–10] being the source of ~70% of all NA in the brain.

Although the LC contains a relatively small number of neurons, the axons of these project and ramify widely[9,11,12] to the spinal cord, the brainstem, the cerebellum, the hypothalamus, the thalamic relay nuclei, the amygdala, the basal telencephalon, and the cortex, although some cortical areas receive more abundant innervation.[12] In principle, such a canvas of neuronal contacts (Fig. 1.2) may synchronously activate a wide array of neural networks in several brain and spinal cord regions. This possibility may be regarded as an anatomical basis for a functional "reset" for many brain networks.[13,14] Indeed, in all these structures, synchronous activation of LC projections[14] leads to synchronized electrical activity, possibly reflected by γ waves on an electroencephalogram.[13] This mode of activity is linked to the most fundamental LC-mediated functions including arousal and the sleep–wake cycle, attention and memory, behavioral flexibility, behavioral inhibition and stress, cognitive control, emotions, neuroplasticity, posture, and balance.[11]

This nucleus develops before birth and it has been postulated that LC efferents are critical for the development of various parts of the brain, especially for neocortex.[9] In rats this nucleus appears to start differentiating during 10–13 days of gestation,[15] which means that LC neurons are born well in advance of many neurons in the LC target brain areas.[9] In

TRAITÉ

D'ANATOMIE

ET

DE PHYSIOLOGIE,

AVEC

DES PLANCHES COLORIÉES

Représentant au naturel les divers organes de l'Homme et des Animaux.

DÉDIÉ AU ROI,

PAR M. VICQ D'AZYR,

TOME PREMIER.

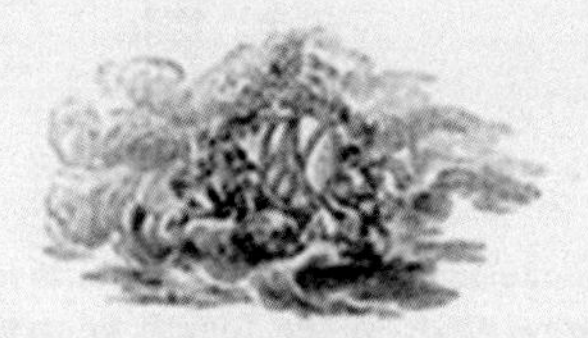

A PARIS,

DE L'IMPRIMERIE DE FRANÇ. AMB. DIDOT L'AÎNÉ.

M. DCC. LXXXVI.

FIGURE 1.1 Cover page of the book by Felix VICQ D'AZYR "Traité d'anatomie et de physiologie, avec des planches coloriées représentant au naturel les divers organes de l'homme et des animaux. Tome Premiere. Paris, Franç. Amb. Didot, 1786." https://hagstromerlibrary.ki.se/books/110

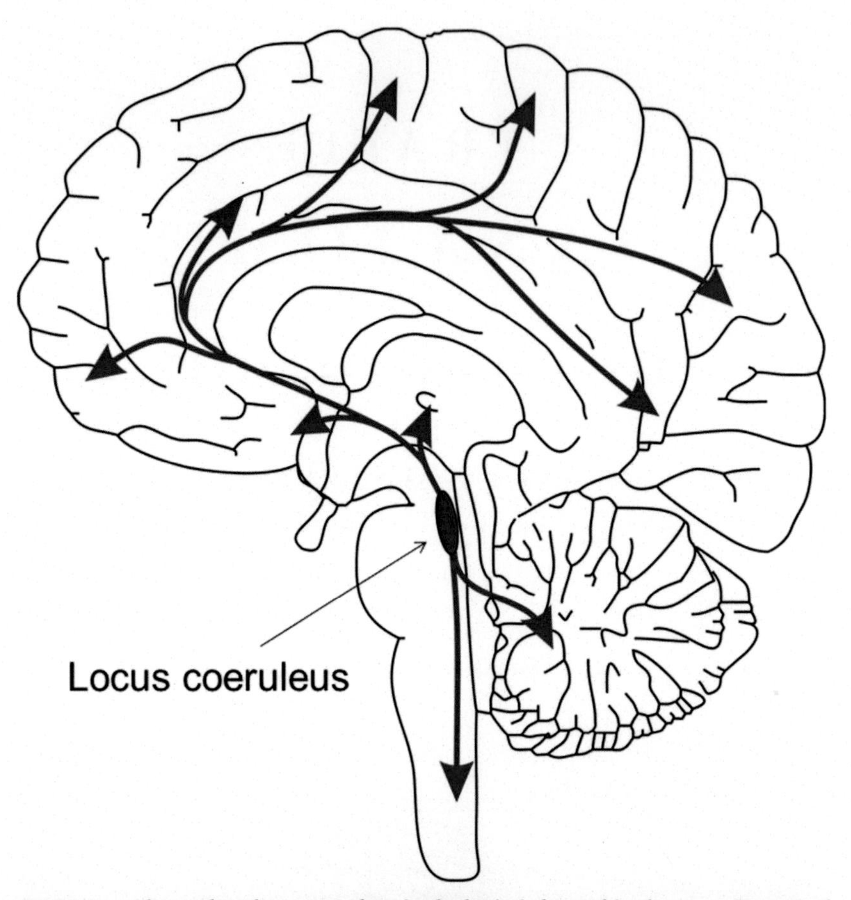

FIGURE 1.2 **The nucleus locus coeruleus in the brain is located in the posterior area of the rostral pons in the lateral floor of the fourth ventricle.** Neurons from this nucleus project axons to most, if not all, areas of the brain and into the spinal cord as denoted by the red arrows. *Adapted from Feinstein DL, Kalinin S, Braun D. Causes, consequences, and cures for neuroinflammation mediated via the locus coeruleus: noradrenergic signaling system.* J Neurochem. *2016;139(Suppl. 2):154–178.*

humans LC nucleus is present during 9–12 weeks of gestation with efferent fibers projecting to the neocortex.[9,16] Based on the morphologic evidence of the presence of LC nerve endings in the neocortex, it has been suggested that NA is involved in the development of the neocortex.[17] Axons with NA are first located in the lower part of the cortical marginal zone, which is the site where the tangential axons of Cajal-Retzius cells extend and give their numerous collaterals. Cajal-Retzius cells, the principal cells of the marginal zone, provide signals for the migration of cells born later in development and for lamination during neocortical development.[18,19] It has been proposed that Cajal-Retzius neurons are the targets of the early NA input,[20] since removal of the NA system after birth resulted in an altered number of Cajal-Retzius cells, revealing a more direct role for NA in development and regulation of neuronal migration and laminar formation in the cerebral cortex.[21]

The growing mass of the developing brain poses a problem for cell-to-cell signaling, since distances between cells are far greater than those reachable by diffusion-mediated signal propagation.[22] Thus two mechanisms appeared to have bypassed this hindrance. First, convection-based signaling, where substances in the extracellular solution are delivered from the source to the targets by the bulk flow. This flow varies diurnally: during the night the "glymphatic" tunnels[23] between cells get enlarged, which increases the flux of cerebrospinal fluid, thus helping to remove the extracellular debris during sleep.[24] Changes in the flux of cerebrospinal fluid are regulated by adrenergic receptors (ARs),[24] which also regulate astroglial morphological plasticity.[25,26] Second, a conductive mechanism mediates communication between cells, which consists of action potential propagation along the ensembles of branching neurons, such as those originating from the LC,[9] innervating brain structures.[11]

Developing tissues, where cells divide and where cells are morphologically plastic, are highly energy demanding, requiring a special form of metabolic adaptation, the aerobic glycolysis. This nonoxidative metabolism of glucose utilization exists in such tissues despite the presence of adequate levels of oxygen, a phenomenon termed "the Warburg effect".[27] Thus in neurodevelopment, cells increase aerobic glycolysis to generate intermediates used, most likely, for biosynthetic pathways. While this form of metabolism is an inefficient way to generate adenosine triphosphate (ATP), the advantage of this process appears to be in providing intermediates for the biosynthesis of lipids, nucleic acids, and amino acids,[28] which are all needed for cell division as well as for their morphological reshaping, both processes defining the developing CNS.[29] This metabolic strategy is considered to generate glycolytic intermediates essential for rapid biomass generation also in cancer cells[30] and thus appears a universal feature of tissue development. Moreover, aerobic glycolysis, the hallmark of which is the production of L-lactate, appears to be regulated in the brain. During alerting, sensory stimulation, exercise, and pathophysiological conditions, L-lactate production and release are up-regulated. Although it is still unclear, how this appears at the cellular level, the process likely depends on LC neurons.[31]

In 1940, relatively abundant vascularization of LC nucleus was described,[32] indicating that LC neurons are metabolically demanding. Indeed, high metabolic rate of LC neurons is mirrored by their relatively high autonomous spiking rate even when glutamate and GABA transmission is blocked.[33] Because of the high exposure to blood circulation, LC neurons are likely to be affected by circulating toxic substances more than neurons in other brain regions. The projections are tightly connected to the brain vascularization; on average, a single LC neuron, with a soma diameter of 45 μm, innervates a 20 m of capillary wall, a length close to that of a tennis court.[34] This relatively large exposure of LC neurons to circulating blood makes them vulnerable to toxin accumulation. Even when present at low concentrations with limited blood-brain barrier (BBB) penetration, toxins could be taken up insufficient quantities by LC terminal axons to be subsequently transported retrogradely to the cell body.[34,35] The close proximity of the LC to the fourth ventricle may additionally expose LC neurons to toxins and viruses present in the cerebrospinal fluid.[36] All this makes LC nucleus highly vulnerable to environmental stress, which may lead to the development of neurological disorders, including Alzheimer's disease (AD), Parkinson's disease (PD), and other diseases.[8,35]

A deficit in LC nucleus was at first proposed to be associated in the etiology of idiopathic PD, but it was later widened into a more general theory viewing PD as member of a family of diseases which are neurological in nature, neurodegenerative in their progressive anatomopathological and functional characteristics and typically associated with the second half of the normal life span.[10] This concept appears to have now gained a much wider acceptance.[8,35] However, the cell-based mechanisms that are instrumental in mediating the loss of LC neurons in neurodegeneration are unclear.

ASTROGLIA AND NEURODEGENERATION

While the term neurodegeneration generally associates with neurons, it is likely that significant part of the pathophysiology in these neurological diseases is mediated by nonneuronal cells, which in some parts of the brain, such as the neocortex, outnumber neurons.[37] Among these nonneuronal cells is neuroglia, represented by astroglia, oligodendroglia, and NG2 glia, which all are of neural origin, and microglia, myeloid cells, which invade CNS early in development. The pathological potential of neuroglia was highlighted by the early neuroanatomists and pathologists, including Rudolph Virchow, Alois Alzheimer, Santiago Ramon-y-Cajal, and many of their colleagues.[38,39] The neuroglial cells provide for multiple functions being, for example the secretory cells of the CNS.[40–42]

In this chapter we are focusing into astrocytes, which populate gray and white matter of the CNS and are, arguably, the most heterogeneous (in form and function) type of neuroglia[43,44]; in particular we shall overview how these cells integrate into the noradrenergic signaling system in health and disease. Astrocytes are responsible for regulating a myriad of processes including synaptogenesis, synapse maturation, neurotransmitter removal from the synaptic cleft, brain microcirculation, brain metabolism and control the formation and maintenance of the BBB.[23,45–58] Moreover, astrocytes provide homeostatic support to the brain and hence these cells are involved in every kind of neuropathology.[59–61] In fact, diseases of the CNS result from homeostatic insufficiency triggered in particular by the disruption of the balance between cell damage and repair. Importantly, astrocytes contribute to various aspects of CNS defense through (1) existing homeostatic mechanisms (which, for example provide for regulation of glutamate and ion movements and hence contain excitotoxicity[62] or protect against reactive oxygen species by astroglia derived glutathione[63]) or (2) by mounting an evolutionary conserved defensive response known as reactive astrogliosis. Reactive astrogliosis protects the CNS, by isolating the damaged area, reconstructing the BBB, and by facilitating remodeling of the neural circuitry after the resolution of pathology.[44,64–67] Astrocytes, however, may also contribute to neuronal damage through failure or reversal of various homeostatic cascades that assume neurotoxic proportions.[60,68]

Considering that the LC-dependent deficit drives neurodegeneration the astrocytic contribution to this pathology is expected to depend on the loss of NA. Thus we need to address how NA operates in normal astrocyte function first. The effects of NA are mediated through α- and β-ARs, which are expressed in neurons, microglia, and astrocytes. When astrocytic α-/β-ARs are simultaneously exposed to NA, a multitude of cytoplasmic second messengers is generated,[69] which regulate a host of downstream cytoplasmic effects including cell shape changes[25] and aerobic glycolysis.[70]

Astroglial β-ARs regulate cell morphology.[71] Stimulation of β-AR with subsequent increase in intracellular cyclic adenosine monophosphate (cAMP) induces in astrocytes stellation, i.e., transformation from a flattened irregular morphology to a stellate process-bearing morphology.[25,72,73] The β-ARs are abundantly expressed by astrocytes in both white and grey matter[74–78] and these receptors are likely to be involved through shape changes in memory formation.[79]

Since the early 1960s the hippocampus was recognized as a fundamental region for memory formation.[80] Subsequently, two distinct memory systems, declarative (explicit) memory for facts and events, for people, places, and objects ("knowing that") and nondeclarative (implicit) memory, the memory for perceptual and motor skills ("knowing how"), have been defined.[81] Both systems rely on similar, if not identical, mechanisms associated with reinforcement of synaptic transmission, which involves morphological changes at the synapse that outlast memory stabilization.[82] This morphology-based mechanism was considered already by Santiago Ramon-y-Cajal, who linked "cerebral gymnastics" with morphological alterations of dendrites and terminals of neurons.[83]

Besides morphological adjustments of neuronal synaptic elements, synaptic transmission may also be directly affected by changes in shape and volume of astroglial processes that tightly enwrap most of CNS synapses.[47,84,85] Local retractions or expansions of astrocytic processes modify the geometry of the extracellular space, affecting neuron–glia interactions.[86] The apposition of astrocyte membrane to the synaptic cleft is an important determinant for the efficient glutamate removal, which defines the properties of synaptic signals.[87] Removal of glutamate from the synaptic cleft consists of diffusion of glutamate in the synaptic cleft and flux into the astrocyte via membrane glutamate transporters; glutamate then diffuses in the cytoplasm to sites where it is metabolized.[85]

Astrocytes display a remarkable structural plasticity under physiologic and pathologic conditions, including reproduction, sensory stimulation, and learning.[25,26,86,88] Distal astrocytic processes can undergo morphological changes in a matter of minutes, thus modifying the geometry and diffusion properties of the extracellular space and relationships with adjacent neuronal elements, especially with synapses. This type of astroglial plasticity has important functional consequences because it modifies extracellular homeostasis of ions and neurotransmitters, thus ultimately modulating neuronal function at the cellular and system levels.[86,88] The mechanisms responsible for morphological changes in astrocytes are not known, but these likely involve ARs, generation of second messenger cAMP[25,26] and cytoskeletal remodeling.[89]

Memory consolidation during the Pavlovian threat conditioning is associated with astrocytic processes to retract from synapses, allowing these synapses to enlarge, suggesting that contact with astroglial processes opposes synapse growth during memory consolidation.[90] In other words, if astrocytic processes enwrap synapses and the latter need to expand during memory formation, astrocytes may hinder this remodeling, which demonstrates how astrocytic structural plasticity enables morphological changes of synapses associated with memory formation. Fig. 1.3 shows a synapse partially unwrapped by astrocyte, which occurred during memory formation.[90] Hebbian memory formation is strongly regulated by NA acting through β-ARs[91] and changes in astrocytic shape have indeed been observed.[90] Moreover, the existence of structural-functional changes of the astrocyte-neuron interactions during memory processes have been detected.[92–94]

Tight association between the synaptic membranes and astrocytes is considered essential for homeostatic control in the synaptic cleft, including rapid removal of glutamate[95] and K^+ from the extracellular space.[57,96] Thus retraction of astrocytic membrane from the synapse during memory formation[90] may facilitate the spillover of neurotransmitter and thus affect synaptic strength.[97] At the same time, memory formation is associated with morphological growth of synaptic elements together with enhanced protein synthesis and rearrangement of receptor proteins, all of which increase the energy consumption.[98]

FIGURE 1.3 **Reconstruction of a tripartite synapse.** Blue: axon; gray: dendrite; orange: astrocyte; red: synaptic cleft; green: mitochondria; yellow: smooth endoplasmic reticulum; black: ribosomes; white: glycogen granules, a marker of astrocytes. Diameter of the presynaptic terminal, adjacent to the synaptic cleft is 5,500 nm. *With permission by Dr. Linneae Ostroff: http://www.cns.nyu.edu/ledoux/SFN2011/L.%20OSTROFF.htm*

How biosynthetic intermediates and energy substrates, needed for ATP synthesis, are delivered to synapses where synaptic plasticity takes place remains an open question. A possibility is that pyruvate is provided to the mitochondria by glycolysis within the neuron. However, the morphology of astrocytes, with extensive end feet plastering blood vessels, is well suited to take up glucose from blood and distribute either glucose itself, or pyruvate or lactate derived from glucose, to astrocytic processes surrounding synapses, possibly by diffusion through gap junctions integrating astroglial syncytia.[99] In support of this mechanism, diffusion of glucose within astrocytes is relatively rapid[100] and may well support glucose delivery via interconnected astrocytes *in situ.* Although synapses are the main energy consumers in the brain, glycogen, the only CNS energy storage system, is present mainly, if not exclusively, in astrocytes (see Fig. 1.3; orange particles in astrocytes). Memory consolidation in young chickens requires glycogenolysis.[101,102] Moreover, the successful consolidation of memory from short-term to long-term memory requires neuronal NA release.[103] Interestingly, noradrenergic signaling regulates the breakdown of glycogen with a rather short time-constant of about 100 s.[70] Therefore it appears that NA, released from neurons, such as those from LC, initiates astrocytic morphological changes and activates astroglial energy metabolism. Thus NA-transmission operating via astrocytes integrates neuronal activity. This requires excitation of astrocytes to initiate morphological changes and at the same time triggers an increase in aerobic metabolism. In support of a key role of NA and LC in excitation-energy coupling, NA is in the adult awake mice brain the main neurotransmitter that triggers a wide synchronous astroglial Ca^{2+} signaling,[78] which represents the universal form of glial excitability.[104]

Thus NA-mediated metabolism support for morphological changes in astrocytes plays an important role in physiologic and in pathologic conditions. Astrocytes may swell under pathologic conditions and contribute to the development of brain edema. Our recent data indicates that in vivo astrocytes swelling could be inhibited by NA.[26] Hypertrophic astrocytes are associated with reactive astrogliosis.[64] At the early stages of neurodegeneration, reduced adrenergic innervation of the brain due to LC degeneration,[105] which may likely result in decreased cAMP signaling and facilitate astrocyte atrophy, as observed in the triple transgenic animal model of AD (3 × Tg-AD).[106] Such astrocytic atrophy may lead to synaptic loss due to insufficient metabolic support for synapses by astrocytes and likely occurs during the presymptomatic stage of neurodegeneration.

PRESYMPTOMATIC STAGE OF NEURODEGENERATION INVOLVES ASTROGLIOSIS

Aging is the main risk factor for the development of neurodegeneration, including that occurring in AD-type pathologies.[107] While there are several stages of AD, the most common one, manifested at a late stage, is linked to the histopathological presence of extracellular deposits of fibrillar β-amyloid peptide (Aβ), and intraneuronal accumulation of aggregates of hyper-phosphorylated Tau protein,[108] which in the healthy brain stabilizes microtubules that transport nutrients and other substances within nerve cells. These proteins have positively charged binding domains that allow them to bind to negatively charged microtubules. Neurodegeneration occurs gradually and dementia may reflect the end stage of an accumulation of pathological changes that start to develop decade(s) before the onset of the clinical symptoms.[109,110]

Mechanisms of the preclinical stages of AD remain obscure. While the current view holds that neurodegeneration in AD reflects neuron-specific deficits, such as is the loss of synapses preceding neuronal death,[111,112] it is likely that preceding or concomitant changes in neuroglia also contribute to this process.[107,110,113–115] The role of neuroglia in dementia was already recognized by Alois Alzheimer, who found pathologically modified glial cells in close contact with damaged neurons.[116–118] In postmortem tissue from patients with AD, astroglial hypertrophy (reflecting reactive astrogliosis accompanied by increased levels of glial fibrillary acidic protein (GFAP) and a calcium binding protein S100B) is often observed, particularly in astrocytes associated with senile plaques.[119–123] Consistent with a recent study in patients with AD,[110] distinct time- and brain-specific morphologic alterations were reported in an animal mouse model of familial AD, i.e., astroglial atrophy was detected in certain brain areas.[106,124–126] Astroglial asthenia preceded the appearance of senile plaques and astroglial reactive hypertrophy and it appeared first in the entorhinal cortex, the region affected at the earliest stage of AD pathology.[126]

Although the pathological developments associated with AD are of particular interest, studies in humans are difficult. Some facets of this pathology can be, to some extent, reproduced in animal models, where the evolution time of AD-like peculiarities is shorter and hence more amenable for experimentation. Thus the mouse models of AD offer certain opportunities for studying AD-related neuropathology. Besides mice, there are many other animal models of AD including nematodes and *Drosophila* flies, rabbits, canines, and nonhuman primates, with each model recapitulating different aspects of AD to some extent (reviewed in Woodruff-Pak[127]).

In addition to the accumulation of peptides, such as Aβ, and morphological alterations of neuroglia, AD progression is also associated with neuroinflammation, because the accumulation of misfolded peptides and proteins provoke an innate immune response in the CNS.[107] This response is in part associated with the Aβ deposits, because hydrophobic proteins may bind apolipoprotein E (ApoE), a major cholesterol carrier in the CNS that is synthesized and released mainly by astrocytes.[128–130] Polymorphic alleles of ApoE are the main genetic determinants of sporadic AD.[131] Why some particular alleles are associated with AD remains unknown, but they may be related to the capacity of the respective ApoE isoforms to prevent neuroinflammation. It has been claimed that the ApoE lipoprotein isoform function, associated with neuroinflammation, arises from differential binding to hydrophobic entities, such as Aβ, regulating their aggregation and clearance in the brain.[132]

ApoE-mediated effects may also be related to a general dysregulation of cell metabolism and fundamental failure in neuronal–glial interaction.[133] The key mechanism of neuronal damage by reactive oxygen species and mitochondrial dysfunction in *Drosophila* is associated with impaired lipid metabolism in glia before or at the onset of neurodegeneration. This likely indicates impairment in lipid traffic, provided by ApoE or similar carriers, between glia and neurons, a primary defect that leads to neurodegeneration. This was further supported by experiments in which human amyloid precursor protein (APP) expressed in *Drosophila* neurons, promoted neurodegeneration; the latter could be attenuated with mimetic ApoE peptides.[134] Before neurodegeneration, lipid droplet accumulation was shown to be present in the astrocytes of an animal model of neurometabolic CNS disorder known as Leigh's disease. The *Ndufs*$^{-/-}$ mutant mice exhibit mitochondrial dysfunction, which further implies impairment of brain lipid metabolism in the early stages of neurodegeneration[133]; the *Ndufs4*s gene encodes a nucleus-encoded accessory subunit of the mitochondrial membrane respiratory chain NADH dehydrogenase (complex I, or NADH:ubiquinone oxidoreductase). The ApoE-mimetic peptides derived from the receptor-binding region of ApoE were selected based on their ability to mimic the functional antiinflammatory and neuroprotective effects of the intact ApoE protein seen in different animal models of neurologic diseases.[135]

ApoE as a cholesterol-carrying protein likely contributes to homeostasis of cholesterol in the brain known to be the most cholesterol-rich organ in the body.[136] Owing to the BBB, there is little uptake of cholesterol from circulating lipoproteins. However, cholesterol synthesis (which occurs mainly in glia[137]) takes place in the brain. In addition to the de novo synthesis, cholesterol homeostasis in the brain also includes its enzymatic conversion to 24(*S*)-hydroxycholesterol,[136] which readily crosses the BBB; this represents the major route for elimination of cholesterol from the CNS. It is appealing to speculate that neurodegeneration may be initiated by defects in components of cell metabolism at the very early stage of AD, during which LC impairments have already started, leading to a reduction in NA availability. Increases in astroglial Ca^{2+} were observed in vivo after stimulation of the LC in anesthetized animals.[138] In awake animals, stimulation of LC neurons triggered (by activation of α_1-ARs) widespread astroglial Ca^{2+} signals, which appeared in almost all astrocytes in the field of study. Simultaneously, through activation of β-ARs, the cAMP-dependent pathways are activated; this in turn instigates rapid degradation of glycogen, which serves as an energy reserve in the brain[70,139] and initiates morphological plasticity of astrocytes.[25] Morphological changes in astrocytes have important implications in memory formation[79] and are affected in AD,[106] likely due to an impaired tone of NA and reduced capacity of astrocytic metabolic support and their capacity to coordinate neural networks via the "reset" mechanism.[14]

DYSREGULATION OF ASTROCYTIC VESICLE DYNAMICS IN NEURODEGENERATION

There are many mechanisms by which astrocytes communicate with the surrounding cells. These include plasma membrane channels, receptors, transporters, and mechanisms that mediate the exchange of molecules by exo- and endocytosis.[22,140–145] Exo- and endocytotic processes require vesicles, from which signaling molecules are released or extracellular material is internalized into vesicles. In the cytoplasm, both exo- and endocytotic vesicles are mobile in the cytoplasm and are subject to an exceedingly complex regulation.[146–148]

Many types of gliosignaling molecules that are stored in membrane-bound vesicles are secreted by astrocytes though the merger of the vesicle and plasma membrane.[40] In addition to molecules stored in the vesicle lumen, ion channels, membrane receptors and transporters, such as major histocompatibility complex II (MHC-II[149]) and excitable amino acid transporter 2 (EAAT2[150]) are delivered to the plasma membrane by vesicle traffic.[148] This traffic is maintained by a system regulated by increases in $[Ca^{2+}]_i$.[148,151] The complexity of vesicle traffic regulation in astrocytes is characterized by two typical, yet opposing, properties of vesicles that contain peptides, such as atrial natriuretic peptide, and those that carry amino acids and are labeled by the vesicular glutamate transporter VGLUT1.[146,151] Namely, glutamatergic vesicle motility is accelerated by an increase in $[Ca^{2+}]_i$,[152] whereas the same increase in $[Ca^{2+}]_i$ slows down peptidergic vesicles and endolysosomes.[153] Such regulation also applies to recycling peptidergic vesicles that have merged with the plasma membrane and subsequently reentered the cytoplasm. The mobility of recycling peptidergic vesicles was studied in cultured astrocytes[154] and in intact brain slices.[155] At rest, peptidergic vesicles linked to cytoskeletal elements move faster and more directionally, than after the exposure of astrocytes to ionomycin to increase $[Ca^{2+}]_i$.[154] The effect of increased $[Ca^{2+}]_i$ was remarkable. The movement of vesicles was almost halted, with only a jitter remaining (that was associated with random diffusional movement). At least some of the peptidergic vesicles carry ATP and a similar attenuation was observed in their mobility when astrocytes were stimulated.[156]

Under pathological conditions this elaborated system of vesicle control appears to be dysregulated.[148] Astrocytes from a mouse model of AD (3 × Tg-AD) isolated at the presymptomatic phase of the disease exhibit attenuated vesicle mobility (Fig. 1.4).[157] Spontaneous mobility of peptidergic and endolysosomal vesicles as well as ATP-evoked, Ca^{2+}-dependent vesicle mobility, were diminished in diseased astrocytes. Similar impairment of peptidergic vesicle trafficking was observed in healthy rat astrocytes transfected with familial AD-associated mutated presenilin 1 ($PS1_{M146V}$). The stimulation-dependent peptide discharge from single vesicles was less efficient in 3xTg-AD and $PS1_{M146V}$-expressing astrocytes than in respective controls. The impaired vesicle dynamics in AD may apply to all peptidergic vesicles including those storing brain-derived neurotrophic factor (BDNF[158]), and ApoE, important for cholesterol transport between astrocytes and neurons.

Reduced secretion may also apply to endolysosomes, which have a similar vesicle dynamics regulation as the peptidergic vesicles.[153] Endolysosomes may store proteolytic enzymes. One such protease is

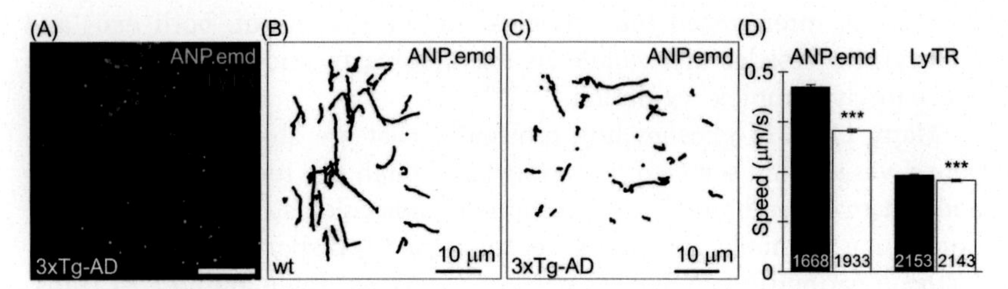

FIGURE 1.4 Slowed vesicle mobility carrying peptides (ANP.emd) and acidic cargo in 3xTg-AD astrocytes. (A) A double fluorescent confocal image of the 3xTg-AD astrocyte expressing ANP.emd stored in individual vesicles observed as bright green fluorescent puncta and LysoTracker-labeled (LyTR) vesicles observed as red fluorescent puncta; scale bar, 10 μm. (B and C) Vesicle tracks ($N = 45$) obtained in a 15-s epoch of imaging (B) representative control wild type (wt) and (C) 3xTg-AD astrocytes expressing ANP.emd. Note less elongated vesicle tracks in the 3xTg-AD astrocyte. (D) Speed of ANP-loaded vesicles and LyTR-labeled vesicles in wt (black bars; mean ± SEM) and 3xTg-AD astrocytes (white bars). Note substantially diminished speed of peptidergic vesicles and modestly diminished speed of LyTR-labeled vesicles in 3xTg-AD astrocytes. The numbers at the bottom of the bars indicate the number of vesicles analyzed. $^{***}P < 0.001$ versus wt (Mann–Whitney U test). *Modified with permission from Stenovec M, Trkov S, Lasič E, et al. Expression of familial Alzheimer disease presenilin 1 gene attenuates vesicle traffic and reduces peptide secretion in cultured astrocytes devoid of pathologic tissue environment. Glia. 2016;64(2):317–329.*

represented by the insulin degrading enzyme (IDE), which when secreted to the extracellular space may degrade Aβ. Although IDE could be secreted from neurons,[159] the major cell type releasing IDE in AD appears to be astrocytes.[160] It has been proposed that in the AD the capacity of astroglial secretion of IDE is reduced, leading to an unchecked accumulation of Aβ, and that this involves a reduction in autophagy-based lysosomal release of IDE.[160] How and why the reduction of lysosomal secretion occurs remains unclear, however it may relate to a general vesicle traffic impairment that has been observed in AD-affected astrocytes.[157]

In AD, astrocytes associated with plaques and Aβ-deposits in the hippocampus appear reactive with morphological hypertrophy and upregulated expression of intermediate filaments.[161] These reactive astrocytes may contribute to the neuroinflammatory response. After exposure to the pro-inflammatory cytokine interferon-γ (IFN-γ), previously immunologically silent astrocytes start to express MHC-II molecules and antigens on their surface and act as nonprofessional antigen-presenting cells (APCs). Intracellular traffic of astrocytic vesicles, including endolysosomes, serves to deliver MHC-II proteins to the plasma membrane.[149,162] The delivery of MHC-II molecules from MHC-II compartments to the cell surface of APCs is mediated via a cytoskeletal network and is most likely accomplished by the fusion of MHC-II-carrying late endolysosomes with the plasma membrane. Actin microfilaments,[163] microtubules,[164,165] and their motor proteins[164,166] have all been shown to mediate trafficking of MHC-II. The role of intermediate filaments in MHC-II trafficking was investigated in IFN-γ-activated reactive (i.e., over-expressing intermediate filament GFAP) astrocytes.[149] Up-regulation of intermediate filaments speed-up vesicle mobility and thus delivery of MHC-II molecules to the cell surface. At elevated $[Ca^{2+}]_i$, reduced mobility of these compartments was observed, which may increase vesicle probability of docking and fusion with the plasma membrane.[153] In astrocytes acting as APCs, this process may serve as an additional regulatory mechanism that controls the onset of late endolysosomal fusion and final delivery of MHC-II molecules to the cell surface.[149] Besides IFN-γ, endogenous suppressors, including NA, have been shown to regulate the expression of MHC-II molecules in astrocytes.[167] Thus when LC neurons degenerate and the amount of NA in the CNS is consequently reduced, this facilitates inflammatory processes, which is consistent with clinical studies.[107]

CONCLUSIONS

Neural reserve of LC and mechanisms mimicking NA action are a strategy to halt neurodegeneration. Neural reserve is a concept arising from the Rush Memory and Aging Project, a longitudinal clinical—pathologic cohort study.[168] In brief, this study, involving 165 individuals, was based on annual evaluations, which included a battery of 19 cognitive tests from which a previously established composite measure of global cognition was derived. Upon death, brain autopsy and uniform neuropathologic examination was performed to determine the density of neurons in the LC and other brainstem nuclei. Measures of neuronal neurofibrillary tangles and Lewy bodies (likely due to the accumulation of α-synuclein[169]) from these nuclei and medial temporal lobe and neocortex were obtained. The results revealed that neuronal densities in each nucleus were approximately normally distributed. Higher neuronal density in each nucleus except the ventral tegmental area was associated with slower rate of cognitive decline. Higher densities of tangles and Lewy bodies in the studied brainstem nuclei were associated with faster cognitive decline even after controlling for pathologic burden elsewhere in the brain. Thus LC neuronal density, brainstem tangles, and brainstem Lewy bodies had independent associations with the rate of cognitive decline. This study confirmed the hypothesis that higher neuronal density in LC is a structural indicator of neural reserve that limits the impact of common neurodegenerative lesions on cognitive function. Therefore a strategy of preserving the viability of neurons in LC and/or mimicking the action of NA in targeted areas is a valid and sensible strategy to mitigate neurodegeneration. This likely includes astroglial integrating signaling and metabolic support functions in the CNS.

ABBREVIATIONS

AD Alzheimer's disease
ApoE apolipoprotein E
APCs antigen-presenting cells
AR adrenergic receptor
ATP adenosine triphosphate
Aβ β-amyloid peptide
BBB blood-brain barrier
BDNF brain-derived neurotrophic factor
cAMP cyclic adenosine monophosphate
CNS central nervous system
EAAT2 excitable amino acid transporter 2
GPCR G-protein coupled receptors
IDE insulin degrading enzyme
IFN-γ interferon-γ
LC locus coeruleus
MHC-II major histocompatibility complex II
NA noradrenaline
PD Parkinson's disease
$PS1_{M146V}$ mutated presenilin 1
VGLUT1 vesicular glutamate transporter

ACKNOWLEDGEMENTS

R.Z. & N.V.'s work is supported by the Slovenian Research Agency (grant nos. P3 310, J3 4051, J3-6789, J3 6790, J3 7605). A.V.'s research was supported by the Alzheimer's Research Trust (United Kingdom).

REFERENCES

1. Tubbs RS, Loukas M, Shoja MM, Mortazavi MM, Cohen-Gadol AA. Felix Vicq d'Azyr (1746–1794): early founder of neuroanatomy and royal French physician. *Childs Nerv Syst.* 2011;27(7):1031–1034.
2. Vicq d'Azyr F. Traité d' anatomie et de physiologie—avec des planches colorës représentant au naturel les divers organes de 'Homme et des Animaux. *FA Didot, Paris.* 1786.
3. Reil JC. Untersuchungen über den Bau des grossen Gehirns im Menschen. *Arch Physiol (Halle).* 1809;9:136–524.
4. Maeda T. The locus coeruleus: history. *J Chem Neuroanat.* 2000;18(1–2):57–64.
5. Wenzel J, Wenzel C. De Penitiori Structura Cerebri Hominis et Brutorum. *Tübingen.* 1812.
6. Jacobsohn L. Uber die Kerne des Menschlichen Hirnstamms. (Meddulla oblongata, Pons und Pedunculus cerebri) Anhang zu den Abhandlungen der Kgl Preuss. *Akad d Wiss Phys-Mathem Kasse ÇPJ Histochem Cytochem.* 1909;306–312.
7. Mouton PR, Pakkenberg B, Gundersen HJ, Price DL. Absolute number and size of pigmented locus coeruleus neurons in young and aged individuals. *J Chem Neuroanat.* 1994;7(3):185–190.
8. Feinstein DL, Kalinin S, Braun D. Causes, consequences, and cures for neuroinflammation mediated via the locus coeruleus: noradrenergic signaling system. *J Neurochem.* 2016;139 (Suppl. 2):154–178.

9. Foote SL, Bloom FE, Aston-Jones G. Nucleus locus ceruleus: new evidence of anatomical and physiological specificity. *Physiol Rev.* 1983;63(3):844–914.
10. Marien MR, Colpaert FC, Rosenquist AC. Noradrenergic mechanisms in neurodegenerative diseases: a theory. *Brain Res Brain Res Rev.* 2004;45(1):38–78.
11. Benarroch EE. The locus ceruleus norepinephrine system: functional organization and potential clinical significance. *Neurology.* 2009;73(20):1699–1704.
12. Chandler DJ, Gao WJ, Waterhouse BD. Heterogeneous organization of the locus coeruleus projections to prefrontal and motor cortices. *Proc Natl Acad Sci USA.* 2014;111(18): 6816–6821.
13. Sara SJ. Locus Coeruleus in time with the making of memories. *Curr Opin Neurobiol.* 2015;35:87–94.
14. Bouret S, Sara SJ. Network reset: a simplified overarching theory of locus coeruleus noradrenaline function. *Trends Neurosci.* 2005;28(11):574–582.
15. Lauder JM, Bloom FE. Ontogeny of monoamine neurons in the locus coeruleus, Raphe nuclei and substantia nigra of the rat. I. Cell differentiation. *J Comp Neurol.* 1974;155(4): 469–481.
16. Olson L, Nystrom B, Seiger A. Monoamine fluorescence histochemistry of human post mortem brain. *Brain Res.* 1973;63:231–247.
17. Latsari M, Dori I, Antonopoulos J, Chiotelli M, Dinopoulos A. Noradrenergic innervation of the developing and mature visual and motor cortex of the rat brain: a light and electron microscopic immunocytochemical analysis. *J Comp Neurol.* 2002;445(2):145–158.
18. D'Arcangelo G, Miao GG, Chen SC, Soares HD, Morgan JI, Curran T. A protein related to extracellular matrix proteins deleted in the mouse mutant reeler. *Nature.* 1995;374(6524): 719–723.
19. Frotscher M. Cajal-Retzius cells, Reelin, and the formation of layers. *Curr Opin Neurobiol.* 1998;8(5):570–575.
20. Marin-Padilla M. Cajal-Retzius cells and the development of the neocortex. *Trends Neurosci.* 1998;21(2):64–71.
21. Naqui SZ, Harris BS, Thomaidou D, Parnavelas JG. The noradrenergic system influences the fate of Cajal-Retzius cells in the developing cerebral cortex. *Brain Res Dev Brain Res.* 1999;113(1-2):75–82.
22. Gucek A, Vardjan N, Zorec R. Exocytosis in astrocytes: transmitter release and membrane signal regulation. *Neurochem Res.* 2012;37(11):2351–2363.
23. Thrane AS, Rangroo Thrane V, Nedergaard M. Drowning stars: reassessing the role of astrocytes in brain edema. *Trends Neurosci.* 2014;37(11):620–628.
24. Xie L, Kang H, Xu Q, et al. Sleep drives metabolite clearance from the adult brain. *Science.* 2013;342(6156):373–377.
25. Vardjan N, Kreft M, Zorec R. Dynamics of β-adrenergic/cAMP signaling and morphological changes in cultured astrocytes. *Glia.* 2014;62(4):566–579.
26. Vardjan N, Horvat A, Anderson JE, et al. Adrenergic activation attenuates astrocyte swelling induced by hypotonicity and neurotrauma. *Glia.* 2016;64(6):1034–1049.
27. Vander Heiden MG, Cantley LC, Thompson CB. Understanding the Warburg effect: the metabolic requirements of cell proliferation. *Science.* 2009;324(5930):1029–1033.
28. Tech K, Gershon TR. Energy metabolism in neurodevelopment and medulloblastoma. *Transl Pediatr.* 2015;4(1):12–19.
29. Goyal MS, Hawrylycz M, Miller JA, Snyder AZ, Raichle ME. Aerobic glycolysis in the human brain is associated with development and neotenous gene expression. *Cell Metab.* 2014;19(1):49–57.

30. Salcedo-Sora JE, Caamano-Gutierrez E, Ward SA, Biagini GA. The proliferating cell hypothesis: a metabolic framework for Plasmodium growth and development. *Trends Parasitol.* 2014;30(4):170–175.
31. Dienel GA, Cruz NF. Aerobic glycolysis during brain activation: Adrenergic regulation and influence of norepinephrine on astrocytic metabolism. *J Neurochem.* 2016;138(1): 14–52.
32. Findley KH, Cobb S. The capillary bed of the locus coeruleus. *J Comp Neurol.* 1940;73:49–58.
33. Sanchez-Padilla J, Guzman JN, Ilijic E, et al. Mitochondrial oxidant stress in locus coeruleus is regulated by activity and nitric oxide synthase. *Nat Neurosci.* 2014;17(6): 832–840.
34. Pamphlett R. Uptake of environmental toxicants by the locus ceruleus: a potential trigger for neurodegenerative, demyelinating and psychiatric disorders. *Med Hypotheses.* 2014;82(1):97–104.
35. Mather M, Harley CW. The Locus Coeruleus: essential for maintaining cognitive function and the aging brain. *Trends Cogn Sci.* 2016;20(3):214–226.
36. Mravec B, Lejavova K, Cubinkova V. Locus (coeruleus) minoris resistentiae in pathogenesis of Alzheimer's disease. *Curr Alzheimer Res.* 2014;11(10):992–1001.
37. Azevedo FA, Carvalho LR, Grinberg LT, et al. Equal numbers of neuronal and nonneuronal cells make the human brain an isometrically scaled-up primate brain. *J Comp Neurol.* 2009;513(5):532–541.
38. Virchow R. Die Cellularpathologie in ihreh Begruendung auf physiologische und pathologische Gewebelehre. First Edition ed. 1858 ed. Berlin: August Hirschwald, p. 440; 1858.
39. Kettenmann H, Verkhratsky A. Neuroglia: the 150 years after. *Trends Neurosci.* 2008;31(12): 653–659.
40. Verkhratsky A, Matteoli M, Parpura V, Mothet JP, Zorec R. Astrocytes as secretory cells of the central nervous system: idiosyncrasies of vesicular secretion. *EMBO J.* 2016;35(3): 239–257.
41. Vardjan N, Parpura V, Zorec R. Loose excitation-secretion coupling in astrocytes. *Glia.* 2016;64(5):655–667.
42. Zorec R, Verkhratsky A, Rodriguez JJ, Parpura V. Astrocytic vesicles and gliotransmitters: slowness of vesicular release and synaptobrevin2-laden vesicle nanoarchitecture. *Neuroscience.* 2016;323:67–75.
43. Verkhratsky A, Butt AM. *Glial Physiology and Pathophysiology.* Chichester: Wiley-Blackwell; 2013:560.
44. Pekny M, Pekna M. Astrocyte reactivity and reactive astrogliosis: costs and benefits. *Physiol Rev.* 2014;94(4):1077–1098.
45. Nedergaard M, Ransom B, Goldman S. New roles for astrocytes: redefining the functional architecture of the brain. *Trends Neurosci.* 2003;26(10):523–530.
46. Stevens B. Neuron-astrocyte signaling in the development and plasticity of neural circuits. *Neurosignals.* 2008;16(4):278–288.
47. Haydon P. GLIA: listening and talking to the synapse. *Nat Rev Neurosci.* 2001;2(3):185–193.
48. Anderson C, Nedergaard M. Astrocyte-mediated control of cerebral microcirculation. *Trends Neurosci.* 2003;26(7):340–344:author reply 344–345
49. Zonta M, Angulo M, Gobbo S, et al. Neuron-to-astrocyte signaling is central to the dynamic control of brain microcirculation. *Nat Neurosci.* 2003;6(1):43–50.
50. Gordon G, Mulligan S, MacVicar B. Astrocyte control of the cerebrovasculature. *Glia.* 2007;55(12):1214–1221.

51. Dong Y, Benveniste E. Immune function of astrocytes. *Glia.* 2001;36(2):180–190.
52. De Keyser J, Zeinstra E, Frohman E. Are astrocytes central players in the pathophysiology of multiple sclerosis? *Arch Neurol.* 2003;60(1):132–136.
53. Abbott N, Rönnbäck L, Hansson E. Astrocyte-endothelial interactions at the blood-brain barrier. *Nat Rev Neurosci.* 2006;7(1):41–53.
54. Ke C, Poon WS, Ng HK, Pang JC, Chan Y. Heterogeneous responses of aquaporin-4 in oedema formation in a replicated severe traumatic brain injury model in rats. *Neurosci Lett.* 2001;301(1):21–24.
55. Nase G, Helm PJ, Enger R, Ottersen OP. Water entry into astrocytes during brain edema formation. *Glia.* 2008;56(8):895–902.
56. Nedergaard M, Verkhratsky A. Artifact versus reality—how astrocytes contribute to synaptic events. *Glia.* 2012;60(7):1013–1023.
57. Verkhratsky A, Nedergaard M. Astroglial cradle in the life of the synapse. *Philos Trans R Soc Lond B Biol Sci.* 2014;369(1654):20130595.
58. Risher WC, Andrew RD, Kirov SA. Real-time passive volume responses of astrocytes to acute osmotic and ischemic stress in cortical slices and in vivo revealed by two-photon microscopy. *Glia.* 2009;57(2):207–221.
59. Seifert G, Schilling K, Steinhäuser C. Astrocyte dysfunction in neurological disorders: a molecular perspective. *Nat Rev Neurosci.* 2006;7(3):194–206.
60. Giaume C, Kirchhoff F, Matute C, Reichenbach A, Verkhratsky A. Glia: the fulcrum of brain diseases. *Cell Death Differ.* 2007;14(7):1324–1335.
61. Verkhratsky A, Sofroniew MV, Messing A, et al. Neurological diseases as primary gliopathies: a reassessment of neurocentrism. *ASN Neuro.* 2012;4(3). Available from: http://dx.doi.org/10.1042/AN20120010.
62. Rose CF, Verkhratsky A, Parpura V. Astrocyte glutamine synthetase: pivotal in health and disease. *Biochem Soc Trans.* 2013;41(6):1518–1524.
63. Dringen R. Metabolism and functions of glutathione in brain. *Prog Neurobiol.* 2000;62(6): 649–671.
64. Wilhelmsson U, Bushong EA, Price DL, et al. Redefining the concept of reactive astrocytes as cells that remain within their unique domains upon reaction to injury. *Proc Natl Acad Sci USA.* 2006;103(46):17513–17518.
65. Sofroniew MV. Astrocyte barriers to neurotoxic inflammation. *Nat Rev Neurosci.* 2015;16(5): 249–263.
66. Anderson MA, Burda JE, Ren Y, et al. Astrocyte scar formation aids central nervous system axon regeneration. *Nature.* 2016;532(7598):195–200.
67. Pekny M, Pekna M, Messing A, et al. Astrocytes: a central element in neurological diseases. *Acta Neuropathol.* 2016;131(3):323–345.
68. Nedergaard M, Dirnagl U. Role of glial cells in cerebral ischemia. *Glia.* 2005;50(4):281–286.
69. Horvat A, Zorec R, Vardjan N. Adrenergic stimulation of single rat astrocytes results in distinct temporal changes in intracellular Ca(2 +) and cAMP-dependent PKA responses. *Cell Calcium.* 2016;59(4):156–163.
70. Prebil M, Vardjan N, Jensen J, Zorec R, Kreft M. Dynamic monitoring of cytosolic glucose in single astrocytes. *Glia.* 2011;59(6):903–913.
71. Hatton GI, Luckman SM, Bicknell RJ. Adrenalin activation of beta 2-adrenoceptors stimulates morphological changes in astrocytes (pituicytes) cultured from adult rat neurohypophyses. *Brain Res Bull.* 1991;26(5):765–769.
72. Shain W, Forman DS, Madelian V, Turner JN. Morphology of astroglial cells is controlled by beta-adrenergic receptors. *J Cell Biol.* 1987;105(5):2307–2314.

73. Bicknell RJ, Luckman SM, Inenaga K, Mason WT, Hatton GI. Beta-adrenergic and opioid receptors on pituicytes cultured from adult rat neurohypophysis: regulation of cell morphology. *Brain Res Bull.* 1989;22(2):379–388.
74. Sutin J, Shao Y. Resting and reactive astrocytes express adrenergic receptors in the adult rat brain. *Brain Res Bull.* 1992;29(3-4):277–284.
75. Zeinstra E, Wilczak N, De Keyser J. [3H]dihydroalprenolol binding to beta adrenergic receptors in multiple sclerosis brain. *Neurosci Lett.* 2000;289(1):75–77.
76. Aoki C. Beta-adrenergic receptors: astrocytic localization in the adult visual cortex and their relation to catecholamine axon terminals as revealed by electron microscopic immunocytochemistry. *J Neurosci.* 1992;12(3):781–792.
77. Catus SL, Gibbs ME, Sato M, Summers RJ, Hutchinson DS. Role of β-adrenoceptors in glucose uptake in astrocytes using β-adrenoceptor knockout mice. *Br J Pharmacol.* 2011;162(8): 1700–1715.
78. Ding F, O'Donnell J, Thrane AS, et al. α1-Adrenergic receptors mediate coordinated Ca2 + signaling of cortical astrocytes in awake, behaving mice. *Cell Calcium.* 2013;54(6): 387–394.
79. Zorec R, Horvat A, Vardjan N, Verkhratsky A. Memory formation shaped by astroglia. *Front Integr Neurosci.* 2015;9:56.
80. Scoville WB, Milner B. Loss of recent memory after bilateral hippocampal lesions. *J Neurol Neurosurg Psychiatry.* 1957;20(1):11–21.
81. Dudai Y, Morris RG. Memorable trends. *Neuron.* 2013;80(3):742–750.
82. Attardo A, Fitzgerald JE, Schnitzer MJ. Impermanence of dendritic spines in live adult CA1 hippocampus. *Nature.* 2015;523(7562):592–596.
83. Cajal SR. The Croonian lecture: La fine structure de centres nerveux. *Proc R Soc Lond.* 1894;444–468.
84. Araque A, Parpura V, Sanzgiri RP, Haydon PG. Tripartite synapses: glia, the unacknowledged partner. *Trends Neurosci.* 1999;22(5):208–215.
85. Reichenbach A, Derouiche A, Kirchhoff F. Morphology and dynamics of perisynaptic glia. *Brain Res Rev.* 2010;63(1-2):11–25.
86. Theodosis DT, Poulain DA, Oliet SH. Activity-dependent structural and functional plasticity of astrocyte-neuron interactions. *Physiol Rev.* 2008;88(3):983–1008.
87. Marcaggi P, Billups D, Attwell D. The role of glial glutamate transporters in maintaining the independent operation of juvenile mouse cerebellar parallel fibre synapses. *J Physiol.* 2003;552(Pt 1):89–107.
88. Oliet SH, Piet R. Anatomical remodelling of the supraoptic nucleus: changes in synaptic and extrasynaptic transmission. *J Neuroendocrinol.* 2004;16(4):303–307.
89. Racchetti G, D'Alessandro R, Meldolesi J. Astrocyte stellation, a process dependent on Rac1 is sustained by the regulated exocytosis of enlargeosomes. *Glia.* 2012;60(3):465–475.
90. Ostroff LE, Manzur MK, Cain CK, Ledoux JE. Synapses lacking astrocyte appear in the amygdala during consolidation of Pavlovian threat conditioning. *J Comp Neurol.* 2014;522(9): 2152–2163.
91. Johansen JP, Diaz-Mataix L, Hamanaka H, et al. Hebbian and neuromodulatory mechanisms interact to trigger associative memory formation. *Proc Natl Acad Sci USA.* 2014;111(51):E5584–E5592.
92. Perez-Alvarez A, Navarrete M, Covelo A, Martin ED, Araque A. Structural and functional plasticity of astrocyte processes and dendritic spine interactions. *J Neurosci.* 2014;34(38): 12738–12744.

93. Bernardinelli Y, Randall J, Janett E, et al. Activity-dependent structural plasticity of perisynaptic astrocytic domains promotes excitatory synapse stability. *Curr Biol.* 2014;24 (15):1679–1688.
94. Lavialle M, Aumann G, Anlauf E, Pröls F, Arpin M, Derouiche A. Structural plasticity of perisynaptic astrocyte processes involves ezrin and metabotropic glutamate receptors. *Proc Natl Acad Sci USA.* 2011;108(31):12915–12919.
95. Bergles DE, Jahr CE. Synaptic activation of glutamate transporters in hippocampal astrocytes. *Neuron.* 1997;19(6):1297–1308.
96. Orkand RK, Nicholls JG, Kuffler SW. Effect of nerve impulses on the membrane potential of glial cells in the central nervous system of amphibia. *J Neurophysiol.* 1966;29(4):788–806.
97. Rusakov DA, Kullmann DM. Extrasynaptic glutamate diffusion in the hippocampus: ultrastructural constraints, uptake, and receptor activation. *J Neurosci.* 1998;18(9):3158–3170.
98. Harris JJ, Jolivet R, Attwell D. Synaptic energy use and supply. *Neuron.* 2012;75(5): 762–777.
99. Rouach N, Koulakoff A, Abudara V, Willecke K, Giaume C. Astroglial metabolic networks sustain hippocampal synaptic transmission. *Science.* 2008;322(5907):1551–1555.
100. Kreft M, Lukšič M, Zorec TM, Prebil M, Zorec R. Diffusion of D-glucose measured in the cytosol of a single astrocyte. *Cell Mol Life Sci.* 2013;70(8):1483–1492.
101. Gibbs ME, Anderson DG, Hertz L. Inhibition of glycogenolysis in astrocytes interrupts memory consolidation in young chickens. *Glia.* 2006;54(3):214–222.
102. Hertz L, Gibbs ME. What learning in day-old chickens can teach a neurochemist: focus on astrocyte metabolism. *J Neurochem.* 2009;109(Suppl 1):10–16.
103. Gibbs ME, Hutchinson DS, Summers RJ. Noradrenaline release in the locus coeruleus modulates memory formation and consolidation; roles for α- and β-adrenergic receptors. *Neuroscience.* 2010;170(4):1209–1222.
104. Verkhratsky A, Orkand RK, Kettenmann H. Glial calcium: homeostasis and signaling function. *Physiol Rev.* 1998;78(1):99–141.
105. Hammerschmidt T, Kummer MP, Terwel D, et al. Selective loss of noradrenaline exacerbates early cognitive dysfunction and synaptic deficits in APP/PS1 mice. *Biol Psychiatry.* 2013;73(5):454–463.
106. Olabarria M, Noristani HN, Verkhratsky A, Rodríguez JJ. Concomitant astroglial atrophy and astrogliosis in a triple transgenic animal model of Alzheimer's disease. *Glia.* 2010;58(7):831–838.
107. Heneka MT, Carson MJ, El Khoury J, et al. Neuroinflammation in Alzheimer's disease. *Lancet Neurol.* 2015;14(4):388–405.
108. Selkoe DJ. Alzheimer's disease: genes, proteins, and therapy. *Physiol Rev.* 2001;81 (2):741–766.
109. Jack Jr. CR, Knopman DS, Jagust WJ, et al. Hypothetical model of dynamic biomarkers of the Alzheimer's pathological cascade. *Lancet Neurol.* 2010;9(1):119–128.
110. Rodriguez-Vieitez E, Saint-Aubert L, Carter SF, et al. Diverging longitudinal changes in astrocytosis and amyloid PET in autosomal dominant Alzheimer's disease. *Brain.* 2016;139(3):922–936.
111. DeKosky ST, Scheff SW. Synapse loss in frontal cortex biopsies in Alzheimer's disease: correlation with cognitive severity. *Ann Neurol.* 1990;27(5):457–464.
112. Scheff SW, Scott SA, DeKosky ST. Quantitation of synaptic density in the septal nuclei of young and aged Fischer 344 rats. *Neurobiol Aging.* 1991;12(1):3–12.

113. Verkhratsky A, Parpura V. Astrogliopathology in neurological, neurodevelopmental and psychiatric disorders. *Neurobiol Dis*. 2015;.
114. Verkhratsky A, Zorec R, Rodríguez JJ, Parpura V. Astroglia dynamics in ageing and Alzheimer's disease. *Curr Opin Pharmacol*. 2016;26:74–79.
115. De Strooper B, Karran E. The cellular phase of Alzheimer's disease. *Cell*. 2016;164(4): 603–615.
116. Alzheimer A. Beiträge zur Kenntnis der pathologischen Neuroglia und ihrer Beziehungen zu den Abbauvorgängen im Nervengewebe. In: Franz Nissl AA, ed. *Histologische und histopathologische Arbeiten über die Grosshirnrinde mit besonderer Berücksichtigung der pathologischen Anatomie der Geisteskrankheiten*. Jena: G. Fischer; 1910:401–562.
117. Strassnig M, Ganguli M. About a peculiar disease of the cerebral cortex: Alzheimer's original case revisited. *Psychiatry (Edgmont)*. 2005;2(9):30–33.
118. Alzheimer A. Über eine eigenartige Erkrankung der Hirnrinde. *Versammlung Südwestdeutscher Irrenärzte in Tübingen am 3. November 1906: Allgemeine Zeitschrift für Psychiatrie und Psychisch-Gerichtliche Medizin*. 1907;64:146–148.
119. Beach TG, McGeer EG. Lamina-specific arrangement of astrocytic gliosis and senile plaques in Alzheimer's disease visual cortex. *Brain Res*. 1988;463(2):357–361.
120. Griffin WS, Stanley LC, Ling C, et al. Brain interleukin 1 and S-100 immunoreactivity are elevated in Down syndrome and Alzheimer disease. *Proc Natl Acad Sci USA*. 1989;86 (19):7611–7615.
121. Nagele RG, Wegiel J, Venkataraman V, Imaki H, Wang KC. Contribution of glial cells to the development of amyloid plaques in Alzheimer's disease. *Neurobiol Aging*. 2004;25 (5):663–674.
122. Mrak RE, Griffin WS. Glia and their cytokines in progression of neurodegeneration. *Neurobiol Aging*. 2005;26(3):349–354.
123. Verkhratsky A, Marutle A, Rodríguez-Arellano JJ, Nordberg A. Glial asthenia and functional paralysis: a new perspective on neurodegeneration and Alzheimer's disease. *Neuroscientist*. 2014;21(5):552–568.
124. Kulijewicz-Nawrot M, Verkhratsky A, Chvátal A, Syková E, Rodríguez JJ. Astrocytic cytoskeletal atrophy in the medial prefrontal cortex of a triple transgenic mouse model of Alzheimer's disease. *J Anat*. 2012;221(3):252–262.
125. Olabarria M, Noristani HN, Verkhratsky A, Rodríguez JJ. Age-dependent decrease in glutamine synthetase expression in the hippocampal astroglia of the triple transgenic Alzheimer's disease mouse model: mechanism for deficient glutamatergic transmission? *Mol Neurodegener*. 2011;6:55.
126. Yeh CY, Vadhwana B, Verkhratsky A, Rodríguez JJ. Early astrocytic atrophy in the entorhinal cortex of a triple transgenic animal model of Alzheimer's disease. *ASN Neuro*. 2011;3(5):271–279.
127. Woodruff-Pak DS. Animal models of Alzheimer's disease: therapeutic implications. *J Alzheimers Dis*. 2008;15(4):507–521.
128. Morikawa M, Fryer JD, Sullivan PM, et al. Production and characterization of astrocyte-derived human apolipoprotein E isoforms from immortalized astrocytes and their interactions with amyloid-beta. *Neurobiol Dis*. 2005;19(1-2):66–76.
129. Peng D, Song C, Reardon CA, Liao S, Getz GS. Lipoproteins produced by ApoE-/- astrocytes infected with adenovirus expressing human ApoE. *J Neurochem*. 2003;86(6):1391–1402.
130. Boyles JK, Pitas RE, Wilson E, Mahley RW, Taylor JM. Apolipoprotein E associated with astrocytic glia of the central nervous system and with nonmyelinating glia of the peripheral nervous system. *J Clin Invest*. 1985;76(4):1501–1513.

131. Yu JT, Tan L, Hardy J. Apolipoprotein E in Alzheimer's disease: an update. *Annu Rev Neurosci.* 2014;37:79–100.
132. Liu CC, Kanekiyo T, Xu H, Bu G. Apolipoprotein E and Alzheimer disease: risk, mechanisms and therapy. *Nat Rev Neurol.* 2013;9(2):106–118.
133. Liu L, Zhang K, Sandoval H, et al. Glial lipid droplets and ROS induced by mitochondrial defects promote neurodegeneration. *Cell.* 2015;160(1-2):177–190.
134. Sarantseva S, Timoshenko S, Bolshakova O, et al. Apolipoprotein E-mimetics inhibit neurodegeneration and restore cognitive functions in a transgenic Drosophila model of Alzheimer's disease. *PLoS One.* 2009;4(12):e8191.
135. Laskowitz DT, Vitek MP. Apolipoprotein E and neurological disease: therapeutic potential and pharmacogenomic interactions. *Pharmacogenomics.* 2007;8(8):959–969.
136. Lütjohann D, Breuer O, Ahlborg G, et al. Cholesterol homeostasis in human brain: evidence for an age-dependent flux of 24S-hydroxycholesterol from the brain into the circulation. *Proc Natl Acad Sci USA.* 1996;93(18):9799–9804.
137. Mauch DH, Nägler K, Schumacher S, et al. CNS synaptogenesis promoted by glia-derived cholesterol. *Science.* 2001;294(5545):1354–1357.
138. Bekar LK, He W, Nedergaard M. Locus coeruleus alpha-adrenergic-mediated activation of cortical astrocytes in vivo. *Cereb Cortex.* 2008;18(12):2789–2795.
139. Kreft M, Bak LK, Waagepetersen HS, Schousboe A. Aspects of astrocyte energy metabolism, amino acid neurotransmitter homoeostasis and metabolic compartmentation. *ASN Neuro.* 2012;4(3).
140. Parpura V, Zorec R. Gliotransmission: exocytotic release from astrocytes. *Brain Res Rev.* 2010;63(1-2):83–92.
141. Osborne KD, Lee W, Malarkey EB, Irving AJ, Parpura V. Dynamic imaging of cannabinoid receptor 1 vesicular trafficking in cultured astrocytes. *ASN Neuro.* 2009;1:5.
142. Kreft M, Stenovec M, Rupnik M, et al. Properties of Ca(2 +)-dependent exocytosis in cultured astrocytes. *Glia.* 2004;46(4):437–445.
143. Zorec R, Araque A, Carmignoto G, Haydon PG, Verkhratsky A, Parpura V. Astroglial excitability and gliotransmission: an appraisal of Ca2 + as a signalling route. *ASN Neuro.* 2012;4:2.
144. Parpura V, Heneka MT, Montana V, et al. Glial cells in (patho)physiology. *J Neurochem.* 2012;121(1):4–27.
145. Parpura V, GrubišiÂc V, Verkhratsky A. Ca(2 +) sources for the exocytotic release of glutamate from astrocytes. *Biochim Biophys Acta.* 2011;1813(5):984–991.
146. Potokar M, Kreft M, Pangrsic T, Zorec R. Vesicle mobility studied in cultured astrocytes. *Biochem Biophys Res Commun.* 2005;329(2):678–683.
147. Potokar M, Kreft M, Li L, et al. Cytoskeleton and vesicle mobility in astrocytes. *Traffic.* 2007;8(1):12–20.
148. Vardjan N, Verkhratsky A, Zorec R. Pathologic potential of astrocytic vesicle traffic: new targets to treat neurologic diseases? *Cell Transplant.* 2015;24(4):599–612.
149. Vardjan N, Gabrijel M, Potokar M, et al. IFN-γ-induced increase in the mobility of MHC class II compartments in astrocytes depends on intermediate filaments. *J Neuroinflammation.* 2012;9:144.
150. Stenovec M, Kreft M, Grilc S, Pangrsic T, Zorec R. EAAT2 density at the astrocyte plasma membrane and Ca(2 +)-regulated exocytosis. *Mol Membr Biol.* 2008;25(3):203–215.
151. Potokar M, Vardjan N, Stenovec M, et al. Astrocytic vesicle mobility in health and disease. *Int J Mol Sci.* 2013;14(6):11238–11258.

152. Stenovec M. Ca2 + -dependent mobility of vesicles capturing anti-VGLUT1 antibodies. *Exp Cell Res*. 2007;313:3809–3818.
153. Potokar M, Stenovec M, Gabrijel M, et al. Intermediate filaments attenuate stimulation-dependent mobility of endosomes/lysosomes in astrocytes. *Glia*. 2010;58(10):1208–1219.
154. Potokar M, Stenovec M, Kreft M, Kreft ME, Zorec R. Stimulation inhibits the mobility of recycling peptidergic vesicles in astrocytes. *Glia*. 2008;56(2):135–144.
155. Potokar M, Kreft M, Lee SY, Takano H, Haydon PG, Zorec R. Trafficking of astrocytic vesicles in hippocampal slices. *Biochem Biophys Res Commun*. 2009;390(4):1192–1196.
156. Pangrsic T. Exocytotic release of ATP from cultured astrocytes. *J Biol Chem*. 2007;282:28749–28758.
157. Stenovec M, Trkov S, Lasič E, et al. Expression of familial Alzheimer disease presenilin 1 gene attenuates vesicle traffic and reduces peptide secretion in cultured astrocytes devoid of pathologic tissue environment. *Glia*. 2016;64(2):317–329.
158. Stenovec M, Lasic E, Bozic M, et al. Ketamine inhibits ATP-evoked exocytotic release of brain-derived neurotrophic factor from vesicles in cultured rat astrocytes. *Mol Neurobiol*. 2015;53(10):6882–6896.
159. Vekrellis K, Ye Z, Qiu WQ, et al. Neurons regulate extracellular levels of amyloid beta-protein via proteolysis by insulin-degrading enzyme. *J Neurosci*. 2000;20(5):1657–1665.
160. Son SM, Cha MY, Choi H, et al. Insulin-degrading enzyme secretion from astrocytes is mediated by an autophagy-based unconventional secretory pathway in Alzheimer disease. *Autophagy*. 2016;12(5):784–800.
161. Verkhratsky A, Parpura V, Pekna M, Pekny M, Sofroniew M. Glia in the pathogenesis of neurodegenerative diseases. *Biochem Soc Trans*. 2014;42(5):1291–1301.
162. Soos JM, Morrow J, Ashley TA, Szente BE, Bikoff EK, Zamvil SS. Astrocytes express elements of the class II endocytic pathway and process central nervous system autoantigen for presentation to encephalitogenic T cells. *J Immunol*. 1998;161(11):5959–5966.
163. Barois N, Forquet F, Davoust J. Actin microfilaments control the MHC class II antigen presentation pathway in B cells. *J Cell Sci*. 1998;111(Pt 13):1791–1800.
164. Wubbolts R, Fernandez-Borja M, Jordens I, et al. Opposing motor activities of dynein and kinesin determine retention and transport of MHC class II-containing compartments. *J Cell Sci*. 1999;112(Pt 6):785–795.
165. Vyas JM, Kim YM, Artavanis-Tsakonas K, Love JC, Van der Veen AG, Ploegh HL. Tubulation of class II MHC compartments is microtubule dependent and involves multiple endolysosomal membrane proteins in primary dendritic cells. *J Immunol*. 2007;178(11):7199–7210.
166. Vascotto F, Lankar D, Faure-André G, et al. The actin-based motor protein myosin II regulates MHC class II trafficking and BCR-driven antigen presentation. *J Cell Biol*. 2007;176(7):1007–1019.
167. Frohman EM, Vayuvegula B, Gupta S, van den Noort S. Norepinephrine inhibits gamma-interferon-induced major histocompatibility class II (Ia) antigen expression on cultured astrocytes via beta-2-adrenergic signal transduction mechanisms. *Proc Natl Acad Sci USA*. 1988;85(4):1292–1296.
168. Wilson RS, Nag S, Boyle PA, et al. Neural reserve, neuronal density in the locus ceruleus, and cognitive decline. *Neurology*. 2013;80(13):1202–1208.
169. Lee HJ, Cho ED, Lee KW, Kim JH, Cho SG, Lee SJ. Autophagic failure promotes the exocytosis and intercellular transfer of alpha-synuclein. *Exp Mol Med*. 2013;45:e22.

Chapter 2

Astroglial Adrenergic Receptor Signaling in Brain Cortex

Leif Hertz[1,✉] **and Ye Chen**[2]

[1]*China Medical University, Shenyang, P.R. China,* [2]*Henry M. Jackson Foundation, Bethesda, MD, United States*

Chapter Outline

Astrocytic and Neuronal Adrenergic Receptor Expression 26
β-Adrenergic Signaling Pathways 28
Signaling Pathways for Astrocytic α-Adrenergic Receptor Subtypes 31
Metabolic Effects of α- and β-Adrenergic Stimulation of Astrocytes 34
Glucose Uptake 34
Glucose Metabolism 34
Glycogen Turnover 38
Importance of Glycogen Turnover for Brain Function 41
β_1-Adrenergic Stimulation of the Astrocytic Na^+,K^+-ATPase 44
Importance of Adrenergic Stimulation During Culturing of Astrocytes 44
Conclusions 47
Abbreviations 47
References 48

ABSTRACT

This paper deals exclusively with gray matter astrocytes because the equally important white matter astrocytes are discussed elsewhere. α- and β-adrenergic receptors have repeatedly been demonstrated immunologically and autoradiographically, by determination of their mRNA or by effects on intracellular Ca^{2+} response. The predominant subtypes are α_{1A}, α_{2A}, and β_1. Signaling by the two latter subtypes is complex and involves transactivation of the epidermal growth factor receptor and phosphorylation of extracellular regulated kinases 1/2 as well as G_s/G_i switch during β_1-adrenergic signaling. Adrenergic signaling stimulates glucose metabolism at many different points as well as Na^+, K^+-ATPase activity. Intermittent stimulation of glycogen synthesis by α_2-adrenergic activation and of glycogenolysis by β-adrenergic activity is essential for learning. Reduction of α-adrenergic activity is probably therapeutic in bipolar disorder, and impairment of adrenergic signaling

✉Correspondence address
E-mail: lhertz538@gmail.com

Noradrenergic Signaling and Astroglia. DOI: http://dx.doi.org/10.1016/B978-0-12-805088-0.00002-5
© 2017 Elsevier Inc. All rights reserved.

may be causatively involved in Alzheimer's disease. Compensation for missing adrenergic stimulation during growth of astrocytic cultures is important.

Keywords: α_1-Adrenergic signaling; α_2-adrenergic signaling; adrenergic receptors in brain; astrocyte; astrocyte culture; Alzheimer's disease; β_1-adrenergic signaling; β_2-adrenergic signaling; bipolar disorder; glucose metabolism

ASTROCYTIC AND NEURONAL ADRENERGIC RECEPTOR EXPRESSION

Almost 50 years ago, noradrenaline (NA), adrenaline, or histamine were found to rapidly increase the concentration of adenosine 3′:5′-cyclic monophosphate (cAMP) in tumoral astrocyte cell lines.[1,2] Although cultured cells are not always reliable models of their in vivo counterparts,[3,4] this finding has been confirmed in cultured astrocytes and freshly dissociated astrocytes.[5,6] β-adrenergic receptor (β-AR) expression in the intact central nervous system (CNS) has been revealed immunologically and autoradiographically.[7–17] Cerebral α_2-ARs were shown by Milner et al.[18] and Glass et al.[19,20] In locus coeruleus (LC), they are expressed on neurons as presynaptic, inhibitory receptors, and on astrocytes.[21] Expression of α_{2A}- and α_{2C}-ARs but not α_{2B}-AR mRNA has been demonstrated in cultured chick astrocytes.[22] The astrocytic α_1-AR has been demonstrated in freshly dissociated brain cells,[13] and rat visual cortex.[23] Astrocytic α_1-adrenergic free cytosolic Ca^{2+} concentration ($[Ca^{2+}]_i$) signals, which are blocked by the specific antagonist prazosin, have been observed in rat brain slices, Bergmann glia, and freshly isolated astrocytes.[24–27]

A pronounced astrocytic expression of mRNA for both α- and β-ARs in freshly isolated cells is shown in Table 2.1.[6] Neuronal mRNA was only found for β_2-ARs and possibly α_2-ARs. The astrocytic and neuronal cell fractions have been obtained using fluorescence-activated cell sorting (FACS)[28] of dissociated cortical brain cells from mice coexpressing a fluorescent drug with either the astrocytic marker glial fibrillary acidic protein (GFAP) or Thy1, a marker mainly of large glutamatergic projection neurons.[29,30] The results may accordingly not be representative for all neurons, and most morphological studies cited above [7–21,23] described both astrocytic and neuronal expression of ARs. Pharmacological evidence for the expression of α_1-, α_2-, and β-ARs in both neurons and astrocytes has also been obtained.[31] Moreover, noradrenergic stimulation affects most, if not all, cell types in the CNS,[32] and Cahoy et al.[3] found approximately similar mRNA expression for both α_2- and β_1-AR expression in their young FACS-isolated forebrain neurons and astrocytes.

Schambra et al.[33] demonstrated neuronal expression of many subtypes of both α_1- and α_2-ARs, although especially the latter, in cerebellum by in situ hybridization. As most cerebellar neurons are GABAergic, this might suggest a high noradrenergic receptor expression on GABAergic neurons,

TABLE 2.1 mRNA Expression of α_1-, α_2-, and β-ARs Subtypes in FACS-Isolated Cell Fractions from 3 Separate Dissociates Obtained from the Adult Murine Brain, Determined by Microarray Analysis of 20 ng of Total RNA

Receptor	Astrocyte	Neuron	Receptor	Astrocyte	Neuron	Receptor	Astrocyte	Neuron
α_{1A}	Present	Consistently absent	α_{2A}	Present	Absent	β_1	Present	Consistently absent
	Present			Present	Present		Present	
	Present			Present	Absent		Present	
α_{1B}	Consistently absent	Consistently absent	α_{2B}	Consistently absent	Consistently absent	β_2	Absent	Present
							Present	Present
							Present	Present
α_{1C}	Consistently absent	Consistently absent	α_{2C}	Consistently absent	Consistently absent	β_3	Consistently absent	Consistently absent

The table indicates presence or absence in three astrocytic or neuronal cell fractions. Three "present" normally indicate that a gene is unequivocally present, two "present" and one "absent" indicates a potential presence.

Modified from Hertz L., Lovatt D., Goldman S.A., Nedergaard M. Adrenoceptors in brain: cellular gene expression and effects on astrocytic metabolism and $[Ca^{2+}]_i$. *Neurochem Int.* 2010;57(4):411–420 and there is a dense noradrenergic-controlled GABAergic cortical network in the prefrontal cortex.[34] Nevertheless, expression of ARs might be more pronounced in cortical astrocytes than in cortical neurons as demonstrated by immunochemical analysis in layer 1 of the developing rat cortex.[8] Because of late astrocytic development, this difference was much more pronounced in 3-week-old than in 2-week-old animals. However, layer 1 may be specially enriched in astrocytes due to presence of glia limitans with its high expression of the astrocyte-specific aquaporin-4, which may play a role in fluid exchange with the subarachnoidal space.[35]

β_3-AR mRNA was not found by Hertz et al.[6] in freshly isolated brain astrocytes, whereas Summers et al.[36] showed β_3-AR expression in rat brain, although the density was much lower than in brown fat. Cultured chick astrocytes express this receptor,[37] which is also found in cultured mouse astrocytes, although less abundantly than the β_1-AR.[38]

Stimulation of LC, the brain stem nucleus that supplies the entire cortex with noradrenergic innervation, results in rapid α_1-AR mediated astrocytic transients in $[Ca^{2+}]_i$.[39] Injection of either the α-AR agonist methoxamine or the β-AR agonist isoproterenol into layer 1 somatosensory cortex elicits Ca^{2+} waves in astrocytes, indicating that astrocytes express both α- and β-ARs and that β-adrenergic signaling influences $[Ca^{2+}]_i$. In awake behaving mice LC activation triggered either by whisker stimulation or air-puffs evoke astrocytic $[Ca^{2+}]_i$ transients that encompass the entire imaged field.[40] Damage of LC and other brain stem nuclei may be an early and perhaps causative event in Alzheimer's disease[41,42] on account of both anti-inflammatory effects[43] (see also chapters by Madrigal and De Keyser) and lack of metabolic/functional activities. The latter may include that the much larger increase in brain glycolysis than in oxidative metabolism, which is characteristic of brain activation, is dependent upon β-adrenergic signaling[44] and accordingly may be reduced or absent in Alzheimer patients.

cAMP and protein kinase A (PKA) measurements employing fluorescence resonance energy transfer (FRET) methods have been used in cultured astrocytes[45] to determine β-AR-mediated increases in cAMP that occurred relatively slowly (compared to the increase in $[Ca^{2+}]_i$ evoked by an α_1-AR agonist). It has also been employed in neurons in intact tissue, although in response to the serotonergic 5-HT_7 receptor.[46]

β-ADRENERGIC SIGNALING PATHWAYS

Stimulation of the G_s-coupled β-ARs activates an adenylyl cyclase that produces cAMP from adenosine triphosphate (ATP). The generated cAMP activates PKA and EPAC1 and 2 (exchange protein directly activated by cAMP).[47] However, it is becoming obvious that in highly differentiated cells signaling by β-ARs may activate a downregulated cAMP/PKA pathway, as elegantly shown in studies comparing young and more differentiated myocardial cells.[48] Accordingly, the downstream pathway can be complex and phosphorylation of extracellular regulated kinases 1 and 2 ($ERK_{1/2}$) has repeatedly been demonstrated after stimulation of β-ARs in different cell types.[49–51]

ERK_1 and ERK_2, which differ slightly in their molecular weights, are members of the mitogen-activated protein kinase (MAPK) family. As reviewed by Cheng et al.,[52] they are important constituents in many pathways regulating glial and neuronal physiology, biochemistry, and development, and they are activated by growth factors, chemokines, cytokines, oxidative stress, or ischemic injury. In astrocytes (and some other cell types), stimulation of some G-protein coupled receptors (GPCRs) leads to a metalloproteinase-mediated release of a growth factor, which "transactivates" the epidermal growth factor receptor (EGFR) and leads to phosphorylation and activation of $ERK_{1/2}$ ($pERK_{1/2}$).[53–56] Such a pathway, activated by β_1-AR stimulation, is shown in Fig. 2.1 and discussed in detail below. pERK can enter the cell nucleus,[55] converting an extracellular stimulus to an intracellular signal that controls gene expression, although nuclear entry of pERK is not a requirement for alteration of gene expression.[57]

Exposure of cultured astrocytes to 10 nM–10 μM isoproterenol and inhibitors of either β_1- or β_2-ARs showed a much more potent effect of isoproterenol on β_2- than on β_1-ARs.[58] Stimulation of either subtype receptor caused phosphorylation of $ERK_{1/2}$. Inhibitors of potential intermediates of the pathway leading to $ERK_{1/2}$ phosphorylation after β_1- and β_2-ARs stimulation by isoproterenol in cultured astrocytes showed that the pathways of the subtypes were very different (Fig. 2.1).

The β_1-adrenergic signaling pathway towards phosphorylation (red arrows) was studied in detail by establishing which inhibitors of specific intermediates could prevent $ERK_{1/2}$ phosphorylation.[58] These inhibitors are described in the legend of Fig. 2.1. A key point in the pathway is that the production of cAMP and stimulation of PKA is followed by a G_s/G_i switch. Such a switch has previously been demonstrated in other cell types.[59,60] G_i signaling leads to a metalloproteinase-induced release of a growth factor via an increase in $[Ca^{2+}]_i$. This growth factor is often heparin-binding epidermal growth factor (HB-EGF),[61] which phosphorylates and activates EGFRs. Such a "transactivation" after β-adrenergic stimulation is also well known in other cell types.[49,54,62] Recruitment of β-arrestin 1 and 2 leads to MEK-dependent $ERK_{1/2}$ phosphorylation. Although the involvement of Ras and Raf was not tested, they are generally part of the growth-factor-induced MAP-kinase signaling that leads to MEK-activated $ERK_{1/2}$ phosphorylation. That the pathway shown in Fig. 2.1 also operates in astrocytes in situ is shown by the ability of each inhibitor that abolishes $ERK_{1/2}$ phosphorylation in cultured astrocytes to prevent ischemia/reperfusion-induced swelling in the brain in vivo, which occurs in astrocytes and depends upon β_1-AR stimulation[63] as

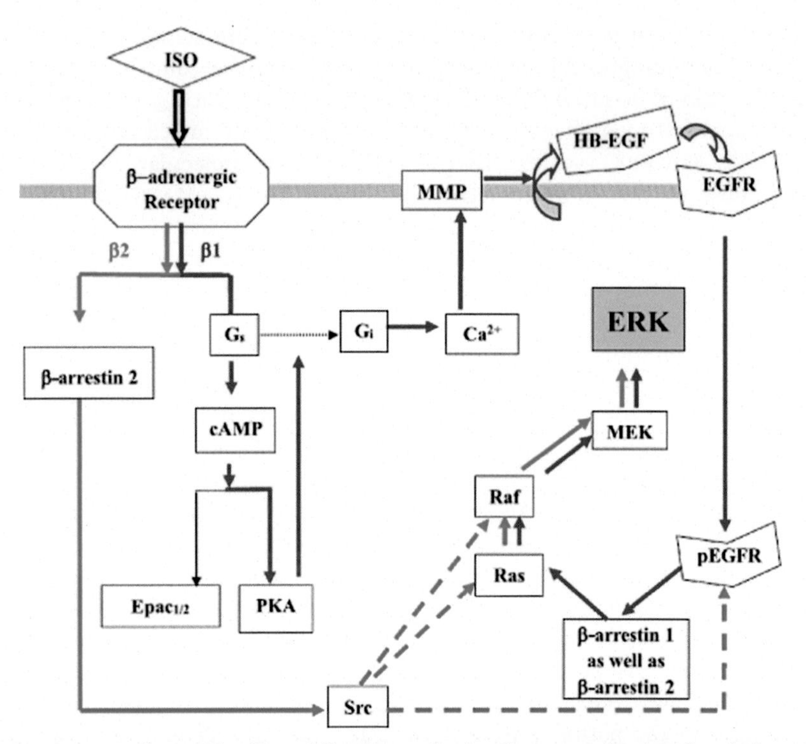

FIGURE 2.1 **Pathways for β_1- and β_2-adrenergic signaling towards phosphorylation of extracellular regulated kinases 1 and 2 ($ERK_{1/2}$) in cultured astrocytes.** Specific inhibitors (H-89, an inhibitor of PKA; PTX, an inhibitor of G_i protein; BAPTA/AM, an intracellular Ca^{2+} chelator; GM6001, an inhibitor of Zn-dependent metalloproteinase; AG1478, an inhibitor of the EGF receptor; U0126, an MEK inhibitor; siRNA against β-arrestins; and PPI, an inhibitor of Src) were used to interrupt one or the other of the pathways, indicating that the inhibited compound or process was part of the pathway. For clarity, the inhibitors are not shown in the figure. The β-adrenergic agonist isoproterenol (ISO) at high concentrations (1 μM or higher) activates β_1-adrenergic signaling (red arrows) and associated EPAC1/2 signaling (black arrow). During β_1-adrenergic signaling PKA-dependent "G_s/G_i switch" induces an increase in $[Ca^{2+}]_i$ by Ca^{2+} release from intracellular stores. This activates Zn-dependent metalloproteinases (MMPs) and leads to shedding of growth factor(s), such as heparin-binding epidermal growth factor (HB-EGF). The released HB-EGF stimulates autophosphorylation of the epidermal growth factor (EGF) receptor in the same and adjacent cells. In turn, EGF receptor stimulation leads to phosphorylation of $ERK_{1/2}$ (shown in blue) via the Ras/Raf/MEK pathway, contingent upon recruitment of β-arrestin 1 and 2. At a low isoproterenol concentration (100 nM) β_2-adrenergic (green arrows) stimulation activates under the conditions used Src via the function of β-arrestin 2. Src stimulates $ERK_{1/2}$ phosphorylation and also phosphorylates the EGF receptor. The $ERK_{1/2}$ phosphorylation is secondary to MEK activation, which probably is induced by direct activation of Raf by Src, although an effect on Ras cannot be excluded. *From Du T, Li B, Li H, Li M, Hertz L, Peng L. Signaling pathways of isoproterenol-induced ERK1/2 phosphorylation in primary cultures of astrocytes are concentration-dependent. J Neurochem. 2010;115(4):1007–1023.*

described in the chapter by Peng et al. As will be discussed in detail later (see "Importance of Adrenergic Stimulation During Culturing of Astrocytes"), our cultures, including those used by the Peng group, are routinely grown in the presence of dibutyryl cyclic AMP (dBcAMP) from the age of 2 weeks, making them very similar to their in vivo counterparts.[64] Whether or not this is important for the G_s/G_i switch, the EGFR transactivation and the resulting phosphorylation of $ERK_{1/2}$ is not known. However, in the above-mentioned section, it will be shown that isoproterenol stimulates glycogenolysis much more in dBcAMP-treated than in untreated cells. As glycogenolysis requires an increase in $[Ca^{2+}]_i$,[65] as will be discussed in detail later, this observation probably shows that at least the G_s/G_i switch is greatly stimulated by the dBcAMP treatment.

Downstream effectors of cAMP production are numerous, including not only PKA but, for example, also hyperpolarization-activated cyclic nucleotide-gated (HCN) channels,[66] Popeye domain-containing (Popdc) proteins, and EPAC.[47,67] HCN channels are widely distributed in brain,[66] and Popdc proteins have been demonstrated in some neurons,[68] but neither has to our knowledge been demonstrated in astrocytes, whereas EPAC is included in the β_1-AR-activated astrocytic pathway shown in Fig. 2.1 and of functional importance for development of the typical astrocytic stellate morphology.[69]

The signaling pathway activated by β_2-adrenergic stimulation in mature cultures of mainly cortical astrocytes (green arrows in Fig. 2.1) is not affected by inhibition of PKA. This may, however, be analogous to the finding[48] that β_2-adrenergic stimulation activates cAMP and PKA only in young and not in differentiated myocardial cells. GPCR-activation under other conditions or at other locations is therefore a distinct possibility. $ERK_{1/2}$ phosphorylation depends on β-arrestin 2 and on Src activation leading to phosphorylation of Ras either directly or via Raf followed by MEK phosphorylation and activation of $ERK_{1/2}$. Src and the EGF receptor coprecipitate and the receptor is phosphorylated at the Src-phosphorylated Y845 site and the Y1045 autophosphorylation site, but not at the Y1173 site phosphorylated by β_1-adrenergic stimulation.[56,58]

β_3-Adrenergic signaling causes $ERK_{1/2}$ phosphorylation in adipocytes through a mechanism depending on G_s/G_i switch and the EGFR but not on PKA.[70]

SIGNALING PATHWAYS FOR ASTROCYTIC α-ADRENERGIC RECEPTOR SUBTYPES

Stimulation of α-ARs leads to activation of phospholipase C (PLC),[71,72] and generation of inositol trisphosphate (IP_3) and diacylglycerol (DAG) from membrane-bound phosphatidylinositide-4,5-bisphosphate (PIP_2). DAG activates protein kinase C (PKC) and stimulation of IP_3 receptors releases bound Ca^{2+}, leading to an increase in $[Ca^{2+}]_i$. This increase is crucial for a multitude of effects in astrocytes, including release of gliotransmitters and as in muscle,[73,74] it activates glycogen synthesis as will be discussed later. IP_3 is dephosphorylated to inositol, which reacts with a DAG metabolite to regenerate PIP_2, but some inositol exits the cells and is replaced by inositol from the food or generated in brain vasculature[75] from glucose. In either case, the inositol has to be accumulated in astrocytes (and other brain cells) to maintain their α-adrenergic signaling.

In cultured astrocytes, inositol uptake can be inhibited by chronic treatment with the lithium ion (Li^+), valproic acid, and carbamazepine.[76] These three drugs all cause astrocytic alkalinization, by either stimulation of the Na^+/H^+ exchanger (NHE), an H^+ extruder, by extracellular Li^+ or increased activity in cultured astrocytes, but not in cultured neurons, of the acid-extruding Na^+/bicarbonate transporter NBCe1 induced by chronic treatment with valproic acid or carbamazepine.[77–79] NBCe1 is also activated by the astrocytic depolarization caused by an increased extracellular K^+ concentration ($[K^+]_o$) in brain.[80] This contributes to K^+ uptake, but to a lesser extent than uptake by the Na^+,K-ATPase.[79]

Alkalinization decreases inositol uptake in astrocytes,[81] and a resulting decrease in phosphatidylinositide-mediated signaling may be major reason for the therapeutic effect of Li^+, valproic and carbamazepine in bipolar disorder.[76,78,82,83] Li^+ also acutely inhibits IP_3 degradation to inositol, an effect that similarly has been suggested as an explanation for its therapeutic effect in bipolar disorder.[84] However, valproic acid and carbamazepine do not share this effect.[78] All three drugs require chronic treatment for their therapeutic effect, and treatment of cultured astrocytes with 1-mM LiCl for 1–2 weeks strongly reduces the increase in $[Ca^{2+}]_i$ by NA, whereas treatment for only 45 min has no effect (Fig. 2.2).[85] Chronic administration of the antidepressant fluoxetine, an agonist at the serotonergic 5-HT_{2B} receptor, similarly inhibits the ability of the α_2-AR agonist dexmedetomidine to increase $[Ca^{2+}]_i$ in cultured astrocytes, and it does so by depletion of Ca^{2+} stores due to inhibition of capacitative Ca^{2+} entry via store-operated channels (SOCEs).[86] However, elevations of $[K^+]_o$ high enough to activate the $Na^+,K^+,2Cl^-$ cotransporter NKCC1 by a depolarization-mediated opening of L-channels for Ca^{2+} caused a larger increase in $[Ca^{2+}]_i$ than in untreated cells. This can be explained by an enhanced expression of the L-channel gene Cav1.2 both in astrocytes from the brain of treated mice and in fluoxetine-treated cultured astrocytes, which augments K^+ uptake via NKCC1 and increases $[Ca^{2+}]_i$.[87] Such an enhanced K^+-induced increase in $[Ca^{2+}]_i$ was not found in astrocytes treated with Li^+, carbamazepine or valproic acid, which in contrast showed a reduced effect due to SOCE inhibition.[88] This inhibition of capacitative Ca^{2+} entry probably also explains the reduced effect of NA on $[Ca^{2+}]_i$ shown in Fig. 2.2.

Downstream α_2-AR signaling in cultured astrocytes, studied by aid of the potent and highly specific α_2-AR agonist dexmedetomidine, leads, like β_1-adrenergic stimulation, to ERK phosphorylation following EGFR transactivation. The pathway reminds of that shown in Fig. 2.1, except that there is no need for a G_s/G_i shift: $ERK_{1/2}$ phosphorylation is inhibited by a G_i inhibitor, GF109203X, a PKC inhibitor, a metalloproteinase inhibitor, or an EGFR inhibitor.[57] It is also inhibited by PP1, an inhibitor of the Src kinase. Moreover, cfos and fosB are induced although $ERK_{1/2}$ did not enter the nucleus, but the induction could be inhibited by a MEK inhibitor, showing downstream effects of $ERK_{1/2}$ phosphorylation. The hypnotic and neuroprotective responses of dexmedetomidine are exerted on the α_{2A}-AR.[89,90] This receptor subtype also reduces forskolin-mediated stimulation of adenylate cyclase activity in CHO cells.[91] This is consistent with interactions between signaling by α_2-AR and by cAMP and PKA in these cells.[92] During surgical stress dexmedetomidine decreases plasma NA.[93]

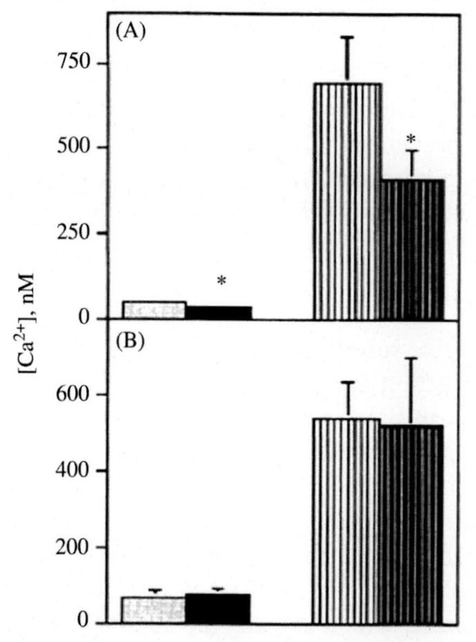

FIGURE 2.2 Free cytosolic calcium concentration ($[Ca^{2+}]_i$) in primary cultures of mouse astrocytes, measured by Ca^{2+}-induced fluorescence changes in INDO-1 at 405 and 485 nm. $[Ca^{2+}]_i$ was determined during control conditions (gray and black columns) and during exposure to 1-μM noradrenaline (columns with vertical lines) in untreated control cultures (black lines on gray background) and in sister cultures (white lines on black background) from the same batches that had been treated for either 7–14 days (A) or 30–45 min (B) with 1-mM lithium chloride. In (A), * indicates statistical significant difference ($P < 0.05$) between treated and untreated cultures, whereas there is no difference between the two groups in (B). In all groups of cultures, there was a significant increase in $[Ca^{2+}]_i$ during exposure to noradrenaline. In the untreated cultures, this response appeared slightly (but not significantly) smaller in (B) than in (A). Results are means ± SEM of 13–14 (A) or 5–6 (B) individual experiments, using cultures obtained from at least two different batches. *From Chen Y., Hertz L. Inhibition of noradrenaline stimulated increase in $[Ca^{2+}]_i$ in cultured astrocytes by chronic treatment with a therapeutically relevant lithium concentration.* Brain Res. *1996;711(1–2):245–248.*

Another α_2-adrenergic agonist, clonidine, causes phosphorylation of AKT and glycogen synthase kinase-3 (GSK-3) in chick astrocytes[22] with important consequences for glycogen synthesis as will be discussed below. Conditioned medium from dexmedetomidine-treated astrocytes, but not dexmedetomidine itself, induces ERK phosphorylation in primary cultures of cerebellar neurons.[57]

None of the studies showing α_1-AR-mediated increase in $[Ca^{2+}]_i$ suggested phosphorylation of $ERK_{1/2}$.[24–27] However, both in the heart and prostate α_1-adrenergic signaling can evoke ERK phosphorylation,[94–97] and α-adrenergic transamination-mediated $ERK_{1/2}$ phosphorylation occurs in aorta.[98]

METABOLIC EFFECTS OF α- AND β-ADRENERGIC STIMULATION OF ASTROCYTES

Glucose Uptake

NA increases glucose content in cultured astrocytes, and although most of this increase is prevented when glycogenolysis is inhibited,[99] part of it is probably due to a stimulation of uptake. β_3-Adrenergic stimulation stimulates glucose uptake in chicken brain and supports learning.[100]

Glucose Metabolism

Glucose is the main substrate for energy production in adult brain. Although a small fraction is degraded via the pentose phosphate shunt, by far most glucose is initially metabolized to pyruvate during glycolysis.[101] Pyruvate is via the pyruvate dehydrogenase complex (PDHc) introduced into the tricarboxylic acid (TCA) cycle, where it condenses with preexisting oxaloacetate to form citrate. For complete oxidative metabolism citrate is via 2 decarboxylations reconverted to oxaloacetate, which then condenses with another molecule of pyruvate. The first decarboxylation metabolizes isocitrate to α-ketoglutarate, which is a key metabolite, because it can be further decarboxylated by the α-ketoglutarate dehydrogenase complex (αKGDHc) to succinate, which is then converted to oxaloacetate. This process goes on as long as pyruvate is available, and the TCA cycle is operating, and it generates large amounts of energy.[101,102] However, α-ketoglutarate can in astrocytes also be converted to glutamate, most of which subsequently is transferred to neurons via the very active glutamate–glutamine cycle for production of the neurotransmitters glutamate and γ-aminobutyric acid (GABA).[103–106] The operation of this cycle is essential for brain function because these transmitters cannot be synthesized in neurons from glucose and do not easily enter the brain from the circulation. Glutamate synthesis is based on carboxylation by the astrocyte-specific pyruvate carboxylase of pyruvate to oxaloacetate, which in the TCA cycle is condensed in the usual manner with pyruvate metabolized via PDHc to form a “new” molecule of citrate, which in the TCA cycle is converted to α-ketoglutarate which is used for synthesis of glutamate.[101,102]

An increase in glutamate synthesis in the chicken brain occurs during the initial phase of learning,[107] and also other forms of increased brain activity are associated with enhanced pyruvate carboxylation or glutamate content.[108–110] A high neuronal glutamate content is important during increased glutamatergic activity, because vesicular filling with glutamate depends upon the cytosolic glutamate concentration.[111] Subsequently, brain glutamate content returns to normal by increased oxidative metabolism, predominantly in astrocytes.[112] Even without any increase in glutamate content, there is a constant de novo formation of glutamate, which is compensated for by corresponding, energy-producing oxidative degradation. This energy is not only used for glutamate uptake but also for other important processes.[112–114] In addition to supplying neurons with glutamate synthesized de novo, the glutamate–glutamine cycle also carries released transmitter glutamate to astrocytes in which most of it is taken up,[114,115] converted to glutamine, and returned to neurons in the cycle, although some might be oxidized.

Noradrenergic signaling enhances virtually all aspects of glucose metabolism (Table 2.2). Glycolysis is potently (EC_{50} 61 nM) increased by α_1-adrenergic stimulation.[116] This probably reflects stimulation of the glycolytic enzyme phosphofructokinase that has been demonstrated in cardiomyocytes.[117] Increased $[K^+]_o$ also increases glycolysis by stimulation of this enzyme.[118]

Metabolism of pyruvate via PDHc has been measured as rate of production of labeled CO_2 from [1-^{14}C]pyruvate, as this carbon atom is released when pyruvate is introduced into the TCA cycle. This process is stimulated by α_2-AR agonists (Fig. 2.3A), shown by the ability of NA and either clonidine[119] or dexmedetomidine[120] to increase $^{14}CO_2$ production from [1-^{14}C]pyruvate. The concentration dependence of the stimulation by dexmedetomidine is similar to its increase of $[Ca^{2+}]_i$.[119] However, the increase in $[Ca^{2+}]_i$ is very small (~100 nM) compared to that evoked by higher concentrations of NA (Fig. 2.3B), and increased $[Ca^{2+}]_i$ is not sufficient for the metabolic response, as phenylephrine also increases $[Ca^{2+}]_i$ but does not stimulate production of labeled CO_2.[119]

The effect of α_2-AR agonists on PDHc is probably exerted via stimulation of pyruvate entry across the inner mitochondrial membrane, mediated by pyruvate carriers.[121,122] In rat renal cortex, this transport is stimulated by the α_1-AR agonist phenylephrine,[123] and with such a mechanism of action pyruvate carboxylation should probably also be activated in astrocytes by α_2-AR stimulation. In hepatocytes, activation of both mitochondrial pyruvate uptake and pyruvate carboxylation has been found after stimulation with phenylephrine[124,125] or NA,[126] which also increases pyruvate carboxylation by an α_1-adrenergic effect.[127] This increase is correlated with elevated $[Ca^{2+}]_i$ and inhibited by phentolamine, an antagonist of α-ARs. The involvement of the α_1-AR in hepatocytes does not disprove a similar α_2-adrenergic mechanism as for PDHc in astrocytes, as it could represent a cell type difference. A similar subtype-dependent difference is seen between astrocytes and neurons in adrenergic stimulation of the Na^+,K^+-ATPase (see below).

NA also stimulates metabolism of [U-^{14}C] aspartate, a precursor of oxaloacetate, in cultured astrocytes but not in cultures of either the glutamatergic cerebellar granule neurons or the GABAergic cortical interneurons.[128] The astrocytic effects also include stimulation of decarboxylation of the TCA cycle intermediates isocitrate to α-ketoglutarate by the isocitrate dehydrogenase and of α-ketoglutarate to succinyl coenzyme A by the αKGDHc during the turn of the TCA cycle.[166]

The effects of noradrenergic agonists on rate of αKGDHc activity were examined in a similar fashion as pyruvate dehydrogenation, measuring

TABLE 2.2 Effects of Adrenergic Stimulation Discussed in This Paper

Receptor	Effect	Species	Reference[a]
α_1	Increase in $[Ca^{2+}]_i$	Mouse, rat	[24–27]
	Increase in glycolysis	Mouse	[116]
	Increase in glutamate uptake	Rat	[129]
α_2	Increase in $[Ca^{2+}]_i$	Mouse	[120]
	Increase in PDHc[b]	Mouse	[120]
	Increase in αKGDHc	Mouse	[116]
	Increase in isocitrate dehydrogenase[c]	Mouse	[128]
	Increase in glycogen synthesis	Mouse, chicken	[22,102]
	Importance for learning	Chicken	[159]
β_1	Increase in Na^+,K^+-ATPase activity	Rat	[213]
	Increase in K^+ uptake	Mouse	[180]
	Increase in glycogenolysis	Mouse, rat	[140,141]
β_2	Increase in glycogenolysis	Chicken	[100]
	Importance for learning	Chicken	[131]
β_3	Increase in glucose uptake	Chicken	[100]
	Importance for learning	Chicken	[100]

[a] *References are examples, not necessarily comprehensive.*
[b] *Probably also pyruvate carboxylase (see text).*
[c] *Effect on aspartate metabolism may not be specific for this enzyme.*

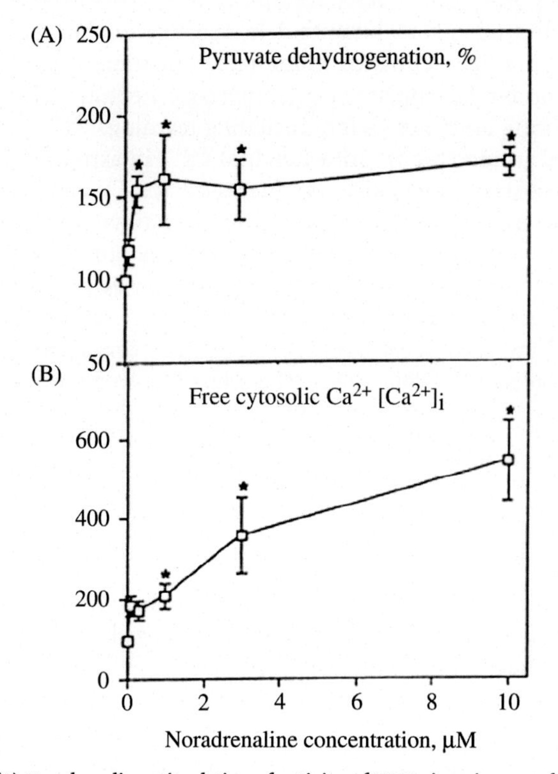

FIGURE 2.3 (A) Noradrenaline stimulation of activity of PDHc in primary cultures of mouse astrocytes. Rate of PDHc operation was measured in individual cultures exposed to noradrenaline concentrations between 0 and 10 μM during a 30-min incubation period by determination of released $^{14}CO_2$ from [1-^{14}C]pyruvate at a pyruvate concentration of 5 mM, of specific activity of incubation medium and of protein content of the culture. Average PDHc activity was determined in each group and expressed as a percentage of that in the absence of noradrenaline. (B) $[Ca^{2+}]_i$ measured by aid of the fluorescent probe Indo-1 and percentage noradrenaline-induced increase determined as in (A). Note that maximum PDHc activity is reached at a noradrenaline concentration giving a relatively small increase in $[Ca^{2+}]_i$. $n = 5-20$ individual experiments. *Significantly different from control ($P < 0.05$). *From Chen Y., Hertz L. Noradrenaline effects on pyruvate decarboxylation: correlation with calcium signaling. J Neurosci Res. 1999;58(4):599–606.*

production of $^{14}CO_2$ from [1-^{14}C]glutamate, an α-ketoglutarate precursor releasing the C-1 atom during the αKGDHc-mediated decarboxylation. Again, both NA and clonidine stimulated metabolism.[116] The α_1-AR agonist phenylephrine also stimulated $^{14}CO_2$ production, but this may be explained by stimulation of glutamate uptake by an α_1-adrenergic effect,[129] which will increase the specific activity of the cellular pools of glutamate and α-ketoglutarate. In contrast to the potent effect of NA on pyruvate dehydrogenation, a >10-fold higher concentration was needed for stimulation of α-ketoglutarate dehydrogenation.[116] Thus, lower concentrations of NA have little effect on α-ketoglutarate dehydrogenation, but facilitate its formation, which might enhance glutamate formation from glucose. This may be important for supplying glutamatergic neurons with transmitter glutamate during brain activation, including learning.[130,131]

NA causes an increase in mitochondrial Ca^{2+} in astrocytes,[132] but stimulation of oxidative metabolism by increased $[Ca^{2+}]_i$ and intramitochondrial Ca^{2+} has mainly been studied in muscle and liver. The mitochondrial dehydrogenases PDHc, isocitrate dehydrogenase and αKGDHc are all stimulated,[133–136] allowing rapid metabolic responses to increased energy demands. These are the same enzymes that are stimulated in astrocytes.[116,119,128] Increase in intramitochondrial Ca^{2+} also stimulates neuronal energy metabolism,[137,138] but the mechanisms involved may be different.

Glycogen Turnover

The regulation of glycogen synthesis and glycogenolysis is important, because glycogenolysis plays a major role both in astrocytic functions and in interactions between astrocytes and neurons that are essential for brain function, as will be described in the next section. Compartmentalization of lactate generated by glycolysis and glycogenolysis has been suggested in cultured astrocytes,[139] and glycogen-derived lactate plays a specially important role in brain function.

Astrocytic glycogenolysis has recently been reviewed,[140–142] and only key points will be repeated here. Several transmitters, including NA and β-AR agonists,[141–144] ATP[142] and arachidonic acid,[145] a downstream product of ATP stimulation,[146] adenosine,[144] vasopressin[142], vasoactive intestinal peptide (VIP),[144] serotonin,[147–149] acting on 5-HT_{2B} receptors,[150] and histamine[144] increase glycogenolysis in brain slices or cultured astrocytes. Many of these transmitters also increase $[Ca^{2+}]_i$ and/or cAMP, although lower concentrations of NA are required to increase $[Ca^{2+}]_i$ (0.5 μM) than to increase cAMP (8 μM).[143] This reflects that an increase in $[Ca^{2+}]_i$ is indispensable for transformation of the inactive glycogen phosphorylase (GP) b to the active GP a, whereas cAMP plays at a facilitatory role, which can be substantial (Fig. 2.4A), although only when $[Ca^{2+}]_i$ is also increased.[73,74,151] The conversion to the active GP b occurs by stimulation of phosphorylase kinase, a complex protein kinase with four subunits, one catalytic and three inhibitory, with increased $[Ca^{2+}]_i$ and cAMP acting on different units to relieve the inhibition.[152] In many tissues, both forms of GP, GP a and GP b, can also be converted to an inactive T-state, which subsequently can be reconverted to the active states.[151]

In the fowl glycogenolysis is stimulated by the $β_2$-AR[153,154] but this $β_2$-AR signaling may be different from that shown in Fig. 2.2, analogous to the change in myocardial cells during ontogenesis and the complexities of $β_1$- and $β_2$-adrenergic signaling.[48] However, in cells and slices from

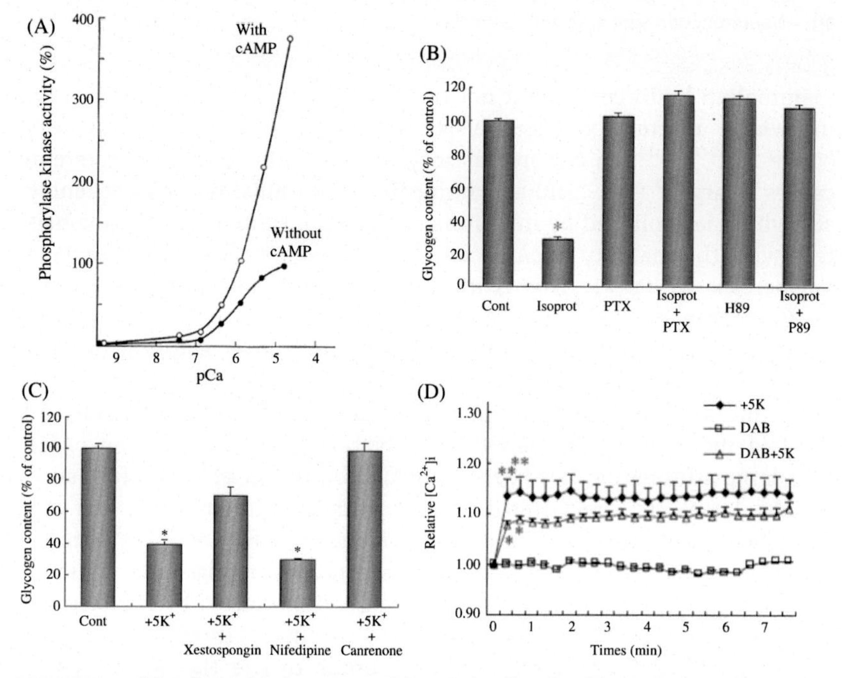

FIGURE 2.4 (A) Relationship between activity of phosphorylase kinase from cardiac muscle and $[Ca^{2+}]_i$ in the presence and absence of cAMP. Closed circles indicate activities without cAMP and open circles indicate those with 3-μM cAMP. Note that the activity is almost zero both in the presence or absence of cAMP, when $[Ca^{2+}]_i$ are between 0.001–0.1-μM, and that Ca^{2+} concentrations between 1 and 10 μM alone activate phosphorylase kinase activity and that this activity is potentiated by cAMP. (B) Effect of isoproterenol on glycogenolysis indicated as reduction of glycogen content in cultured mouse astrocytes incubated in tissue culture medium for 20 min under control conditions (Cont) or in the presence of 1-μM isoproterenol (Isoprot), with or without 15-min pretreatment with the G_i inhibitor PTX or the PKA inhibitor H89. (C) Effects on cultured astrocytes [treated as in (B)] of addition of 5 mM K^+ to a final extracellular concentration of 10 mM (+5 K^+), with or without 15 min pretreatment with xestospongin, an inhibitor of the IP_3 receptor known to be part of the signaling pathway by endogenous ouabains, nifedipine, an inhibitor of L-channel opening but not of the ouabain signaling pathway, or the ouabain inhibitor canrenone. In both (B) and (C), S.E.M. values are indicated by vertical bars, and * indicates significant difference ($P < 0.05$) from control conditions. (D) Glycogenolysis is required for increase of $[Ca^{2+}]_i$ in cultured astrocytes induced by addition of 5-mM K^+ and measured by aid of a Ca^{2+}-sensing drug. After incubation of fura-2-loaded cells in saline solution for 30 min and subsequent wash, the cells, grown on a glass coverslip, were perfused either in similar solution or in a solution to which an additional 5-mM KCl had been added at zero time (with a corresponding reduction of NaCl concentration). In some experiments, 10-mM DAB, an inhibitor of glycolgenolysis, was added 2 min before the addition of K^+. Results are averages of 29 to 77 cells on two–three individual coverslips. S.E.M. values are indicated by vertical bars. All results after addition of K^+ are significantly different ($P < 0.05$) from control conditions. *Statistically significant ($P < 0.05$) difference from drug-free group at the same time period. *(A) From Ozawa E. Regulation of phosphorylase kinase by low concentrations of Ca ions upon muscle contraction: the connection between metabolism and muscle contraction and the connection between muscle physiology and Ca-dependent signal transduction. Proc Jpn Acad Ser B Phys Biol Sci. 2011;87(8):486–508, (B) and (C) from Xu J., Song D., Bai Q., Cai L., Hertz L., Peng L. Basic mechanism leading to stimulation of glycogenolysis by isoproterenol, EGF, elevated extracellular K^+ concentrations, or GABA. Neurochem Res. 2014;39(4):661–667, and (D) from Hertz L., Xu J., Song D., et al. Astrocytic glycogenolysis: mechanisms and functions. Metab Brain Dis. 2015;30(1):317–333.*

mammalian brain cortex (but not necessarily in subcortical structures or in white matter) glycogenolysis is caused by β_1-AR stimulation (Table 2.2).[140,141,143] In cultured astrocytes, the β-AR agonist isoproterenol causes a large[141] (Fig. 2.4B) and immediate[155] stimulation of glycogenolysis, which is abolished by inhibitors of the G_s/G_i switch and of G_i, consistent with the inability of cAMP to elicit glycogenolysis in the absence of an increase in $[Ca^{2+}]_i$. This is in agreement with previous observations by Ververken et al.[156] that cAMP in brain slices is only very slightly increased and still rapidly rising half a minute after addition of 10 or 100 μM NA, whereas glycogen phosphorylase a is already fully activated. The same authors observed that highly elevated $[K^+]_o$ (25 mM) caused a fourfold increase in phosphorylase activation with no effect on cAMP. This large stimulation has been confirmed by Hof et al.[157] in brain slices and by Xu et al.[140] and Hertz et al.[141] in cultured astrocytes, and it is inhibited by nifedipine, an inhibitor of L-channels for Ca^{2+}.[140,141] Smaller increases in $[K^+]_o$ (e.g., 5 mM) cause less stimulation (Fig. 2.4C), which is unaffected by L-channel inhibitors[141,157] but counteracted by inhibitors of signaling by endogenous ouabains,[141] a signaling pathway that is activated in cultured astrocytes when K^+ binds to the Na^+,K^+-ATPase.[158] This signaling induces an increase in $[Ca^{2+}]_i$, which is slightly inhibited by 1,4-dideoxy-1,4-imino-D-arabinitol DAB (Fig. 2.4D).

In both mammals and fowl, glycogen synthesis in brain and brain cells is stimulated by α_2-adrenergic signaling[22,101,159] and by insulin,[22,160,161] although the insulin effect may be masked by a concomitant decrease in blood sugar. The main source for glycogen synthesis is glucose-6-phosphate formed by hexokinase from glucose, although glycogen can also be formed from pyruvate, alanine or TCA cycle constituents by gluconeogenesis.[101] Glucose-6-phosphate is introduced into glycogenin, an oligosaccharide primer, via glucose-1-phosphate and the nucleotide sugar uridine diphosphate (UDP)-glucose and subsequently further converted to the mature glycogen molecule.[151,162] The incorporation of UDP-glucose into glycogenin is catalyzed by glycogen synthase (GS), which like GP is found in an inactive (GS b) and active (GS a) form. However, in contrast to GP in which the active form is phosphorylated, activation of GS occurs by dephosphorylation, explaining why PKA at least in some tissues can activate GP but inactivate GS.[151]

Hutchinson et al.[22] showed that increase of glycogen synthesis in primary cultures of astrocytes (assessed by measurement of the incorporation of D-[U-^{14}C]glucose) can be mediated by insulin or by stimulation of α_2-ARs and that it occurs by a phosphatidylinositol-3 kinase (PI3K)-dependent mechanism. The effect of insulin was not enhanced by the α_2-AR agonist clonidine, suggesting that a common signaling pathway is used by these two agents. The inhibitors of PI3K, wortmannin and 2-(4-morpolinyl)-8-phenyl-4*H*-1-benzopyran-4-one (LY294002), both significantly inhibited total ^{14}C incorporation in response to low concentrations (1 μM) of insulin, NA or clonidine. In cultured mouse astrocytes, simulation of the 5-HT_{2B} receptor by low concentrations of fluoxetine has shown that EGF receptor activation leads to stimulation (phosphorylation) of AKT, an intermediate between PI3K and GSK-3.[150,163] This serine/threonine protein kinase exists as relatively similar α and β forms (GSK-3α and GSK-3β). Its activation (phosphorylation) inhibits the glycogen synthase, but the inhibition is relieved by α_2-adrenergic stimulation as discussed below. As both the 5-HT_{2B} and the α_2-ARs are PKC-linked, the latter is also likely to activate the PI3K-AKT-GSK-3 pathway, and in chick astrocytes AKT is phosphorylated on Ser 473, GSK-3α on Ser 21, and GSK-3β on Ser9 by clonidine.[22] These events can explain α_2-AR-mediated stimulation of glycogen synthesis, because phosphorylated AKT inhibits glycogen GSK-3β by phosphorylation.[164] This stimulates glycogen synthesis, as GSK-3 decreases GS activity.[165,166] Consistent with these findings in cultured astrocytes, Plenge[167] found an acute increase in brain glycogen after administration of Li^+ that is known to inhibit GSK-3. It is likely that the 5-HT_{2B} agonist fluoxetine also activates glycogen synthesis, as the AKT pathway is stimulated, as discussed above, and 5-HT_{2B} stimulation or acute fluoxetine administration has been found to decrease the levels of phosphorylated GSK-3.[168,169] The cannabinoid 1 receptor (CB1) also increases glycogen synthesis in rat primary astrocytes by a G_i-PI3K-mediated pathway.[170] This receptor is abundantly expressed in astrocytes in vivo and functions in a very complex manner.[171]

Importance of Glycogen Turnover for Brain Function

The demonstration by Swanson[172] that stimulation of the rat face and vibrissae accelerated utilization of glycogen in brain regions receiving sensory input from these regions changed the concept of glycogen from being a rather uninteresting emergency substrate to a metabolic substrate of key importance for brain function. As glycogen and its GS-mediated degradation in brain parenchyma are virtually restricted to astrocytes,[173,174] many of these functions, including metabolic support of signaling pathways and processes, energize the astrocytes' own function.[175] These include glutamate formation from glucose by combined PDHc and pyruvate carboxylase activity[107,176] and astrocytic uptake of glutamate[177] and K^+,[158,178–180] release of gliotransmitter ATP[142,181] and metabolic support for Ca^{2+} homeostasis.[182] However, Sickmann et al.[177] reported that neuronal glutamatergic function was also affected by inhibition of glycogenolysis. This is of special interest in connection with studies of inhibition of learning by inhibition of glycogenolysis. As discussed below, both astrocytes and neurons are affected.

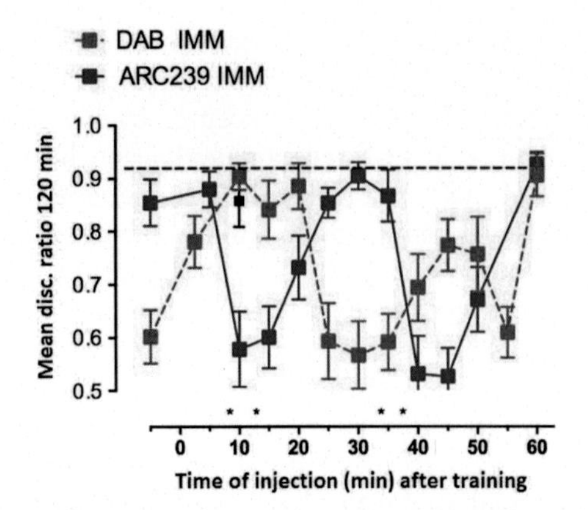

FIGURE 2.5 Effect of memory in day old chickens by inhibition of glycogen turnover, inhibiting either its synthesis with injection into the intermediate medium mesopallium, an equivalent of the mammalian brain cortex with the $\alpha_{2B/C}$-AR antagonist ARC239 or glycogenolysis with DAB at different times after training. The times when memory is sensitive to ARC are immediately preceding or during glycogen synthesis, and sensitivity to DAB is immediately preceding or during glycogenolysis. The figure thus indicates a series of intermittent glycogen breakdown and resynthesis. *From Gibbs M.E. Role of glycogenolysis in memory and learning: regulation by noradrenaline, serotonin and ATP. Front Integr Neurosci. 2015;9:70.*

Gibbs et al.[183,184] were the first to show that glycogen turnover is indispensable for memory by demonstrating that one-trial aversive learning in day-old chickens is abolished by administration of the glycogenolysis inhibitor DAB. Later Duran et al.[185] showed that knock-out of brain GS abolishes learning of new motor and cognitive skills. In the chickens intermittent glycogen synthesis (mediated by α_2-adrenergic signaling) and glycogenolysis (mediated by serotonin during the first glycogenolytic period and by β_2-adrenergic signaling during the second)[131] occurs during the first hour of the learning process as shown by the effect of inhibiting either process at different times (Fig. 2.5). With the injection times used by Gibbs and her coworkers, the inhibitory effect of DAB on learning was exclusively due to inhibition of glutamate synthesis,[107,186] an astrocytic effect, as the inhibition could be completely reversed by administration of glutamine or other glutamate precursors. However, inhibition of release of gliotransmitter ATP also inhibits learning in the same learning paradigm.[187]

That inhibition of glycogenolysis also abolishes neuronal contributions to hippocampus-mediated learning in rat and does so by inhibiting release of glycogenolysis-derived lactate has been shown by Suzuki et al. and Steinman et al.[188,189] At the same time DAB also impedes long-term potentiation (LTP) and expression of cofilin and other memory-related genes, and both these effects and the memory impairment can be rescued by administration of lactate, known to exit astrocytes via monocarboxylate transporters (MCT) 1 and 4. This exit causes a small increase in extracellular lactate (from ~1 to ~1.5 mM),[189] and MCT1 and 4 are upregulated by learning but MCT2, the MCT mediating lactate entry into neurons is not.[188,190] The upregulation strongly suggests an essential effect of signaling by extracellular lactate, probably that activating effect neurons by astrocyte-released L-lactate, but not D-lactate, with an EC_{50} of ~0.5 mM, in a glycogenolysis-dependent manner which stimulates processes needed for memory consolidation.[191,192]

In spite of the lack of MCT2 upregulation, learning memory in rats is inhibited after MCT2 blockade, implying additional importance of neuronal lactate uptake.[188,189] The lack of upregulation of MCT2 and the low magnitude of the rise in lactate suggest, however, that the uptake is very limited.[193] A small lactate uptake might be especially useful in the glutamatergic dendritic spines, which rarely have mitochondria,[194] express AMPA glutamate receptors, are enriched in MCT2 colocalized with GluA2/3[194] and express the actin depolymerizing protein cofilin, which is required for spine enlargement during early LTP.[195] In the absence of mitochondria MCT2-mediated lactate accumulation and subsequent oxidation to pyruvate would rapidly produce ATP needed for the remodeling of actin without interfering with glucose utilization by competition for NAD^+.[196] LTP dependence on glycogenolysis has been confirmed in brain slices from young rats,[197] and in astrocyte-neuronal cocultures the amplitudes of miniature excitatory postsynaptic currents (mEPSC) are ~2 times larger than in pure neuronal cultures, a difference abolished by inhibition of glycogenolysis.[198] Although NA, together with many other transmitters, also is involved in memory retrieval[199] and astrocytes are involved in this process,[200] no information seems to be available about the role of glycogenolysis during retrieval.

Release of gliotransmitters is necessary for fear memory consolidation in the basolateral amygdala,[201] and administration of DAB into this structure in rats reduces a cocaine-induced conditioned place preference memory.[202] The latter effect could be counteracted by lactate administration in the lateral amygdala, which also altered expression of both $ERK_{1/2}$ and the transcription factor *Zif*268, which is not only a synaptic plasticity-related gene[202] in neurons but also is expressed in glia.[203] Although $ERK_{1/2}$ is expressed in both neurons and glia its involvement in NA-mediated signaling in astrocytes discussed above is of great importance (Fig. 2.1), and stimulation of the dopamine D_2 receptors (which occurs during cocaine administration)[204,205] seems to activate $ERK_{1/2}$ in a similar manner.[206] An abundant amount of studies show effects of cocaine on astrocytes or the importance of glycogenolysis to maintain cocaine memory.[202,207–210] As extracellular lactate enters astrocytes two-to-four fold faster than neurons,[211] it is likely that at least some of the administered lactate rather than exclusively being accumulated into neurons[202] is taken up into astrocytes, where it might counteract some of the many effects of inhibition of glycogenolysis discussed above, and LC stimulation by extracellular signaling[191,192,212] might also be important effect in astrocytes.

β_1-ADRENERGIC STIMULATION OF THE ASTROCYTIC NA^+,K^+-ATPASE

Both the astrocytic and the neuronal Na^+,K^+-ATPases are stimulated by NA, but only when $[K^+]_o$ is not simultaneously elevated or decreased (Fig. 2.6A).[213] This also applies to K^+ uptake in astrocytes mediated by either a 5-mM increase in $[K^+]_o$ or 10-μM isoproterenol, with no additive effect of the two stimuli (Fig. 2.6B)[180] due to mechanisms discussed elsewhere.[214] For this reason, NA is normally of no importance for the initial astrocytic clearance of increased extracellular K^+ known to occur after neuronal excitation,[215–220] but it is important when the K^+-stimulation is inhibited[221,222] as discussed in depth in the chapter by Peng et al. However, noradrenergic stimulation of astrocytic uptake of Na^+, K^+, $2Cl^-$ and water by the Na^+,K^+-ATPase-dependent[223] cotransporter NKCC1 is crucial for the postexcitatory undershoot.[105,180]

Stimulation of the astrocytic Na^+,K^+-ATPase is β-adrenergic[213] and exerted on the β_1-AR (Table 2.2),[63] whereas the neuronal stimulation is α_1-adrenergic.[224] β_1-Adrenergic stimulation of uptake of the K^+ analogue rubidium has also been shown in pig hearts.[225] Baskey et al. found no stimulation of the astrocytic Na^+,K^+-ATPase by 100 μM adrenaline,[224] which must cause a huge increase in astrocytic $[Ca^{2+}]_i$ (Fig. 2.3B). Too high $[Ca^{2+}]_i$ inhibits adrenergic stimulation of decarboxylations in the astrocytic TCA cycle.[116,119]

IMPORTANCE OF ADRENERGIC STIMULATION DURING CULTURING OF ASTROCYTES

Primary cultures of astrocytes including our own[64,226–228] are generally prepared from neonatal mice or rats. They therefore fail to benefit from effects of noradrenergic innervation reaching the cerebral hemispheres around the time of birth.[229] dBcAMP, a cAMP analog that penetrates the cell membrane[230] causes extension of processes[231–234] (Fig. 2.7A), which is EPAC-dependent,[69] and it also induces many functional changes, including the ability of elevated extracellular K^+ concentrations to activate glycogenolysis (as it does in the brain in vivo[157]) and enhancement of isoproterenol-stimulated glycogenolysis[235,236] (Fig. 2.7B) as well as depolarization-mediated L-channel opening.[237] Moreover, the α_2-AR agonists dexmedetomidine and clonidine only increase $[Ca^{2+}]_i$ in dBcAMP-treated cells.[72] Our cultures are therefore routinely treated with dBcAMP from the age of two weeks[228] making them more similar to their in vivo counterparts.[64]

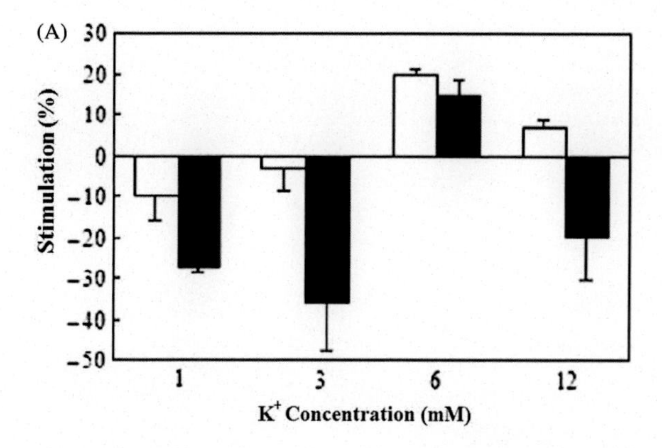

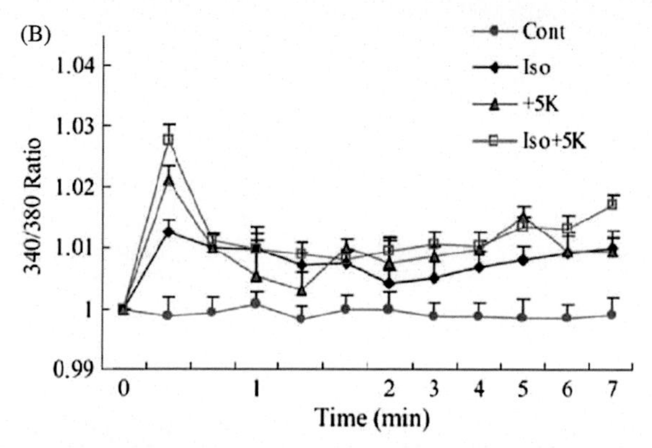

FIGURE 2.6 (A) Na^+,K^+-ATPase stimulation (indicated in the upward direction) and inhibition (indicated in the downward direction) by 10-μM noradrenaline at 1, 3, 6, or 12 mM extracellular K^+ in homogenates of primary cultures of mouse cerebral cortical astrocytes (open bars) or neurons (solid bars). The activity in the same homogenates in the absence of any transmitter equals 0%. Results are averages ± SEM for four to seven different homogenates. (B) K^+ uptake into astrocytes measured as increase in intracellular K^+ concentration, determined in arbitrary units based on enhanced fluorescence of the K^+ indicator benzofuran isophthalate (PBFI) in the presence of pluronic acid. After incubation of PBFI-AM-loaded cells in saline solution and a subsequent wash, the cells were, from zero time, incubated either in a solution to which 1-μM isoproterenol had been added, or in a solution to which 5-mM KCl and 1-μM isoproterenol had been jointly added. The results for the control group represent mean ± SEM of 42 cells, for those with added KCl mean ± SEM of 43 cells, for those with added isoproterenol mean ± SEM of 53 cells, and for those with joint addition of KCl and isoproterenol mean ± SEM of 66 cells. *(A) From Hajek I., Subbarao K.V., Hertz L. Acute and chronic effects of potassium and noradrenaline on Na + , K + -ATPase activity in cultured mouse neurons and astrocytes. Neurochem Int. 1996;28(3):335–342, (B) from Hertz L., Gerkau N.J., Xu J., et al. Roles of astrocytic Na(+),K(+)-ATPase and glycogenolysis for K(+) homeostasis in mammalian brain. J Neurosci Res. 2015;93(7):1019–1030.*

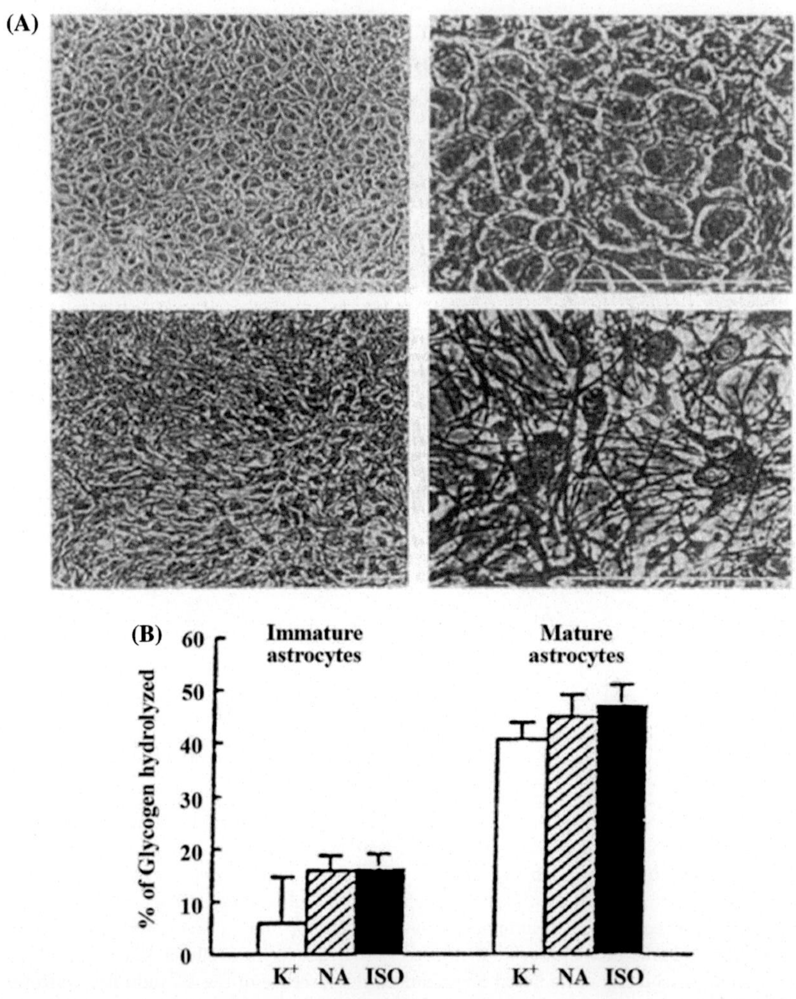

FIGURE 2.7 (A) Phase contrast micrographs of 4-week-old primary cultures of astrocytes from neonatal mouse brain grown in the absence (upper row) or presence (lower row) of a cAMP analog dBcAMP (0.25 mM) from the age of 2 weeks. Two different magnifications are shown, with bars in right lower corner indicating 200 μm. (B) An increase of extracellular K^+ concentration (to 50 mM) has no significant effect on glycogenolysis (measured by trapping of 6-[^{3}H]glycogen in K^+-treated and untreated cultures) in astrocytes that had not been grown with dBcAMP ("Immature") but causes a substantial increase in cells cultured with dBcAMP for at least one week ("Mature"). Noradrenaline (NA) and isoproterenol (ISO) (25 μM) have also much larger effect in the dBcAMP-treated than in the untreated cells. *(A) From Hertz L., Juurlink B.H.J., Fosmark H., Schousboe A. Methodological appendix. In: Pfeiffer S.E., ed. Neuroscience Approached Through Cell Culture. Vol 1. Boca Raton, Florida: CRC Press; 1982:175–186, (B) From Hertz L., Code W.E. Calcium channel signalling in astrocytes. In: Paoletti R., Godfraind T., Vankoullen P.M., eds. Calcium Antagonists: Pharmacology and Clinical Research. Boston: Kluwer; 1993:205–213.*

CONCLUSIONS

Noradrenergic signaling, acting on different receptor subtypes, exerts profound effects both on mature and developing gray matter astrocytes. At least in the mature cells, signaling is complex and acts on many different processes, including glucose metabolism and K^+ uptake. The effects on glycogen turnover have been especially thoroughly studied during learning, where glycogenolysis supports many astrocytic processes, and release of glycogenolysis-produced lactate exerts extracellular signaling effects and may sustain function of dendritic spines. NA and elevated $[K^+]_o$ have many effects in common, and both depend on elevated $[Ca^{2+}]_i$. However, the effects are not identical as also indicated by the fluoxetine-induced decrease in SOCE-mediated Ca^{2+} entry which will decrease effects of adrenergic agonists but not those of increases in $[K^+]_o$.

ABBREVIATIONS

5-HT$_{2B}$ serotonin receptor 2B
αKGDHc α-ketoglutarate dehydrogenase complex
ATP adenosine triphosphate
AR adrenergic receptor
$[Ca^{2+}]_i$ free cytosolic Ca^{2+} concentration
cAMP adenosine 3′:5′-cyclic monophosphate
CB1 cannabinoid receptor 1
CNS central nervous system
DAB 1,4-dideoxy-1,4-imino-D-arabinitol
DAG diacylglycerol
dBcAMP dibutyryl cyclic AMP
EGF epidermal growth factor
EGFR epidermal growth factor receptor
EPAC exchange factor directly activated by cAMP
ERK$_{1/2}$ extracellular regulated kinases 1 and 2
FACS fluorescence-activated cell sorting
FRET fluorescence resonance energy transfer
GABA γ-aminobutyric acid
GFAP glial fibrillary acidic protein
GP glycogen phosphorylase
GPCRs G-protein coupled receptors
GS glycogen synthase
GSK glycogen synthase kinase
GSK-3 glycogen synthase kinase-3
HB-EGF heparin-binding epidermal growth factor
HCN hyperpolarization-activated cyclic nucleotide-gated
IP$_3$ inositol trisphosphate
ISO isoproterenol or isoprenaline
$[K^+]_o$ extracellular K^+ concentration

LC locus coeruleus
Li$^+$ lithium ion
LTP long-term potentiation
MAPK mitogen-activated protein kinase
MCT monocarboxylate transporters
mEPSC miniature excitatory postsynaptic currents
NA noradrenaline or norepinephrine
NHE Na^+/H^+ exchanger
NKCC1 $Na^+,K^+,2Cl^-$ cotransporter 1
PKA protein kinase A
Popdc Popeye domain-containing
PDHc pyruvate dehydrogenase complex
PKC protein kinase C
PI3K phosphatidylinositol-3 kinase
PIP$_2$ phosphatidylinositide-4,5-bisphosphate
PLC phospholipase C
SOCEs store-operated channels
Na^+,K^+-ATPase sodium–potassium adenosine triphosphatase
TCA tricarboxylic acid
UDP uridine diphosphate
VIP vasoactive intestinal peptide

REFERENCES

1. Clark RB, Perkins JP. Regulation of adenosine 3′:5′-cyclic monophosphate concentration in cultured human astrocytoma cells by catecholamines and histamine. *Proc Natl Acad Sci U S A.* 1971;68(11):2757–2760.
2. Gilman AG, Nirenberg M. Regulation of adenosine 3′,5′-cyclic monophosphate metabolism in cultured neuroblastoma cells. *Nature.* 1971;234(5328):356–358.
3. Cahoy JD, Emery B, Kaushal A, et al. A transcriptome database for astrocytes, neurons, and oligodendrocytes: a new resource for understanding brain development and function. *J Neurosci.* 2008;28(1):264–278.
4. Kimelberg HK. Functions of mature mammalian astrocytes: a current view. *Neuroscientist.* 2010;16(1):79–106.
5. Hertz L, Chen Y, Gibbs ME, Zang P, Peng L. Astrocytic adrenoceptors: a major drug target in neurological and psychiatric disorders? *Curr Drug Targets CNS Neurol Disord.* 2004;3(3): 239–267.
6. Hertz L, Lovatt D, Goldman SA, Nedergaard M. Adrenoceptors in brain: cellular gene expression and effects on astrocytic metabolism and [Ca(2 +)]i. *Neurochem Int.* 2010;57(4): 411–420.
7. Aoki C. Beta-adrenergic receptors: astrocytic localization in the adult visual cortex and their relation to catecholamine axon terminals as revealed by electron microscopic immunocytochemistry. *J Neurosci.* 1992;12(3):781–792.
8. Aoki C. Differential timing for the appearance of neuronal and astrocytic beta-adrenergic receptors in the developing rat visual cortex as revealed by light and electron-microscopic immunocytochemistry. *Vis Neurosci.* 1997;14(6):1129–1142.
9. Aoki C, Joh TH, Pickel VM. Ultrastructural localization of beta-adrenergic receptor-like immunoreactivity in the cortex and neostriatum of rat brain. *Brain Res.* 1987;437(2):264–282.

10. Aoki C, Lubin M, Fenstemaker S. Columnar activity regulates astrocytic beta-adrenergic receptor-like immunoreactivity in V1 of adult monkeys. *Vis Neurosci.* 1994;11(1):179–187.
11. Aoki C, Pickel VM. Ultrastructural relations between beta-adrenergic receptors and catecholaminergic neurons. *Brain Res Bull.* 1992;29(3-4):257–263.
12. Aoki C, Pickel VM. C-terminal tail of beta-adrenergic receptors: immunocytochemical localization within astrocytes and their relation to catecholaminergic neurons in N. tractus solitarii and area postrema. *Brain Res.* 1992;571(1):35–49.
13. Shao Y, Sutin J. Expression of adrenergic receptors in individual astrocytes and motor neurons isolated from the adult rat brain. *Glia.* 1992;6(2):108–117.
14. Shao YP, Sutin J. Noradrenergic facilitation of motor neurons: localization of adrenergic receptors in neurons and nonneuronal cells in the trigeminal motor nucleus. *Exp Neurol.* 1991;114(2):216–227.
15. Sutin J, Shao Y. Resting and reactive astrocytes express adrenergic receptors in the adult rat brain. *Brain Res Bull.* 1992;29(3–4):277–284.
16. Liu Y, Jia W, Strosberg AD, Cynader M. Development and regulation of beta adrenergic receptors in kitten visual cortex: an immunocytochemical and autoradiographic study. *Brain Res.* 1993;632(1–2):274–286.
17. Nakadate K, Imamura K, Watanabe Y. Effects of monocular deprivation on the expression pattern of alpha-1 and beta-1 adrenergic receptors in the kitten visual cortex. *Neurosci Res.* 2001;40(2):155–162.
18. Milner TA, Lee A, Aicher SA, Rosin DL. Hippocampal alpha2a-adrenergic receptors are located predominantly presynaptically but are also found postsynaptically and in selective astrocytes. *J Comp Neurol.* 1998;395(3):310–327.
19. Glass MJ, Colago EE, Pickel VM. Alpha-2A-adrenergic receptors are present on neurons in the central nucleus of the amygdala that project to the dorsal vagal complex in the rat. *Synapse.* 2002;46(4):258–268.
20. Glass MJ, Huang J, Aicher SA, Milner TA, Pickel VM. Subcellular localization of alpha-2A-adrenergic receptors in the rat medial nucleus tractus solitarius: regional targeting and relationship with catecholamine neurons. *J Comp Neurol.* 2001;433(2):193–207.
21. Lee A, Rosin DL, Van Bockstaele EJ. alpha2A-adrenergic receptors in the rat nucleus locus coeruleus: subcellular localization in catecholaminergic dendrites, astrocytes, and presynaptic axon terminals. *Brain Res.* 1998;795(1–2):157–169.
22. Hutchinson DS, Catus SL, Merlin J, Summers RJ, Gibbs ME. alpha(2)-Adrenoceptors activate noradrenaline-mediated glycogen turnover in chick astrocytes. *J Neurochem.* 2011;117(5):915–926.
23. Nakadate K, Imamura K, Watanabe Y. Cellular and subcellular localization of alpha-1 adrenoceptors in the rat visual cortex. *Neuroscience.* 2006;141(4):1783–1792.
24. Duffy S, MacVicar BA. Adrenergic calcium signaling in astrocyte networks within the hippocampal slice. *J Neurosci.* 1995;15(8):5535–5550.
25. Kirischuk S, Tuschick S, Verkhratsky A, Kettenmann H. Calcium signalling in mouse Bergmann glial cells mediated by alpha1-adrenoreceptors and H1 histamine receptors. *Eur J Neurosci.* 1996;8(6):1198–1208.
26. Kulik A, Haentzsch A, Luckermann M, Reichelt W, Ballanyi K. Neuron-glia signaling via alpha(1) adrenoceptor-mediated Ca(2 +) release in Bergmann glial cells in situ. *J Neurosci.* 1999;19(19):8401–8408.
27. Thorlin T, Eriksson PS, Ronnback L, Hansson E. Receptor-activated Ca2 + increases in vibrodissociated cortical astrocytes: a nonenzymatic method for acute isolation of astrocytes. *J Neurosci Res.* 1998;54(3):390–401.

28. Lovatt D, Sonnewald U, Waagepetersen HS, et al. The transcriptome and metabolic gene signature of protoplasmic astrocytes in the adult murine cortex. *J Neurosci.* 2007;27(45):12255–12266.
29. Feng G, Mellor RH, Bernstein M, et al. Imaging neuronal subsets in transgenic mice expressing multiple spectral variants of GFP. *Neuron.* 2000;28(1):41–51.
30. Seki M, Nawa H, Morioka T, et al. Establishment of a novel enzyme-linked immunosorbent assay for Thy-1; quantitative assessment of neuronal degeneration. *Neurosci Lett.* 2002;329(2):185–188.
31. Atkinson BN, Minneman KP. Multiple adrenergic receptor subtypes controlling cyclic AMP formation: comparison of brain slices and primary neuronal and glial cultures. *J Neurochem.* 1991;56(2):587–595.
32. O'Donnell J, Zeppenfeld D, McConnell E, Pena S, Nedergaard M. Norepinephrine: a neuromodulator that boosts the function of multiple cell types to optimize CNS performance. *Neurochem Res.* 2012;37(11):2496–2512.
33. Schambra UB, Mackensen GB, Stafford-Smith M, Haines DE, Schwinn DA. Neuron specific alpha-adrenergic receptor expression in human cerebellum: implications for emerging cerebellar roles in neurologic disease. *Neuroscience.* 2005;135(2):507–523.
34. Roychowdhury S, Zwierzchowski AN, Garcia-Oscos F, Olguin RC, Delgado RS, Atzori M. Layer- and area-specificity of the adrenergic modulation of synaptic transmission in the rat neocortex. *Neurochem Res.* 2014;39(12):2377–2384.
35. Hubbard JA, Hsu MS, Seldin MM, Binder DK. Expression of the astrocyte water channel aquaporin-4 in the mouse brain. *ASN Neuro.* 2015;7(5). Available from: http://dx.doi.org/10.1177/1759091415605486.
36. Summers RJ, Papaioannou M, Harris S, Evans BA. Expression of beta 3-adrenoceptor mRNA in rat brain. *Br J Pharmacol.* 1995;116(6):2547–2548.
37. Hutchinson DS, Summers RJ, Gibbs ME. Beta2- and beta3-adrenoceptors activate glucose uptake in chick astrocytes by distinct mechanisms: a mechanism for memory enhancement? *J Neurochem.* 2007;103(3):997–1008.
38. Catus SL, Gibbs ME, Sato M, Summers RJ, Hutchinson DS. Role of beta-adrenoceptors in glucose uptake in astrocytes using beta-adrenoceptor knockout mice. *Br J Pharmacol.* 2011;162(8):1700–1715.
39. Bekar LK, He W, Nedergaard M. Locus coeruleus alpha-adrenergic-mediated activation of cortical astrocytes in vivo. *Cereb Cortex.* 2008;18(12):2789–2795.
40. Ding F, O'Donnell J, Thrane AS, et al. alpha1-Adrenergic receptors mediate coordinated Ca2 + signaling of cortical astrocytes in awake, behaving mice. *Cell Calcium.* 2013;54(6):387–394.
41. Braak H, Del Tredici K. Amyloid-beta may be released from non-junctional varicosities of axons generated from abnormal tau-containing brainstem nuclei in sporadic Alzheimer's disease: a hypothesis. *Acta Neuropathol.* 2013;126(2):303–306.
42. Hertz L, Chen Y, Waagepetersen HS. Effects of ketone bodies in Alzheimer's disease in relation to neural hypometabolism, beta-amyloid toxicity, and astrocyte function. *J Neurochem.* 2015;134(1):7–20.
43. Feinstein DL, Kalinin S, Braun D. Causes, consequences, and cures for neuroinflammation mediated via the locus coeruleus: noradrenergic signaling system. *J Neurochem.* 2016;. Available from: http://dx.doi.org/10.1111/jnc.13447.
44. Schmalbruch IK, Linde R, Paulson OB, Madsen PL. Activation-induced resetting of cerebral metabolism and flow is abolished by beta-adrenergic blockade with propranolol. *Stroke.* 2002;33(1):251–255.

45. Horvat A, Zorec R, Vardjan N. Adrenergic stimulation of single rat astrocytes results in distinct temporal changes in intracellular Ca(2 +) and cAMP-dependent PKA responses. *Cell Calcium*. 2016;59(4):156–163.
46. Goaillard JM, Vincent P. Serotonin suppresses the slow afterhyperpolarization in rat intralaminar and midline thalamic neurones by activating 5-HT(7) receptors. *J Physiol*. 2002;541(Pt 2):453–465.
47. Lezoualc'h F, Fazal L, Laudette M, Conte C. Cyclic AMP sensor EPAC proteins and their role in cardiovascular function and disease. *Circ Res*. 2016;118(5):881–897.
48. Steinberg SF. beta(2)-Adrenergic receptor signaling complexes in cardiomyocyte caveolae/lipid rafts. *J Mol Cell Cardiol*. 2004;37(2):407–415.
49. Maudsley S, Pierce KL, Zamah AM, et al. The beta(2)-adrenergic receptor mediates extracellular signal-regulated kinase activation via assembly of a multi-receptor complex with the epidermal growth factor receptor. *J Biol Chem*. 2000;275(13):9572–9580.
50. Luttrell LM, Lefkowitz RJ. The role of beta-arrestins in the termination and transduction of G-protein-coupled receptor signals. *J Cell Sci*. 2002;115(Pt 3):455–465.
51. Lefkowitz RJ. A brief history of G-protein coupled receptors (Nobel Lecture). *Angew Chem Int Ed Engl*. 2013;52(25):6366–6378.
52. Cheng P, Alberts I, Li X. The role of ERK1/2 in the regulation of proliferation and differentiation of astrocytes in developing brain. *Int J Dev Neurosci*. 2013;31(8):783–789.
53. Pierce KL, Tohgo A, Ahn S, Field ME, Luttrell LM, Lefkowitz RJ. Epidermal growth factor (EGF) receptor-dependent ERK activation by G protein-coupled receptors: a co-culture system for identifying intermediates upstream and downstream of heparin-binding EGF shedding. *J Biol Chem*. 2001;276(25):23155–23160.
54. Peng L. Transactivation in astrocytes as a novel mechanism of neuroprotection. In: Hertz L, ed. *Non-neuronal Cells of the Nervous System: Function and Dysfunction*. Vol II. Amsterdam: Elsevier; 2004:503–518.
55. Peng L, Li B, Du T, et al. Astrocytic transactivation by alpha2A-adrenergic and 5-HT2B serotonergic signaling. *Neurochem Int*. 2010;57(4):421–431.
56. Peng L, Du T, Xu J, et al. Adrenergic and V1-ergic agonists/antagonists affecting recovery from brain trauma in the Lund project act on astrocytes. *Curr Signal Trans Ther*. 2012;7:43–55.
57. Li B, Du T, Li H, et al. Signalling pathways for transactivation by dexmedetomidine of epidermal growth factor receptors in astrocytes and its paracrine effect on neurons. *Br J Pharmacol*. 2008;154(1):191–203.
58. Du T, Li B, Li H, Li M, Hertz L, Peng L. Signaling pathways of isoproterenol-induced ERK1/2 phosphorylation in primary cultures of astrocytes are concentration-dependent. *J Neurochem*. 2010;115(4):1007–1023.
59. Daaka Y, Luttrell LM, Lefkowitz RJ. Switching of the coupling of the beta2-adrenergic receptor to different G proteins by protein kinase A. *Nature*. 1997;390(6655):88–91.
60. Baillie GS, Sood A, McPhee I, et al. beta-Arrestin-mediated PDE4 cAMP phosphodiesterase recruitment regulates beta-adrenoceptor switching from Gs to Gi. *Proc Natl Acad Sci U S A*. 2003;100(3):940–945.
61. Zhang M, Shan X, Gu L, Hertz L, Peng L. Dexmedetomidine causes neuroprotection via astrocytic α2-adrenergic receptor stimulation and HB-EGF release. *J Anesthesiol Clin Sci*. 2013;2:6.
62. Kim J, Eckhart AD, Eguchi S, Koch WJ. Beta-adrenergic receptor-mediated DNA synthesis in cardiac fibroblasts is dependent on transactivation of the epidermal growth factor

receptor and subsequent activation of extracellular signal-regulated kinases. *J Biol Chem.* 2002;277(35):32116–32123.

63. Song D, Xu J, Hertz L, Peng L. Regulatory volume increase in astrocytes exposed to hypertonic medium requires beta1-adrenergic Na(+) /K(+)-ATPase stimulation and glycogenolysis. *J Neurosci Res.* 2015;93(1):130–139.
64. Hertz L., Chen Y., Song D. Astrocyte cultures mimicking brain astrocytes in gene expression, signaling, metabolism and K uptake and showing astrocytic gene expression overlooked by immunohistochemistry and in situ hybridization. *Neurochem Res.* 2017; 42 (1):254-271.
65. Zorec R, Horvat A, Vardjan N, Verkhratsky A. Memory formation shaped by astroglia. *Front Integr Neurosci.* 2015;9:56.
66. Benarroch EE. HCN channels: function and clinical implications. *Neurology.* 2013;80(3): 304–310.
67. Brand T, Poon KL, Simrick S, Schindler RFR. The popeye domain containing genes and cAMP signaling. *J Cardiovasc Dev Dis.* 2014;1(1):121–133.
68. Brand T. The Popeye domain-containing gene family. *Cell Biochem Biophys.* 2005;43(1): 95–103.
69. Vardjan N, Kreft M, Zorec R. Dynamics of beta-adrenergic/cAMP signaling and morphological changes in cultured astrocytes. *Glia.* 2014;62(4):566–579.
70. Soeder KJ, Snedden SK, Cao W, et al. The beta3-adrenergic receptor activates mitogen-activated protein kinase in adipocytes through a Gi-dependent mechanism. *J Biol Chem.* 1999;274(17):12017–12022.
71. Abdel-Latif AA. Calcium-mobilizing receptors, polyphosphoinositides, generation of second messengers and contraction in the mammalian iris smooth muscle: historical perspectives and current status. *Life Sci.* 1989;45(9):757–786.
72. Enkvist MO, Hamalainen H, Jansson CC, et al. Coupling of astroglial alpha 2-adrenoreceptors to second messenger pathways. *J Neurochem.* 1996;66(6):2394–2401.
73. Ozawa E. Activation of muscular phosphorylase b kinase by a minute amount of Ca ion. *J Biochem.* 1972;71(2):321–331.
74. Ozawa E. Regulation of phosphorylase kinase by low concentrations of Ca ions upon muscle contraction: the connection between metabolism and muscle contraction and the connection between muscle physiology and Ca-dependent signal transduction. *Proc Jpn Acad Ser B Phys Biol Sci.* 2011;87(8):486–508.
75. Wong YH, Kalmbach SJ, Hartman BK, Sherman WR. Immunohistochemical staining and enzyme activity measurements show myo-inositol-1-phosphate synthase to be localized in the vasculature of brain. *J Neurochem.* 1987;48(5):1434–1442.
76. Wolfson M, Bersudsky Y, Zinger E, Simkin M, Belmaker RH, Hertz L. Chronic treatment of human astrocytoma cells with lithium, carbamazepine or valproic acid decreases inositol uptake at high inositol concentrations but increases it at low inositol concentrations. *Brain Res.* 2000;855(1):158–161.
77. Song D, Du T, Li B, et al. Astrocytic alkalinization by therapeutically relevant lithium concentrations: implications for myo-inositol depletion. *Psychopharmacology (Berl).* 2008;200(2):187–195.
78. Song D, Li B, Yan E, et al. Chronic treatment with anti-bipolar drugs causes intracellular alkalinization in astrocytes, altering their functions. *Neurochem Res.* 2012;37(11):2524–2540.
79. Song D, Man Y, Li B, Xu J, Hertz L, Peng L. Comparison between drug-induced and K (+)-induced changes in molar acid extrusion fluxes (JH(+)) and in energy consumption rates in astrocytes. *Neurochem Res.* 2013;38(11):2364–2374.

80. Brookes N, Turner RJ. K(+)-induced alkalinization in mouse cerebral astrocytes mediated by reversal of electrogenic Na(+)-HCO3-cotransport. *Am J Physiol.* 1994;267(6 Pt 1): C1633–C1640.
81. Fu H, Li B, Hertz L, Peng L. Contributions in astrocytes of SMIT1/2 and HMIT to myo-inositol uptake at different concentrations and pH. *Neurochem Int.* 2012;61(2):187–194.
82. Hertz L, Song D, Li B, et al. Signal transduction in astrocytes during chronic or acute treatment with drugs (SSRIs, antibipolar drugs, GABA-ergic drugs, and benzodiazepines) ameliorating mood disorders. *J Signal Transduct.* 2014;2014:593934.
83. Peng L, Li B, Verkhratsky A. Targeting astrocytes in bipolar disorder. *Expert Rev Neurother.* 2016;16(6):762–776.
84. Berridge MJ. Inositol trisphosphate and diacylglycerol as second messengers. *Biochem J.* 1984;220(2):345–360.
85. Chen Y, Hertz L. Inhibition of noradrenaline stimulated increase in [Ca2 +]i in cultured astrocytes by chronic treatment with a therapeutically relevant lithium concentration. *Brain Res.* 1996;711(1–2):245–248.
86. Li B, Dong L, Fu H, Wang B, Hertz L, Peng L. Effects of chronic treatment with fluoxetine on receptor-stimulated increase of [Ca2 +]i in astrocytes mimic those of acute inhibition of TRPC1 channel activity. *Cell Calcium.* 2011;50(1):42–53.
87. Du T, Liang C, Li B, Hertz L, Peng L. Chronic fluoxetine administration increases expression of the L-channel gene Cav1.2 in astrocytes from the brain of treated mice and in culture and augments K(+)-induced increase in [Ca(2 +)]i. *Cell Calcium.* 2014;55(3):166–174.
88. Yan E, Li B, Gu L, Hertz L, Peng L. Mechanisms for L-channel-mediated increase in [Ca (2 +)]i and its reduction by anti-bipolar drugs in cultured astrocytes combined with its mRNA expression in freshly isolated cells support the importance of astrocytic L-channels. *Cell Calcium.* 2013;54(5):335–342.
89. Mizobe T, Maghsoudi K, Sitwala K, Tianzhi G, Ou J, Maze M. Antisense technology reveals the alpha2A adrenoceptor to be the subtype mediating the hypnotic response to the highly selective agonist, dexmedetomidine, in the locus coeruleus of the rat. *J Clin Invest.* 1996;98(5):1076–1080.
90. Ma D, Hossain M, Rajakumaraswamy N, et al. Dexmedetomidine produces its neuroprotective effect via the alpha 2A-adrenoceptor subtype. *Eur J Pharmacol.* 2004;502 (1–2):87–97.
91. Pohjanoksa K, Jansson CC, Luomala K, Marjamaki A, Savola JM, Scheinin M. Alpha2-adrenoceptor regulation of adenylyl cyclase in CHO cells: dependence on receptor density, receptor subtype and current activity of adenylyl cyclase. *Eur J Pharmacol.* 1997;335(1): 53–63.
92. Takesono A, Zahner J, Blumer KJ, Nagao T, Kurose H. Negative regulation of alpha2-adrenergic receptor-mediated Gi signalling by a novel pathway. *Biochem J.* 1999;343 (Pt 1):77–85.
93. Tang C, Huang X, Kang F, et al. Intranasal dexmedetomidine on stress hormones, inflammatory markers, and postoperative analgesia after functional endoscopic sinus surgery. *Mediators Inflamm.* 2015;2015:939431.
94. Marshall I, Burt RP, Chapple CR. Signal transduction pathways associated with alpha1-adrenoceptor subtypes in cells and tissues including human prostate. *Eur Urol.* 1999;36 (Suppl 1):42–47:discussion 65
95. Huang Y, Wright CD, Merkwan CL, et al. An alpha1A-adrenergic-extracellular signal-regulated kinase survival signaling pathway in cardiac myocytes. *Circulation.* 2007;115(6): 763–772.

96. Rohde S, Sabri A, Kamasamudran R, Steinberg SF. The alpha(1)-adrenoceptor subtype- and protein kinase C isoform-dependence of Norepinephrine's actions in cardiomyocytes. *J Mol Cell Cardiol.* 2000;32(7):1193–1209.
97. Jensen BC, O'Connell TD, Simpson PC. Alpha-1-adrenergic receptors in heart failure: the adaptive arm of the cardiac response to chronic catecholamine stimulation. *J Cardiovasc Pharmacol.* 2014;63(4):291–301.
98. Ulu N, Gurdal H, Landheer SW, et al. alpha1-Adrenoceptor-mediated contraction of rat aorta is partly mediated via transactivation of the epidermal growth factor receptor. *Br J Pharmacol.* 2010;161(6):1301–1310.
99. Prebil M, Vardjan N, Jensen J, Zorec R, Kreft M. Dynamic monitoring of cytosolic glucose in single astrocytes. *Glia.* 2011;59(6):903–913.
100. Gibbs ME, Hutchinson DS, Summers RJ. Role of beta-adrenoceptors in memory consolidation: beta3-adrenoceptors act on glucose uptake and beta2-adrenoceptors on glycogenolysis. *Neuropsychopharmacology.* 2008;33(10):2384–2397.
101. Hertz L, Dienel GA. Energy metabolism in the brain. *Int Rev Neurobiol.* 2002;51:1–102.
102. Hertz L, Peng L, Dienel GA. Energy metabolism in astrocytes: high rate of oxidative metabolism and spatiotemporal dependence on glycolysis/glycogenolysis. *J Cereb Blood Flow Metab.* 2007;27(2):219–249.
103. van den Berg CJ, Garfinkel D. A stimulation study of brain compartments. Metabolism of glutamate and related substances in mouse brain. *Biochem J.* 1971;123(2): 211–218.
104. Sibson NR, Dhankhar A, Mason GF, Rothman DL, Behar KL, Shulman RG. Stoichiometric coupling of brain glucose metabolism and glutamatergic neuronal activity. *Proc Natl Acad Sci U S A.* 1998;95(1):316–321.
105. Hertz L. The glutamate–glutamine (GABA) cycle: importance of late postnatal development and potential reciprocal interactions between biosynthesis and degradation. *Front Endocrinol (Lausanne).* 2013;4:59.
106. Hertz L, Rothman DL. Glucose, lactate, β-hydroxybutyrate, acetate, GABA, and succinate as substrates for synthesis of glutamate and GABA in the glutamine-glutamate/GABA cycle. In: Sonnewald U, Schousboe A, eds. *Adv Neurobiol.* Springer pp 9-42; 2016.
107. Gibbs ME, Lloyd HG, Santa T, Hertz L. Glycogen is a preferred glutamate precursor during learning in 1-day-old chick: biochemical and behavioral evidence. *J Neurosci Res.* 2007;85(15):3326–3333.
108. Oz G, Berkich DA, Henry PG, et al. Neuroglial metabolism in the awake rat brain: CO_2 fixation increases with brain activity. *J Neurosci.* 2004;24(50):11273–11279.
109. Serres S, Raffard G, Franconi JM, Merle M. Close coupling between astrocytic and neuronal metabolisms to fulfill anaplerotic and energy needs in the rat brain. *J Cereb Blood Flow Metab.* 2008;28(4):712–724.
110. Mangia S, Giove F, Dinuzzo M. Metabolic pathways and activity-dependent modulation of glutamate concentration in the human brain. *Neurochem Res.* 2012;37(11):2554–2561.
111. Wilson NR, Kang J, Hueske EV, et al. Presynaptic regulation of quantal size by the vesicular glutamate transporter VGLUT1. *J Neurosci.* 2005;25(26):6221–6234.
112. McKenna MC. Glutamate pays its own way in astrocytes. *Front Endocrinol (Lausanne).* 2013;4:191.
113. Karaca M, Frigerio F, Migrenne S, et al. GDH-dependent glutamate oxidation in the brain dictates peripheral energy substrate distribution. *Cell Rep.* 2015;13(2):365–375.
114. Robinson MB, Jackson JG. Astroglial glutamate transporters coordinate excitatory signaling and brain energetics. *Neurochem Int.* 2016;98:56–71.

115. Danbolt NC, Furness DN, Zhou Y. Neuronal vs glial glutamate uptake: resolving the conundrum. *Neurochem Int*. 2016;98:29–45.
116. Subbarao KV, Hertz L. Stimulation of energy metabolism by alpha-adrenergic agonists in primary cultures of astrocytes. *J Neurosci Res*. 1991;28(3):399–405.
117. Salvi S. Protecting the myocardium from ischemic injury: a critical role for alpha(1)-adrenoreceptors? *Chest*. 2001;119(4):1242–1249.
118. Sugden PH, Newsholme EA. The effects of ammonium, inorganic phosphate and potassium ions on the activity of phosphofructokinases from muscle and nervous tissues of vertebrates and invertebrates. *Biochem J*. 1975;150(1):113–122.
119. Chen Y, Hertz L. Noradrenaline effects on pyruvate decarboxylation: correlation with calcium signaling. *J Neurosci Res*. 1999;58(4):599–606.
120. Chen Y, Zhao Z, Code WE, Hertz L. A correlation between dexmedetomidine-induced biphasic increases in free cytosolic calcium concentration and energy metabolism in astrocytes. *Anesth Analg*. 2000;91(2):353–357.
121. Vanderperre B, Bender T, Kunji ER, Martinou JC. Mitochondrial pyruvate import and its effects on homeostasis. *Curr Opin Cell Biol*. 2015;33:35–41.
122. McCommis KS, Finck BN. Mitochondrial pyruvate transport: a historical perspective and future research directions. *Biochem J*. 2015;466(3):443–454.
123. Oliver J, Sola MM, Salto R, Vargas AM. Stimulation of mitochondrial pyruvate transport in rat renal cortex by phenylephrine. *Life Sci*. 1990;47(5):401–406.
124. Thomas AP, Halestrap AP. The role of mitochondrial pyruvate transport in the stimulation by glucagon and phenylephrine of gluconeogenesis from L-lactate in isolated rat hepatocytes. *Biochem J*. 1981;198(3):551–560.
125. Adam PA, Haynes Jr. RC. Control of hepatic mitochondrial CO_2 fixation by glucagon, epinephrine, and cortisol. *J Biol Chem*. 1969;244(23):6444–6450.
126. Martin AD, Titheradge MA. Hormonal stimulation of gluconeogenesis through increased mitochondrial metabolic flux. *Biochem Soc Trans*. 1983;11(1):78–81.
127. Garrison JC, Borland MK. Regulation of mitochondrial pyruvate carboxylation and gluconeogenesis in rat hepatocytes via an alpha-adrenergic, adenosine 3′:5′-monophosphate-independent mechanism. *J Biol Chem*. 1979;254(4):1129–1133.
128. Subbarao KV, Hertz L. Noradrenaline induced stimulation of oxidative metabolism in astrocytes but not in neurons in primary cultures. *Brain Res*. 1990;527(2):346–349.
129. Hansson E, Ronnback L. Regulation of glutamate and GABA transport by adrenoceptors in primary astroglial cell cultures. *Life Sci*. 1989;44(1):27–34.
130. Gibbs ME, Hertz L. Importance of glutamate-generating metabolic pathways for memory consolidation in chicks. *J Neurosci Res*. 2005;81(2):293–300.
131. Gibbs ME. Role of glycogenolysis in memory and learning: regulation by noradrenaline, serotonin and ATP. *Front Integr Neurosci*. 2015;9:70.
132. Simpson PB, Mehotra S, Langley D, Sheppard CA, Russell JT. Specialized distributions of mitochondria and endoplasmic reticulum proteins define Ca2 + wave amplification sites in cultured astrocytes. *J Neurosci Res*. 1998;52(6):672–683.
133. McCormack JG, Denton RM. The role of mitochondrial Ca2 + transport and matrix Ca2 + in signal transduction in mammalian tissues. *Biochim Biophys Acta*. 1990;1018(2–3):287–291.
134. Rutter GA, Burnett P, Rizzuto R, et al. Subcellular imaging of intramitochondrial Ca2 + with recombinant targeted aequorin: significance for the regulation of pyruvate dehydrogenase activity. *Proc Natl Acad Sci USA*. 1996;93(11):5489–5494.
135. Gaspers LD, Thomas AP. Calcium-dependent activation of mitochondrial metabolism in mammalian cells. *Methods*. 2008;46(3):224–232.

136. Griffiths EJ, Rutter GA. Mitochondrial calcium as a key regulator of mitochondrial ATP production in mammalian cells. *Biochim Biophys Acta*. 2009;1787(11):1324–1333.
137. Duchen MR. Ca(2 +)-dependent changes in the mitochondrial energetics in single dissociated mouse sensory neurons. *Biochem J*. 1992;283(Pt 1):41–50.
138. Llorente-Folch I, Rueda CB, Pardo B, Szabadkai G, Duchen MR, Satrustegui J. The regulation of neuronal mitochondrial metabolism by calcium. *J Physiol*. 2015;593(16): 3447–3462.
139. Sickmann HM, Schousboe A, Fosgerau K, Waagepetersen HS. Compartmentation of lactate originating from glycogen and glucose in cultured astrocytes. *Neurochem Res*. 2005;30(10):1295–1304.
140. Xu J, Song D, Bai Q, Cai L, Hertz L, Peng L. Basic mechanism leading to stimulation of glycogenolysis by isoproterenol, EGF, elevated extracellular K + concentrations, or GABA. *Neurochem Res*. 2014;39(4):661–667.
141. Hertz L, Xu J, Song D, et al. Astrocytic glycogenolysis: mechanisms and functions. *Metab Brain Dis*. 2015;30(1):317–333.
142. Hertz L, Xu J, Peng L. Glycogenolysis and purinergic signaling. *Adv Neurobiol*. 2014;11: 31–54.
143. Quach TT, Duchemin AM, Rose C, Schwartz JC. [3H]glycogenolysis in brain slices mediated by beta-adrenoceptors: comparison of physiological response and [3H]dihydroalprenolol binding parameters. *Neuropharmacology*. 1988;27(6):629–635.
144. Sorg O, Magistretti PJ. Characterization of the glycogenolysis elicited by vasoactive intestinal peptide, noradrenaline and adenosine in primary cultures of mouse cerebral cortical astrocytes. *Brain Res*. 1991;563(1-2):227–233.
145. Sorg O, Pellerin L, Stolz M, Beggah S, Magistretti PJ. Adenosine triphosphate and arachidonic acid stimulate glycogenolysis in primary cultures of mouse cerebral cortical astrocytes. *Neurosci Lett*. 1995;188(2):109–112.
146. Xia M, Zhu Y. Signaling pathways of ATP-induced PGE2 release in spinal cord astrocytes are EGFR transactivation-dependent. *Glia*. 2011;59(4):664–674.
147. Darvesh AS, Gudelsky GA. Activation of 5-HT2 receptors induces glycogenolysis in the rat brain. *Eur J Pharmacol*. 2003;464(2-3):135–140.
148. Chen Y, Peng L, Zhang X, Stolzenburg JU, Hertz L. Further evidence that fluoxetine interacts with a 5-HT2C receptor in glial cells. *Brain Res Bull*. 1995;38(2):153–159.
149. Quach TT, Rose C, Duchemin AM, Schwartz JC. Glycogenolysis induced by serotonin in brain: identification of a new class of receptor. *Nature*. 1982;298(5872):373–375.
150. Hertz L, Rothman DL, Li B, Peng L. Chronic SSRI stimulation of astrocytic 5-HT2B receptors change multiple gene expressions/editings and metabolism of glutamate, glucose and glycogen: a potential paradigm shift. *Front Behav Neurosci*. 2015;9:25.
151. Zois CE, Harris AL. Glycogen metabolism has a key role in the cancer microenvironment and provides new targets for cancer therapy. *J Mol Med (Berl)*. 2016;94(2):137–154.
152. Brushia RJ, Walsh DA. Phosphorylase kinase: the complexity of its regulation is reflected in the complexity of its structure. *Front Biosci*. 1999;4:D618–D641.
153. Fernandez-Lopez A, Revilla V, Candelas MA, Gonzalez-Gil J, Diaz A, Pazos A. A comparative study of alpha2- and beta-adrenoceptor distribution in pigeon and chick brain. *Eur J Neurosci*. 1997;9(5):871–883.
154. Gibbs ME, Bowser DN, Hutchinson DS, Loiacono RE, Summers RJ. Memory processing in the avian hippocampus involves interactions between beta-adrenoceptors, glutamate receptors, and metabolism. *Neuropsychopharmacology*. 2008;33(12):2831–2846.

155. Subbarao KV, Hertz L. Effect of adrenergic agonists on glycogenolysis in primary cultures of astrocytes. *Brain Res.* 1990;536(1–2):220–226.
156. Ververken D, Van Veldhoven P, Proost C, Carton H, De Wulf H. On the role of calcium ions in the regulation of glycogenolysis in mouse brain cortical slices. *J Neurochem.* 1982;38(5):1286–1295.
157. Hof PR, Pascale E, Magistretti PJ. K + at concentrations reached in the extracellular space during neuronal activity promotes a Ca2 + -dependent glycogen hydrolysis in mouse cerebral cortex. *J Neurosci.* 1988;8(6):1922–1928.
158. Xu J, Song D, Xue Z, Gu L, Hertz L, Peng L. Requirement of glycogenolysis for uptake of increased extracellular K + in astrocytes: potential implications for K + homeostasis and glycogen usage in brain. *Neurochem Res.* 2013;38(3):472–485.
159. Gibbs ME, Hutchinson DS. Rapid turnover of glycogen in memory formation. *Neurochem Res.* 2012;37(11):2456–2463.
160. Mellerup ET, Rafaelsen OJ. Brain glycogen after intracisternal insulin injection. *J Neurochem.* 1969;16(5):777–781.
161. Dringen R, Hamprecht B. Glucose, insulin, and insulin-like growth factor I regulate the glycogen content of astroglia-rich primary cultures. *J Neurochem.* 1992;58(2):511–517.
162. Adeva-Andany MM, Gonzalez-Lucan M, Donapetry-Garcia C, Fernandez-Fernandez C, Ameneiros-Rodriguez E. Glycogen metabolism in humans. *BBA Clin.* 2016;5:85–100.
163. Hertz L, Li B, Song D, et al. Astrocytes as a 5-HT2B-mediated, SERT-independent SSRI target, slowly altering depression-associated genes and function. *Curr Signal Trans Ther.* 2012;7:65–80.
164. Cross DA, Alessi DR, Cohen P, Andjelkovich M, Hemmings BA. Inhibition of glycogen synthase kinase-3 by insulin mediated by protein kinase B. *Nature.* 1995;378(6559):785–789.
165. Embi N, Rylatt DB, Cohen P. Glycogen synthase kinase-3 from rabbit skeletal muscle. Separation from cyclic-AMP-dependent protein kinase and phosphorylase kinase. *Eur J Biochem.* 1980;107(2):519–527.
166. De Sarno P, Li X, Jope RS. Regulation of Akt and glycogen synthase kinase-3 beta phosphorylation by sodium valproate and lithium. *Neuropharmacology.* 2002;43(7): 1158–1164.
167. Plenge P. Acute lithium effects on rat brain glucose metabolism—in vivo. *Int Pharmacopsychiatry.* 1976;11(2):84–92.
168. Li X, Zhu W, Roh MS, Friedman AB, Rosborough K, Jope RS. In vivo regulation of glycogen synthase kinase-3beta (GSK3beta) by serotonergic activity in mouse brain. *Neuropsychopharmacology.* 2004;29(8):1426–1431.
169. Polter A, Beurel E, Yang S, et al. Deficiency in the inhibitory serine-phosphorylation of glycogen synthase kinase-3 increases sensitivity to mood disturbances. *Neuropsychopharmacology.* 2010;35(8):1761–1774.
170. Sanchez C, Galve-Roperh I, Rueda D, Guzman M. Involvement of sphingomyelin hydrolysis and the mitogen-activated protein kinase cascade in the Delta9-tetrahydrocannabinol-induced stimulation of glucose metabolism in primary astrocytes. *Mol Pharmacol.* 1998;54(5):834–843.
171. Scheller A, Kirchhoff F. Endocannabinoids and Heterogeneity of Glial Cells in Brain Function. *Front Integr Neurosci.* 2016;10:24.
172. Swanson RA, Morton MM, Sagar SM, Sharp FR. Sensory stimulation induces local cerebral glycogenolysis: demonstration by autoradiography. *Neuroscience.* 1992;51(2):451–461.
173. Ibrahim MZ. Glycogen and its related enzymes of metabolism in the central nervous system. *Adv Anat Embryol Cell Biol.* 1975;52(1):3–89.

174. Pfeiffer-Guglielmi B, Broer S, Broer A, Hamprecht B. Isozyme pattern of glycogen phosphorylase in the rat nervous system and rat astroglia-rich primary cultures: electrophoretic and polymerase chain reaction studies. *Neurochem Res*. 2000;25(11):1485–1491.

175. Muller MS. Functional impact of glycogen degradation on astrocytic signalling. *Biochem Soc Trans*. 2014;42(5):1311–1315.

176. Sickmann HM, Waagepetersen HS, Schousboe A, Benie AJ, Bouman SD. Brain glycogen and its role in supporting glutamate and GABA homeostasis in a type 2 diabetes rat model. *Neurochem Int*. 2012;60(3):267–275.

177. Sickmann HM, Walls AB, Schousboe A, Bouman SD, Waagepetersen HS. Functional significance of brain glycogen in sustaining glutamatergic neurotransmission. *J Neurochem*. 2009;109(Suppl 1):80–86.

178. Dinuzzo M, Mangia S, Maraviglia B, Giove F. The role of astrocytic glycogen in supporting the energetics of neuronal activity. *Neurochem Res*. 2012;37(11):2432–2438.

179. DiNuzzo M, Mangia S, Maraviglia B, Giove F. Regulatory mechanisms for glycogenolysis and K+ uptake in brain astrocytes. *Neurochem Int*. 2013;63(5):458–464.

180. Hertz L, Gerkau NJ, Xu J, et al. Roles of astrocytic Na(+),K(+)-ATPase and glycogenolysis for K(+) homeostasis in mammalian brain. *J Neurosci Res*. 2015;93(7):1019–1030.

181. Xu J, Song D, Bai Q, et al. Role of glycogenolysis in stimulation of ATP release from cultured mouse astrocytes by transmitters and high K+ concentrations. *ASN Neuro*. 2014;6(1):e00132.

182. Muller MS, Fox R, Schousboe A, Waagepetersen HS, Bak LK. Astrocyte glycogenolysis is triggered by store-operated calcium entry and provides metabolic energy for cellular calcium homeostasis. *Glia*. 2014;62(4):526–534.

183. Gibbs ME, Anderson DG, Hertz L. Inhibition of glycogenolysis in astrocytes interrupts memory consolidation in young chickens. *Glia*. 2006;54(3):214–222.

184. Gibbs ME, Hutchinson D, Hertz L. Astrocytic involvement in learning and memory consolidation. *Neurosci Biobehav Rev*. 2008;32(5):927–944.

185. Duran J, Saez I, Gruart A, Guinovart JJ, Delgado-Garcia JM. Impairment in long-term memory formation and learning-dependent synaptic plasticity in mice lacking glycogen synthase in the brain. *J Cereb Blood Flow Metab*. 2013;33(4):550–556.

186. Hertz L, O'Dowd BS, Ng KT, Gibbs ME. Reciprocal changes in forebrain contents of glycogen and of glutamate/glutamine during early memory consolidation in the day-old chick. *Brain Res*. 2003;994(2):226–233.

187. Gibbs ME, Shleper M, Mustafa T, Burnstock G, Bowser DN. ATP derived from astrocytes modulates memory in the chick. *Neuron Glia Biol*. 2011;7(2–4):177–186.

188. Suzuki A, Stern SA, Bozdagi O, et al. Astrocyte-neuron lactate transport is required for long-term memory formation. *Cell*. 2011;144(5):810–823.

189. Steinman M, Gao V, Alberini CM. The role of lactate-mediated metabolic coupling between astrocytes and neurons in long-term memory formation. *Front Integr Neurosci J*. 2016;:http://dx.doi.org/103389/fnint201600010.

190. Tadi M, Allaman I, Lengacher S, Grenningloh G, Magistretti PJ. Learning-induced gene expression in the hippocampus reveals a role of neuron-astrocyte metabolic coupling in long term memory. *PLoS ONE*. 2015;10(10):e0141568.

191. Tang F, Lane S, Korsak A, et al. Lactate-mediated glia-neuronal signalling in the mammalian brain. *Nat Commun*. 2014;5:3284.

192. DiNuzzo M. Astrocyte-neuron interactions during learning may occur by lactate signaling rather than metabolism. *Front Integr Neurosci*. 2016;10:2.

193. Hertz L, Chen Y. All 3 types of glial cells are important for memory formation. *Front Integr Neurosci.* 2016;:In press.2016; 10:31. eCollection 2016
194. Bergersen LH, Magistretti PJ, Pellerin L. Selective postsynaptic co-localization of MCT2 with AMPA receptor GluR2/3 subunits at excitatory synapses exhibiting AMPA receptor trafficking. *Cereb Cortex.* 2005;15(4):361–370.
195. Rust MB. ADF/cofilin: a crucial regulator of synapse physiology and behavior. *Cell Mol Life Sci: CMLS.* 2015;72(18):3521–3529.
196. Hertz L. The astrocyte-neuron lactate shuttle: a challenge of a challenge. *J Cereb Blood Flow Metab.* 2004;24(11):1241–1248.
197. Drulis-Fajdasz D, Wojtowicz T, Wawrzyniak M, Wlodarczyk J, Mozrzymas JW, Rakus D. Involvement of cellular metabolism in age-related LTP modifications in rat hippocampal slices. *Oncotarget.* 2015;6(16):14065–14081.
198. Mozrzymas J, Szczesny T, Rakus D. The effect of glycogen phosphorolysis on basal glutaminergic transmission. *Biochem Biophys Res Commun.* 2011;404(2):652–655.
199. Barros DM, Mello e Souza T, De David T, et al. Simultaneous modulation of retrieval by dopaminergic D(1), beta-noradrenergic, serotonergic-1A and cholinergic muscarinic receptors in cortical structures of the rat. *Behav Brain Res.* 2001;124(1):1–7.
200. Zhang Z, Gong N, Wang W, Xu L, Xu TL. Bell-shaped D-serine actions on hippocampal long-term depression and spatial memory retrieval. *Cereb Cortex.* 2008;18(10):2391–2401.
201. Stehberg J, Moraga-Amaro R, Salazar C, et al. Release of gliotransmitters through astroglial connexin 43 hemichannels is necessary for fear memory consolidation in the basolateral amygdala. *FASEB J.* 2012;26(9):3649–3657.
202. Boury-Jamot B, Carrard A, Martin JL, Halfon O, Magistretti PJ, Boutrel B. Disrupting astrocyte-neuron lactate transfer persistently reduces conditioned responses to cocaine. *Mol Psychiatry.* 2016;21(8):1070–1076.
203. Beck H, Semisch M, Culmsee C, Plesnila N, Hatzopoulos AK. Egr-1 regulates expression of the glial scar component phosphacan in astrocytes after experimental stroke. *Am J Pathol.* 2008;173(1):77–92.
204. Shelton K, Bogyo K, Schick T, Ettenberg A. Pharmacological modulation of lateral habenular dopamine D2 receptors alters the anxiogenic response to cocaine in a runway model of drug self-administration. *Behav Brain Res.* 2016;310:42–50.
205. Liu K, Steketee JD. The role of adenylyl cyclase in the medial prefrontal cortex in cocaine-induced behavioral sensitization in rats. *Neuropharmacology.* 2016;111:70–77.
206. Luo Y, Kokkonen GC, Wang X, Neve KA, Roth GS. D2 dopamine receptors stimulate mitogenesis through pertussis toxin-sensitive G proteins and Ras-involved ERK and SAP/JNK pathways in rat C6-D2L glioma cells. *J Neurochem.* 1998;71(3):980–990.
207. Peng L, Hertz L. Long-lasting abolishment of noradrenaline induced stimulation of oxidative metabolism after chronic exposure of developing mouse astrocytes to cocaine. *Brain Res.* 1992;581(2):334–338.
208. Zhang Y, Xue Y, Meng S, et al. Inhibition of Lactate Transport Erases Drug Memory and Prevents Drug Relapse. *Biol Psychiatry.* 2016;79(11):928–939.
209. Turner JR, Ecke LE, Briand LA, Haydon PG, Blendy JA. Cocaine-related behaviors in mice with deficient gliotransmission. *Psychopharmacology (Berl).* 2013;226(1):167–176.
210. Cadet JL, Bisagno V. Glial-neuronal ensembles: partners in drug addiction-associated synaptic plasticity. *Front Pharmacol.* 2014;5:204.
211. Gandhi GK, Cruz NF, Ball KK, Dienel GA. Astrocytes are poised for lactate trafficking and release from activated brain and for supply of glucose to neurons. *J Neurochem.* 2009;111(2):522–536.

212. Bergersen LH. Lactate transport and signaling in the brain: potential therapeutic targets and roles in body—brain interaction. *J Cereb Blood Flow Metab*. 2015;35(2):176—185.
213. Hajek I, Subbarao KV, Hertz L. Acute and chronic effects of potassium and noradrenaline on Na + , K + -ATPase activity in cultured mouse neurons and astrocytes. *Neurochem Int*. 1996;28(3):335—342.
214. Hertz L, Peng L, Song D. The astrocytic Na + ,K + -ATPase—its stimulation by increased extracellular K + or b-adrenergic activation and its ouabain-mediated signaling. In: Chakraborti S, Dhalla NS, eds. *Regulation of Na + ,K + - ATPase*. Springer; 2016:195—221.
215. Ransom CB, Ransom BR, Sontheimer H. Activity-dependent extracellular K + accumulation in rat optic nerve: the role of glial and axonal Na + pumps. *J Physiol*. 2000;522(Pt 3):427—442.
216. Somjen GG, Kager H, Wadman WJ. Computer simulations of neuron-glia interactions mediated by ion flux. *J Comput Neurosci*. 2008;25(2):349—365.
217. Dufour S, Dufour P, Chever O, Vallee R, Amzica F. In vivo simultaneous intra- and extracellular potassium recordings using a micro-optrode. *J Neurosci Methods*. 2011;194(2): 206—217.
218. Macaulay N, Zeuthen T. Glial K(+) clearance and cell swelling: key roles for cotransporters and pumps. *Neurochem Res*. 2012;37(11):2299—2309.
219. Larsen BR, Assentoft M, Cotrina ML, et al. Contributions of the Na(+)/K(+)-ATPase, NKCC1, and Kir4.1 to hippocampal K(+) clearance and volume responses. *Glia*. 2014;62(4): 608—622.
220. Larsen BR, Stoica A, MacAulay N. Managing brain extracellular K(+) during neuronal activity: the physiological role of the Na(+)/K(+)-ATPase subunit isoforms. *Front Physiol*. 2016;7:141.
221. Song D, Xu J, Du T, et al. Inhibition of brain swelling after ischemia-reperfusion by beta-adrenergic antagonists: correlation with increased K + and decreased Ca2 + concentrations in extracellular fluid. *Biomed Res Int*. 2014;2014:873590.
222. Hertz L, Xu J, Chen Y, et al. Antagonists of the vasopressin V1 receptor and of the beta (1)-adrenoceptor inhibit cytotoxic brain edema in stroke by effects on astrocytes—but the mechanisms differ. *Curr Neuropharmacol*. 2014;12(4):308—323.
223. Pedersen SF, O'Donnell ME, Anderson SE, Cala PM. Physiology and pathophysiology of Na + /H + exchange and Na + -K + -2Cl-cotransport in the heart, brain, and blood. *Am J Physiol Regul Integr Comp Physiol*. 2006;291(1):R1—R25.
224. Baskey G, Singh A, Sharma R, Mallick BN. REM sleep deprivation-induced noradrenaline stimulates neuronal and inhibits glial Na-K ATPase in rat brain: in vivo and in vitro studies. *Neurochem Int*. 2009;54(1):65—71.
225. Kupriyanov VV, Xiang B, Sun J, Jilkina O. Effect of adrenergic stimulation on Rb(+) uptake in normal and ischemic areas of isolated pig hearts: (87)Rb MRI study. *Magn Reson Med*. 2002;48(1):15—20.
226. Hertz L, Schousboe A, Boechler N, Mukerji S, Fedoroff S. Kinetic characteristics of the glutamate uptake into normal astrocytes in cultures. *Neurochem Res*. 1978;3(1):1—14.
227. Hertz L, Juurlink BHJ, Fosmark H, Schousboe A. Methodological appendix. In: Pfeiffer SE, ed. *Neuroscence Approached Through Cell Culture*. Vol 1. Boca Raton, Florida: CRC Press; 1982:175—186.
228. Juurlink B.H.J., Hertz L. Astrocytes. In: Boulton A.A., Baker G.B., Walz W., eds. *Neuromethods in cell cultures*. New York, NY: Humana Clifton; 1992:269—321. Also on the

Internet in Springer Protocols (Neuromethods, vol. 223. Practical cell Culture techniques, Boulton A.A., Baker G.B., Walz W., eds).

229. Foote SL, Bloom FE, Aston-Jones G. Nucleus locus ceruleus: new evidence of anatomical and physiological specificity. *Physiol Rev.* 1983;63(3):844–914.
230. Facci L, Skaper SD, Levin DL, Varon S. Dissociation of the stellate morphology from intracellular cyclic AMP levels in cultured rat brain astroglial cells: effects of ganglioside GM1 and lysophosphatidylserine. *J Neurochem.* 1987;48(2):566–573.
231. Shapiro DL. Morphological and biochemical alterations in foetal rat brain cells cultured in the presence of monobutyryl cyclic AMP. *Nature.* 1973;241(5386):203–204.
232. Lim R, Mitsunobu K, Li WK. Maturation–stimulation effect of brain extract and dibutyryl cyclic AMP on dissociated embryonic brain cells in culture. *Exp Cell Res.* 1973;79(1):243–246.
233. Hertz L. Dibutyryl cyclic AMP treatment of astrocytes in primary cultures as a substitute for normal morphogenic and 'functiogenic' transmitter signals. In: Lauder J, Privat A, Giacobini E, Timiras P, Vernadakis A, eds. *Molecular Aspects of Developmenmt and Aging of the Nervous System.* 265. NY: Plenjm Press; 1990:227–243.
234. Meier E, Hertz L, Schousboe A. Neurotransmitters as developmental signals. *Neurochem Int.* 1991;19(1/2):1–15.
235. Hertz L, Code WE. Calcium channel signalling in astrocytes. In: Paoletti R, Godfraind T, Vankoullen PM, eds. *Calcium Antagonists: Pharmacology and Clinical Research.* Boston: Kluwer; 1993:205–213.
236. Subbarao KV, Stolzenburg JU, Hertz L. Pharmacological characteristics of potassium-induced, glycogenolysis in astrocytes. *Neurosci Lett.* 1995;196(1–2):45–48.
237. Hertz L, Bender AS, Woodbury DM, White HS. Potassium-stimulated calcium uptake in astrocytes and its potent inhibition by nimodipine. *J Neurosci Res.* 1989;22(2):209–215.

Chapter 3

White Matter Astrocytes: Adrenergic Mechanisms

Maria Papanikolaou and Arthur Morgan Butt[✉]
University of Portsmouth, Portsmouth, United Kingdom

Chapter Outline

Introduction 64
White Matter Glia Ensure Rapid Neuronal Signaling Over Long Distances 65
Neuroglial Communication in White Matter: The "Nodal Synapse" 66
Adrenergic Mechanisms in White Matter 68
Adrenergic Signaling in White Matter Physiology 70
Adrenergic Signaling in White Matter Pathology 71
Adrenergic Signaling Regulates Blood Flow 72
Conclusions 73
Abbreviations 73
References 74

ABSTRACT
White matter (WM) tracts contain bundles of myelinated axons and the glia that support them, namely oligodendrocytes that form the myelin and astrocytes, which provide structural and homeostatic support, together with small populations of microglia and oligodendrocyte precursor cells. Neurotransmitter signaling is prominent in WM, and there is a significant role for adrenergic signaling, together with glutamatergic and purinergic mechanisms. The site of axoglial signaling in myelinated axons is the node of Ranvier, which is the site of action potential propagation and displays many properties of synapses. Neurotransmitters released during axonal electrical activity activate receptors on perinodal astrocytes, to stimulate their release of gliotransmitters that integrate function within axon–glial–vascular networks. One of the key functions of astroglial adrenergic signaling in WM is to maintain energy supply to axons and oligodendrocytes, ensuring efficient and rapid information transfer throughout the central nervous system and maintaining WM integrity in times of energy stress.

Keywords: Glia; axon; astrocyte; oligodendrocyte; adrenergics; white matter

[✉]Correspondence address
E-mail: arthur.butt@port.ac.uk

Noradrenergic Signaling and Astroglia. DOI: http://dx.doi.org/10.1016/B978-0-12-805088-0.00003-7
© 2017 Elsevier Inc. All rights reserved.

INTRODUCTION

White matter (WM) tracts are bundles of myelinated axons that transmit electrical signals rapidly in the form of action potentials from one part of the central nervous system (CNS) gray matter (GM) to another. WM appears opaque or dense in anatomical sections and in brain scans due to myelin, the fatty insulating layer around axons that is produced by oligodendrocytes. WM is a prominent feature of the human cortex and spinal cord (Fig. 3.1), and major WM tracts include the following:

- corpus callosum—largest WM tract in the brain and crucially interconnecting the two cerebral hemispheres,
- posterior funiculus (dorsal columns) of the spinal cord—the longest WM tract connecting the brain with the periphery,
- internal capsule—in a unique location where all the motor and sensory fibers travel to and from the cortex and damage to this area is the most common in stroke,
- optic tracts—transmit all visual information from the eyes to the brain,
- cerebellar and brainstem WM—control life-sustaining functions, such as motor control and respiration,
- fornix—connects the hippocampus with subcortical structures, which is essential for memory and cognitive function.

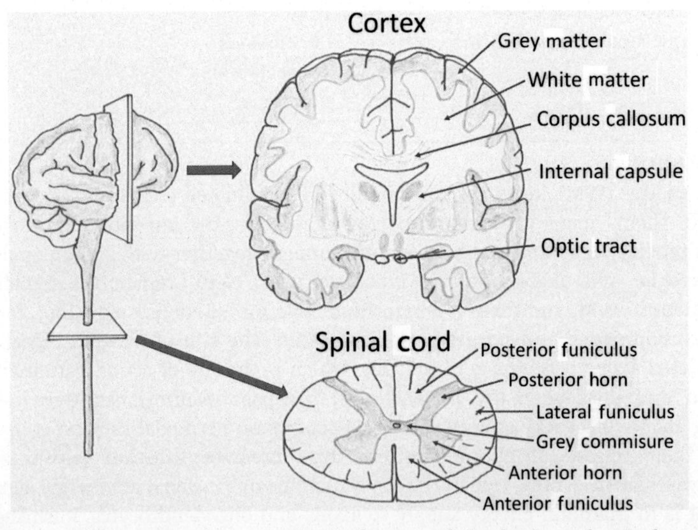

FIGURE 3.1 **White matter is a prominent feature of the brain and spinal cord.** The white matter (WM) contains bundles of myelinated axons that interconnect neurons located in the widely dispersed gray matter areas. The corpus callosum is the largest WM tract in the brain and interconnects the two cortical hemispheres, and the posterior funiculus (dorsal columns) of the spinal cord is the longest WM tract connecting the brain with the periphery. Other important WM tracts include the internal capsule, which contains the major motor and sensory fibers that travel to and from the cortex, and the optic tracts, which transmit all visual information from the eyes to the brain.

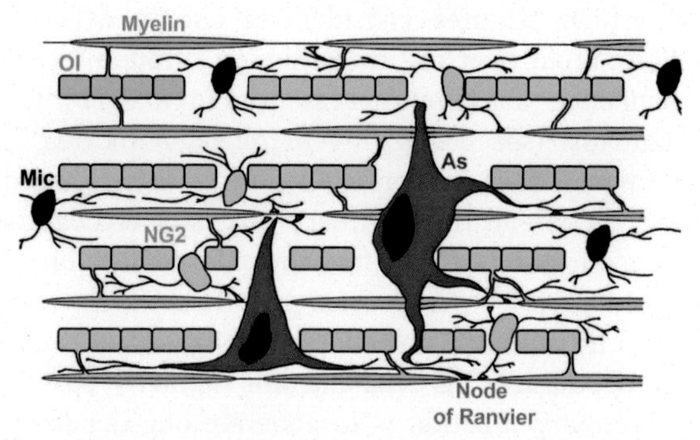

FIGURE 3.2 **White matter contains myelinated axons and the glia that support them.** The two main cell types in white matter (WM) are astrocytes (As; red) and oligodendrocytes (Ol; green). Oligodendrocytes are located in rows of cells and extend processes to form myelin sheaths that insulate axons and provide for rapid electrical transmission of action potentials. Astrocytes send processes both perpendicular to the axonal axis to provide the structural support of WM and to nodes of Ranvier, the sites of axonal action potential propagation. Small but significant numbers of NG2-glia (blue) and microglia (Mic; black) are scattered throughout WM; the function of NG2-glia is the life-long generation of oligodendrocytes and microglia are the immune cells of the CNS.

Hence, rapid signal transmission in WM is critical for all CNS functions and underlies the massive computing power of the human brain. Consequently, WM pathology has potentially devastating effects, such as occurs in demyelinating diseases multiple sclerosis (MS) and leukodystrophies,[1,2] as well as stroke, traumatic injury, dementias, and Alzheimer's disease, autism and neuropsychiatric diseases.[3–5] The main cells in WM are the myelinating oligodendrocytes, together with astrocytes and small populations of microglia and NG2-glia (Fig. 3.2). Although WM is largely devoid of neuronal somata, it does exhibit significant neurotransmitter signaling in the absence of classical neuronal synaptic signaling.[6] Most evidence of neurotransmitter signaling in WM is for glutamate and adenosine triphosphate (ATP) signaling,[6–8] which involves release from axons and astrocytes, but there is increasing evidence that adrenergic signaling plays a key role in regulating glycogenolysis in astrocytes and in protecting axons and myelin during energetic stress.[9–14] This chapter will provide an overview of neurotransmission in WM and focus on the pathophysiological mechanisms of astroglial adrenergic signaling.

WHITE MATTER GLIA ENSURE RAPID NEURONAL SIGNALING OVER LONG DISTANCES

WM tracts comprise myelinated axons and the glia that support them, namely oligodendrocytes, astrocytes, microglia, and NG2-glia, together with the vasculature (Fig. 3.2).[15] The function of oligodendrocytes is to produce myelin, a living structure and vast expansion of the oligodendrocyte plasmalemma, which requires considerable energy expenditure by the cells,[16] to facilitate rapid electrical transmission through its insulatory benefits. There is evidence that myelin is essential for axonal integrity through multiple mechanisms (see below).[17] WM oligodendrocytes lie in regular interfascicular rows of five or more cells and are interspersed with astrocytes.[18,19] Astrocytes extend their primary processes orthogonally and radially to separate axon bundles into fascicles and contact blood vessels and axons at nodes of Ranvier.[20,21] Astrocytes provide the WM with its structure and are central to homeostatic and metabolic support of the tissue.[9,21,22] In addition, small populations of NG2-glia are regularly spaced throughout WM, and their primary function is to generate oligodendrocytes throughout life, required to myelinate new connections and to replace myelin lost through natural "wear and tear" or through demyelination.[23–25] Finally, microglia are the resident immune cells of the CNS and have multiple roles in WM pathology[26]; "resting" microglia are constantly moving their fine processes to monitor for brain disturbances and become "activated" in response to tissue damage. Overall, each glial cell type performs essential specialist functions to ensure the integrity of rapid neuronal signaling within the brain and over long distances along the spinal cord in health and disease. Astrocytes are at the very center of this complex and via their cellular contacts and repertoire of receptors, ion channels, transporters, and signal transduction mechanisms serve to integrate the various WM elements into an axon–glial–vascular network.[27,28]

NEUROGLIAL COMMUNICATION IN WHITE MATTER: THE "NODAL SYNAPSE"

Neuroglial communication is a recognized feature in GM and has given rise to the concept of the tripartite synapse,[29] in which astrocytes are partners at neuronal synapses, including monoaminergic synapses.[30] In addition, microglia and NG2-glia extend processes to synapses and are actively involved in their function.[31,32] There are no synapses in WM, and the sites of axoglial signaling are at nodes of Ranvier,[6] which represent the only area of naked axolemma in myelinated axons. These sites can be considered analogous to synapses in GM and express many of the same neurotransmitter receptors, ion channels, transporters, and structural elements.[33,34] It is significant, therefore, that astrocytes extend minute collaterals that engage nodes of Ranvier and respond to axonal electrical activity as they do at synapses in GM (Fig. 3.3).[28,35] In addition, noradrenergic axons are found throughout WM, and noradrenaline (NA) is probably released from axonal varicosities,[36] which would activate adrenergic receptors (ARs) on glia through "volume transmission" by diffusion through the extracellular space, or by direct "synaptic-like" signaling via cellular contacts with the varicosities, as has been shown for

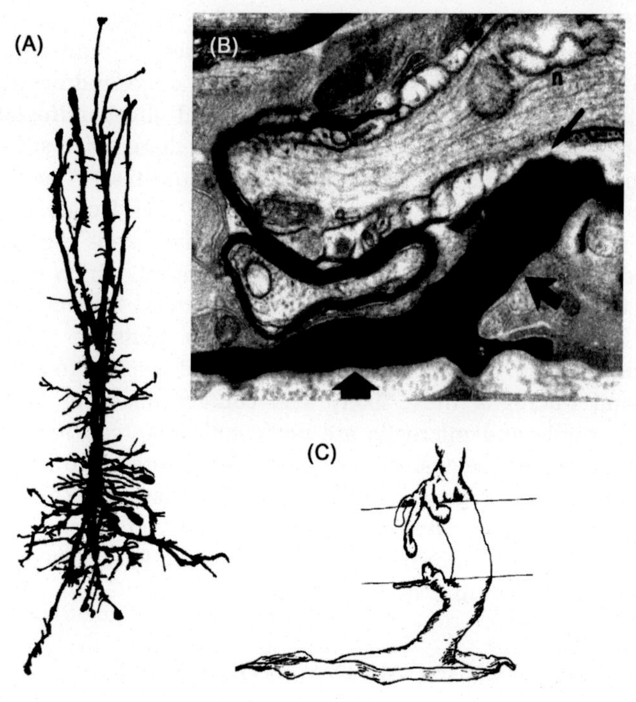

FIGURE 3.3 **Astrocytes contact nodes of Ranvier.** Camera lucida drawing (A), electron microscope (EM) (B), and 3D ultrastructural reconstruction of perinodal processes (C) of an astrocyte in the adult mouse optic nerve following intracellular electrophysiological recording and dye-filling with horseradish peroxidase. (A) Astrocytes extend primary processes that traverse the white matter (WM) tract to provide physical structural support for the tissue and extend collateral processes that form contacts with blood vessels, glial cells, and axons. (B) EM of fine collaterals located along astrocyte primary processes that form perinodal processes to contact axons at nodes of Ranvier, the site of axonal electrical activity. (C) 3D reconstruction of serial EM sections illustrates astroglial fillipodia forming multiple axonal contacts at a node of Ranvier. Each astrocyte contacts many hundreds of nodes in addition to contacting blood vessels via perivascular endfeet, forming a communications pathway between the vasculature and the sites of axonal electrical activity. *Adapted from Butt et al. J Neurocytol* **23 (8),** *1994, 486-499.*

glutamate signaling onto NG2-glia in WM.[37,38] In addition to nodes of Ranvier, myelinated axons are organized into distinct paranodal, juxtaparanodal, and internodal domains. The paranodal and juxtaparanodal regions are highly specialized sites of contact between the oligodendroglial myelin sheath and axon; by comparison, the internodal axolemma beneath the compacted myelin sheath is relatively unspecialized,[34] but represents the largest area for bidirectional communication between axons and oligodendrocytes.[18,19] By this means, oligodendrocytes sense axonal electrical activity[39] and are able to deliver proteins and metabolic substrates to axons.[14,40–42] Importantly, astrocytes are at the center of these WM communication pathways, contacting axons, oligodendrocytes, and the vasculature, via their cellular processes,[28,35] and through gap junction connections with each other and with oligodendrocytes, principally utilizing astroglial connexin 43 (Cx43) and oligodendroglial Cx47.[43] Thus, astrocytes form WM axon—glial—vascular domains to supply metabolic support directly to axons and to oligodendrocytes,[44,45] and oligodendrocytes in turn are able to deliver pyruvate and lactate to axons.[40] Moreover, glutamate released by electrically active axons can directly act on oligodendroglial NMDA-type glutamate receptors, which mobilize the glucose transporter GLUT1 in oligodendrocytes and enhance uptake of glucose and glycolytic support of axons.[14] In addition, NG2-glia also extend processes to contact nodes of Ranvier and respond to axonal activity,[23,27] and microglia most likely also contact nodes of Ranvier.[46] It appears NG2-glia and microglia are not coupled via gap junctions with each other or to astrocytes, but they are incorporated into an integrated WM network by purinergic signaling in response to ATP released by astrocytes.[7,8,27,28] Thus, astrocytes sit at the center of a panglial network that serves to ensure information transfer in the form of axonal action potential propagation in CNS WM (Fig. 3.4).

ADRENERGIC MECHANISMS IN WHITE MATTER

NA acts on ARs, which are G protein-coupled and divided into α_1-AR, α_2-AR β_{1-3}-AR. The α_1-ARs are coupled to phospholipase C, and their activation results in formation of IP_3 (inositol triphosphate) and subsequent Ca^{2+} release from the endoplasmic reticulum (ER), whereas α_2-ARs inhibit cAMP (cyclic adenosine monophosphate) production, and β-ARs stimulate cAMP production. Overall, ARs are less expressed in WM compared with GM, [47] although ARs are enriched in the splenium of the human corpus callosum,[48] and adrenergic fibers in the indusium griseum and longitudinal striae of the corpus callosum are considered components of the limbic system.[49] In the optic nerve, expression of β_2-AR has been detected in rabbits, rats, and humans, in which they are expressed predominantly by astrocytes.[50] More recently, β_2-ARs were observed to be downregulated in astrocytes identified by glial fibrillary acidic protein in human MS brain samples, in both astrogliotic plaques and normal appearing WM.[51–53] In addition, prominent α_2-AR immunoreactivity has been demonstrated in astrocytes and myelinated axons in the rodent optic nerve and spinal dorsal columns, in which they modulate WM responses to ischemia.[12] Astroglial expression of α_1-AR in WM has not been demonstrated directly, and autoradiographic localization indicated diffuse expression of α_1-AR in human optic nerves, consistent with axonal expression.[54] Studies using transgenic reporter mice did not detect expression of α_{1A}-AR or α_{1B}-AR in astrocytes, whereas α_{1A}-ARs were detected in NG2-glia, but not mature

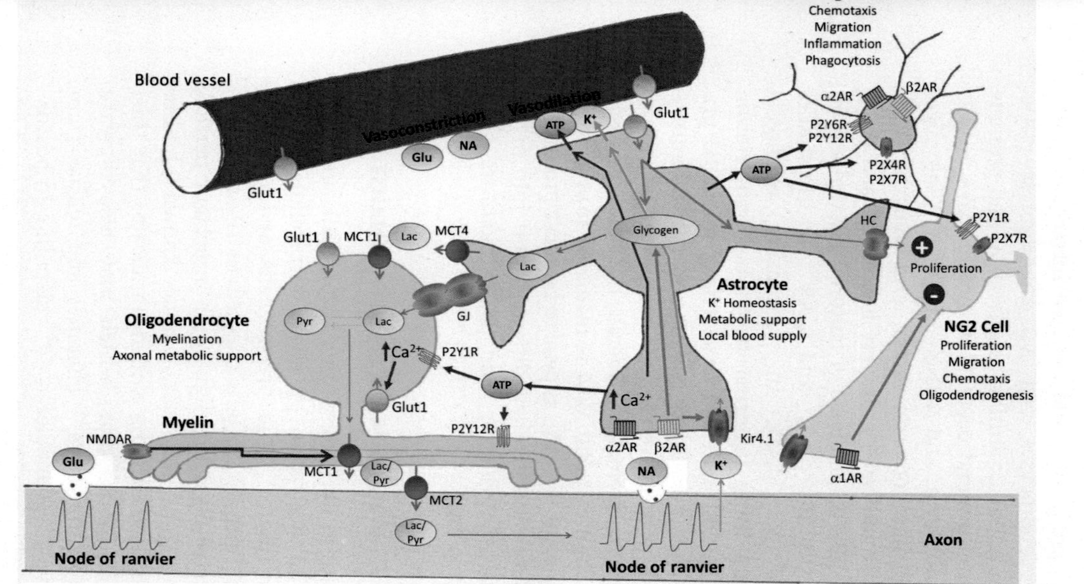

FIGURE 3.4 **Astrocytes form white matter networks.** Noradrenaline signaling (NA) in astrocytes provides a communications and metabolic axon–glial–vascular network within white matter (WM), via astroglial contacts with axons, all glial cells and the vasculature. (1) Activation of astroglial β_2-AR stimulates uptake of K^+ released during axonal electrical activity, which if it built up would cause hyperexcitability and ultimately conduction block. (2) Activation of astroglial β_2-AR stimulates glycogenolysis and the production of lactate, which astrocytes provide as an energy source for axons and oligodendrocytes, via gap junctions (GJ), and by release through monocarboxylate transporters (MCT4) and subsequent uptake by oligodendrocytes (MCT1) and axons (MCT2). Astrocytes also release glucose via hemichannels to provide metabolic support for NG2-glia, which is necessary for their proliferation and regeneration of oligodendrocytes. In addition, glutamate released during action potential propagation can act directly on oligodendrocyte NMDA-type receptors to stimulate their metabolic support of axons. (3) Activation of α_2-AR induces astroglial Ca^{2+} signaling, which has been shown to trigger their release of ATP to activate P2 receptors on oligodendrocytes, NG2-glia and microglia, incorporating them into a panglial network. Activation of P2YR in oligodendrocytes evokes raised Ca^{2+}, which has been shown to increase activity of the glucose transporters (Glut1), providing a further mechanism by which astrocytes may enhance metabolic support for oligodendrocytes. In microglia, P2 receptors in combination with β-ARs regulate their chemotactic, migratory, inflammatory, and phagocytic behaviors, whereas P2 receptors and α_{1A}-ARs in NG2-glia regulate their proliferation and differentiation into myelinating oligodendrocytes. (4) Astrocytes regulate local blood flow through the release of ATP and K^+. Axonal activity and NA release would be sensed by astrocytes via α_2-AR on their perinodal processes, and this evokes an intracellular Ca^{2+} signal that is transmitted to their perivascular endfeet to trigger the release of ATP and K^+, which are potent vasodilators; this would counterbalance the actions of NA and glutamate, which are potent vasoconstrictors and are released during axonal activity. In this way, release of NA would couple astroglial functions to axonal activity and via the release of ATP incorporate oligodendrocytes, NG2-glia, microglia, and the vasculature into an axon–glial–vascular network that is essential for WM function and integrity.

oligodendrocytes, and α_{1B}-ARs were weakly expressed by NG2-glia and oligodendrocytes.[55,56] In vitro studies indicate NG2-glia, and oligodendrocytes express all three α_1-ARs,[57] although α_{1A}-AR may be the most prominent physiologically, promoting IP_3 production and elevated intracellular Ca^{2+} levels.[58,59] In addition, oligodendrocytes and NG2-glia express β-ARs in vitro,[60] and they have been shown to inhibit proliferation and promote differentiation of NG2-glia into oligodendrocytes.[61] As noted above, α_1-ARs have been detected in oligodendrocytes and NG2-glia in vivo,[55,56] but oligodendroglial expression of β-AR has not been shown, and analysis of the RNA-Seq transcriptome database indicated only very low expression levels of AR subtypes by NG2-glia.[62] RT-PCR analysis of cultured microglia indicated that they express mRNA for α_1-, α_2-, β_1-, and β_2-AR[63]; in vitro studies indicate a prominent role for β_1-AR in preventing microglial activation.[64–66] A recent study using quantitative reverse transcription polymerase chain reaction (RT-PCR) indicated that resting microglia primarily express β_2-AR and switch expression to α_{2A}-AR under proinflammatory conditions[67]; NA caused process retraction via β_2-AR in resting microglia and α_{2A}-AR in activated microglia.[67] Overall, the balance of evidence is that adrenergic signaling in adult WM is mediated primarily by astroglial α_2-AR and β_2-AR, whereas α_{1A}-ARs appear to be most prominent in NG2-glia, in which they may regulate their differentiation into myelinating oligodendrocytes, and microglial β_1- and β_2-AR modulate their responses to WM disruption (Fig. 3.4).

ADRENERGIC SIGNALING IN WHITE MATTER PHYSIOLOGY

WM-glia in the developing rodent optic nerve and corpus callosum have been shown to respond to AR activation by raised intracellular Ca^{2+}, often termed glial Ca^{2+}-signaling.[68,69] The precise cells and receptors involved were not identified, but α_1-ARs are coupled to IP_3 and Ca^{2+} release from ER stores and have been highlighted as important participants in astrocyte Ca^{2+} signaling.[70,71] However, it is not known whether WM astrocytes express α_1-ARs, and existing evidence indicates that they predominantly express the α_2-AR subtype, which may also increase intracellular Ca^{2+} through opening of Ca^{2+} channels and release from intracellular stores. Alternatively, NA-mediated Ca^{2+} signaling in postnatal WM may represent predominantly the responses of NG2-glia, which are numerous and express α_1-ARs.[55,56] Notwithstanding this, NA is implicated in ischemic injury in developing and adult WM,[12,72] and modulation of α_2-AR activation provided robust protection against ischemia-hypoxia in the optic nerve.[12] In addition, there is evidence that β_1-ARs modulate excitability of premyelinated optic nerve axons, although this effect appears to be lost when axons become fully myelinated.[73] Further studies are required to determine whether astroglial AR are important in sustaining axonal excitability in adult WM, but the loss of astroglial β_2-AR in MS demyelinated lesions is implicated in impaired axonal metabolism and degeneration.[9] Overall, the balance of evidence is that the adrenergic system exerts a potent modulatory effect in WM, most likely via multiple pathways (Fig. 3.4). The source of NA in WM is unclear, but it is probably released from axonal varicosities that are present along noradrenergic axons throughout WM,[36] a mechanism that has been shown for glutamate release in premyelinated axons in the optic nerve and corpus callosum.[37,38,74] Neurotransmitters can also be released by reversed uptake through transporters, and immunolocalization of NA transporters has been demonstrated on astrocyte processes and the myelin sheath in rat optic nerves, and there is evidence that they are responsible for the release of NA in hypoxia/ischemia.[12] Activation of astroglial β-AR regulates their essential K^+ homeostatic function,[75,76] and astroglial metabolic support via glycogenolysis.[77] Astrocytes can also release neurotransmitters through vesicles, hemichannels, P2X7 receptors, anion channels, and stretch-activated receptors.[78] Moreover, astroglial α_1-ARs have been shown to activate exocytosis of the gliotransmitter ATP, which contributes to modulation of synaptic plasticity in neurons.[79] It is likely that an equivalent mechanism exists in WM, whereby NA evokes Ca^{2+} signals in WM astrocytes and stimulates their release of ATP, which activates purine (P2X and P2Y) receptors on oligodendrocytes, NG2-glia, and microglia.[8,27,28] Although the precise physiological functions of adrenergic signaling in WM require further investigation, it is likely that adrenergic signaling regulates the homeostatic and metabolic functions of astrocytes and coupled with astroglial purinergic signaling it serves as a powerful signaling pathway integrating axon and glial functions (Fig. 3.4).

ADRENERGIC SIGNALING IN WHITE MATTER PATHOLOGY

There is growing appreciation of the importance of astroglial AR in WM pathology. Modulation of astroglial α_2-ARs has been shown to have a potent neuroprotective effect in both perinatal and adult WM ischemia.[12,80] In addition, β-ARs regulate essential astroglial homeostatic and metabolic support for axons through their uptake of K^+ during action potential propagation[22,81] and by the production and release of lactate.[40,44,45] Notably, there is a deficiency of astrocytic β_2-AR in WM of MS patients[9] (see the chapter by De Keyser), which would potentially disrupt astroglial K^+ homeostasis and glycogenolysis, thereby interfering with their support of axonal electrical conduction and energy supply. In addition, oligodendrocytes in the WM of the rat cerebellum and corpus callosum take up lactate released by astrocytes via monocarboxylate transporter 1 (MCT1), and this can support oligodendrocytes and myelination in ischemia.[82] In turn, oligodendrocytes can provide lactate via

MCT1 to support axon survival and function.[83] Thus, astroglial β_2-AR are likely to be important for both oligodendroglial and axonal metabolism during energetic stress, and loss of astroglial β_2-AR is potentially a factor in oligodendrocyte and axon demise in MS and other neuropathologies. However, a recent study indicated that the axonal protective effects of glial β_2-AR activation are not mediated by enhanced lactate production,[10] suggesting that other mechanisms are also important. Both α- and β-ARs mediate astroglial activation in vivo following stimulation of adrenergic fibers,[84,85] and pharmacological inhibition of β-AR suppresses astroglial scar formation.[86] In experimental autoimmune encephalomyelitis (EAE), the mouse model of MS, depletion of NA levels in the CNS exacerbated clinical scores and reduced astrocyte activation. NG2-glia respond to CNS insults by increased proliferation, which is geared primarily towards the regeneration of myelinating oligodendrocytes, and interact with inflammatory cells to promote repair.[87,88] These responses are modulated by α_{1A}- and β_2-AR in NG2-glia,[55,61] in combination with $P2Y_1R$ and $P2X_7R$, which mediate oligodendrocyte regeneration and remyelination.[7] In addition, NA exerts potent anti-inflammatory effects on microglia via β_1- and β_2-ARs,[89] reducing production of IL-1β and TNF-α in microglia activated by lipopolysaccharide (LPS)[64–66]; adrenergic signaling is likely to be coupled to purinergic signaling, acting via multiple purinoceptors, with prominent roles for $P2Y_{12}R$ and $P2X_4R$, which mediate chemotactic responses in resting and activated microglia, respectively, and $P2X_7R$, which drive microglial activation, proliferation, and inflammatory responses.[7,8] Microglial inflammatory responses are prominent in WM pathology and, for example, IL-1β and TNF-α mimic the effects of LPS and induce widespread oligodendroglial damage in the rat optic nerve[90,91]; hence, evidence that activation of microglial β_1- and β_2-AR reduces production of IL-1β and TNF-α[64–66] suggests that they may be a useful therapeutic target for protecting oligodendrocytes and myelin. Overall, the evidence is that NA signaling acting in combination with purinergic signaling on astrocytes, NG2-glia, and microglia has an overall protective effect in WM pathology (Fig. 3.4).

ADRENERGIC SIGNALING REGULATES BLOOD FLOW

Astrocytes play a key role in coupling local blood flow to neuronal activity by multiple mechanisms involving Ca^{2+} signals in astrocyte endfeet (Fig. 3.4).[92] NA is a potent vasoconstrictor and has been shown to mediate its effects at least in part by increasing Ca^{2+} in astrocyte endfeet.[93] In contrast, K^+ is a potent vasodilator, and K^+ released during neuronal activity is transmitted to astroglial perivascular endfeet in the form of an astrocytic Ca^{2+} signal, where it triggers the local release of K^+ via astrocytic big potassium (BK) channels to activate inward rectifying potassium channels on vascular smooth muscle cells and cause vasodilation.[94] In addition, astrocytes respond to axonal activity at nodes of Ranvier via their perinodal processes and by Ca^{2+} signaling transmit this information to their perivascular endfeet, where release of ATP causes vasodilation.[95] As noted above, ARs play key roles in regulating astroglial K^+ channels and can regulate astroglial release of ATP and are likely to be key players in regulation of WM perfusion. Indeed, WM hypoperfusion is a factor in MS, and this may be directly related to the observed downregulation of astroglial β_2-AR, which would interfere with astroglial ATP signaling and K^+ transport, resulting in disruption of the regulation of local blood supply.[13]

CONCLUSIONS

It is now clear that adrenergic signaling is a prominent feature of WM, across a wide range of species including humans. This is notable, because WM is characterized in general by a lack of neurons and synapses, and so adrenergic signaling has physiological functions other than neuron-to-neuron communication. Astroglial ARs are implicated in multiple physiological functions, but their key role in WM may be in regulating metabolic support to axons and oligodendrocytes, which is essential for WM function and integrity. There is clear evidence that astroglial ARs are important in WM pathology, in particular those involving energy deprivation, such as stroke, periventricular leukomalacia, MS, vascular dementia, and traumatic CNS injury. Most evidence points to astroglial α_2-AR and β_2-AR being of greatest importance in WM, regulating the key astroglial homeostatic functions of K^+ regulation, glycogenolysis, and regulation of local blood flow. Notably, WM function depends on integrated communication between the different elements—axons, oligodendrocytes, astrocytes, NG2-glia, microglia, and the vasculature. Astrocytes are at the very center of this network and via their cellular contacts and repertoire of receptors, ion channels, transporters, and signal transduction mechanisms serve to integrate the various WM elements into an axon—glial—vascular network (Fig. 3.4). The coexistence of astroglial adrenergic signaling with other neurotransmitter mechanisms in WM serves to ensure rapid and efficient information transfer through WM, in health and disease.

ABBREVIATIONS

AD Alzheimer's disease
AR adrenergic receptor
ATP adenosine triphosphate
cAMP cyclic adenosine monophosphate

BK "big potassium" channels, calcium-sensitive large conductance K^+ channels
CNS central nervous system
Cspg4 chondroitin sulphate proteoglycan 4 (NG2)
Cx connexin, e.g., Cx43
EAE experimental autoimmune encephalomyelitis
ER endoplasmic reticulum
GFAP glial fibrillary acidic protein
GJ gap junction
Glu glutamate
GluT1 glucose transporter 1
GM gray matter
HC hemichannel
IP3 inositol triphosphate
Kir inward rectifying potassium channel
Lac lactate
MCT monocarboxylate transporter (e.g., MCT1, MCT2, MCT4)
MS multiple sclerosis
NA noradrenaline or norepinephrine
NG2 neuron glia 2 chondroitin sulphate proteoglycan (Cspg4)
OPC oligodendrocyte precursor cell
P2XR purinergic ionotropic receptors (seven genes identified, P2rx1-7)
P2YR purinergic G protein-coupled receptor (10 genes identified, P2ry1, 2, 4, 6, 8, 10–14)
WM white matter

REFERENCES

1. Lassmann H. Mechanisms of white matter damage in multiple sclerosis. *Glia*. 2014;62(11):1816–1830.
2. Back SA, Rosenberg PA. Pathophysiology of glia in perinatal white matter injury. *Glia*. 2014;62(11):1790–1815.
3. Fern RF, Matute C, Stys PK. White matter injury: ischemic and nonischemic. *Glia*. 2014;62(11):1780–1789.
4. Haroutunian V, Katsel P, Roussos P, Davis KL, Altshuler LL, Bartzokis G. Myelination, oligodendrocytes, and serious mental illness. *Glia*. 2014;62(11):1856–1877.
5. Kou Z, VandeVord PJ. Traumatic white matter injury and glial activation: from basic science to clinics. *Glia*. 2014;62(11):1831–1855.
6. Butt AM, Fern RF, Matute C. Neurotransmitter signaling in white matter. *Glia*. 2014;62(11):1762–1779.
7. Rivera A, Vanzulli I, Butt AM. A central role for ATP signaling in glial interactions in the CNS. *Curr Drug Targets*. 2016;17:1829–1833.
8. Butt AM. ATP: a ubiquitous gliotransmitter integrating neuron-glial networks. *Semin Cell Dev Biol*. 2011;22(2):205–213.
9. Cambron M, D'Haeseleer M, Laureys G, Clinckers R, Debruyne J, De Keyser J. White-matter astrocytes, axonal energy metabolism, and axonal degeneration in multiple sclerosis. *J Cereb Blood Flow Metab*. 2012;32(3):413–424.

10. Laureys G, Valentino M, Demol F, et al. Beta(2)-adrenergic receptors protect axons during energetic stress but do not influence basal glio-axonal lactate shuttling in mouse white matter. *Neuroscience*. 2014;277:367–374.
11. De Keyser J, Laureys G, Demol F, Wilczak N, Mostert J, Clinckers R. Astrocytes as potential targets to suppress inflammatory demyelinating lesions in multiple sclerosis. *Neurochem Int*. 2010;57(4):446–450.
12. Nikolaeva MA, Richard S, Mouihate A, Stys PK. Effects of the noradrenergic system in rat white matter exposed to oxygen-glucose deprivation in vitro. *J Neurosci*. 2009;29(6):1796–1804.
13. De Keyser J, Steen C, Mostert JP, Koch MW. Hypoperfusion of the cerebral white matter in multiple sclerosis: possible mechanisms and pathophysiological significance. *J Cereb Blood Flow Metab*. 2008;28(10):1645–1651.
14. Saab AS, Tzvetavona ID, Trevisiol A, et al. Oligodendroglial NMDA receptors regulate glucose import and axonal energy metabolism. *Neuron*. 2016;91(1):119–132.
15. Butt AM, Pugh M, Hubbard P, James G. Functions of optic nerve glia: axoglial signalling in physiology and pathology. *Eye (Lond)*. 2004;18(11):1110–1121.
16. Harris JJ, Attwell D. The energetics of CNS white matter. *J Neurosci*. 2012;32(1):356–371.
17. Hirrlinger J, Nave KA. Adapting brain metabolism to myelination and long-range signal transduction. *Glia*. 2014;62(11):1749–1761.
18. Butt AM, Ransom BR. Visualization of oligodendrocytes and astrocytes in the intact rat optic nerve by intracellular injection of lucifer yellow and horseradish peroxidase. *Glia*. 1989;2(6):470–475.
19. Ransom BR, Butt AM, Black JA. Ultrastructural identification of HRP-injected oligodendrocytes in the intact rat optic nerve. *Glia*. 1991;4(1):37–45.
20. Butt AM, Colquhoun K, Tutton M, Berry M. Three-dimensional morphology of astrocytes and oligodendrocytes in the intact mouse optic nerve. *J Neurocytol*. 1994;23(8):469–485.
21. Butt AM, Ransom BR. Morphology of astrocytes and oligodendrocytes during development in the intact rat optic nerve. *J Comp Neurol*. 1993;338(1):141–158.
22. Bay V, Butt AM. Relationship between glial potassium regulation and axon excitability: a role for glial Kir4.1 channels. *Glia*. 2012;60(4):651–660.
23. Butt AM, Duncan A, Hornby MF, et al. Cells expressing the NG2 antigen contact nodes of Ranvier in adult CNS white matter. *Glia*. 1999;26(1):84–91.
24. Greenwood K, Butt AM. Evidence that perinatal and adult NG2-glia are not conventional oligodendrocyte progenitors and do not depend on axons for their survival. *Mol Cell Neurosci*. 2003;23(4):544–558.
25. Young KM, Psachoulia K, Tripathi RB, et al. Oligodendrocyte dynamics in the healthy adult CNS: evidence for myelin remodeling. *Neuron*. 2013;77(5):873–885.
26. Verney C, Monier A, Fallet-Bianco C, Gressens P. Early microglial colonization of the human forebrain and possible involvement in periventricular white-matter injury of preterm infants. *J Anat*. 2010;217(4):436–448.
27. Hamilton N, Vayro S, Wigley R, Butt AM. Axons and astrocytes release ATP and glutamate to evoke calcium signals in NG2-glia. *Glia*. 2010;58(1):66–79.
28. Hamilton N, Vayro S, Kirchhoff F, et al. Mechanisms of ATP- and glutamate-mediated calcium signaling in white matter astrocytes. *Glia*. 2008;56(7):734–749.
29. Perez-Alvarez A, Araque A. Astrocyte-neuron interaction at tripartite synapses. *Curr Drug Targets*. 2013;14(11):1220–1224.

30. Quesseveur G, Gardier AM, Guiard BP. The monoaminergic tripartite synapse: a putative target for currently available antidepressant drugs. *Curr Drug Targets*. 2013;14(11):1277–1294.
31. Wake H, Moorhouse AJ, Miyamoto A, Nabekura J. Microglia: actively surveying and shaping neuronal circuit structure and function. *Trends Neurosci*. 2013;36(4):209–217.
32. Bergles DE, Jabs R, Steinhauser C. Neuron-glia synapses in the brain. *Brain Res Rev*. 2010;63(1–2):130–137.
33. Micu I, Plemel JR, Lachance C, et al. The molecular physiology of the axo-myelinic synapse. *Exp Neurol*. 2016;276:41–50.
34. Poliak S, Peles E. The local differentiation of myelinated axons at nodes of Ranvier. *Nat Rev Neurosci*. 2003;4(12):968–980.
35. Butt AM, Duncan A, Berry M. Astrocyte associations with nodes of Ranvier: ultrastructural analysis of HRP-filled astrocytes in the mouse optic nerve. *J Neurocytol*. 1994;23(8):486–499.
36. Chiti Z, Teschemacher AG. Exocytosis of norepinephrine at axon varicosities and neuronal cell bodies in the rat brain. *FASEB J*. 2007;21(10):2540–2550.
37. Kukley M, Capetillo-Zarate E, Dietrich D. Vesicular glutamate release from axons in white matter. *Nat Neurosci*. 2007;10(3):311–320.
38. Ziskin JL, Nishiyama A, Rubio M, Fukaya M, Bergles DE. Vesicular release of glutamate from unmyelinated axons in white matter. *Nat Neurosc*. 2007;10(3):321–330.
39. Yamazaki Y, Fujiwara H, Kaneko K, et al. Short- and long-term functional plasticity of white matter induced by oligodendrocyte depolarization in the hippocampus. *Glia*. 2014;62(8):1299–1312.
40. Funfschilling U, Supplie LM, Mahad D, et al. Glycolytic oligodendrocytes maintain myelin and long-term axonal integrity. *Nature*. 2012;485(7399):517–521.
41. Fruhbeis C, Frohlich D, Kuo WP, et al. Neurotransmitter-triggered transfer of exosomes mediates oligodendrocyte-neuron communication. *PLoS Biol*. 2013;11(7):e1001604.
42. Duncan A, Ibrahim M, Berry M, Butt AM. Transfer of horseradish peroxidase from oligodendrocyte to axon in the myelinating neonatal rat optic nerve: artefact or transcellular exchange? *Glia*. 1996;17(4):349–355.
43. Nagy JI, Ionescu AV, Lynn BD, Rash JE. Coupling of astrocyte connexins Cx26, Cx30, Cx43 to oligodendrocyte Cx29, Cx32, Cx47: implications from normal and connexin32 knockout mice. *Glia*. 2003;44(3):205–218.
44. Niu J, Li T, Yi C, et al. Connexin-based channels contribute to metabolic pathways in the oligodendroglial lineage. *J Cell Sci*. 2016;129(9):1902–1914.
45. Brown AM, Ransom BR. Astrocyte glycogen and brain energy metabolism. *Glia*. 2007;55(12):1263–1271.
46. Howell OW, Rundle JL, Garg A, Komada M, Brophy PJ, Reynolds R. Activated microglia mediate axoglial disruption that contributes to axonal injury in multiple sclerosis. *J Neuropathol Exp Neurol*. 2010;69(10):1017–1033.
47. Nahimi A, Jakobsen S, Munk OL, et al. Mapping alpha2 adrenoceptors of the human brain with 11C-yohimbine. *J Nucl Med*. 2015;56(3):392–398.
48. Penke L, Munoz Maniega S, Houlihan LM, et al. White matter integrity in the splenium of the corpus callosum is related to successful cognitive aging and partly mediates the protective effect of an ancestral polymorphism in ADRB2. *Behav Genet*. 2010;40(2):146–156.

49. Di Ieva A, Fathalla H, Cusimano MD, Tschabitscher M. The indusium griseum and the longitudinal striae of the corpus callosum. *Cortex.* 2015;62:34–40.
50. Mantyh PW, Rogers SD, Allen CJ, et al. Beta 2-adrenergic receptors are expressed by glia in vivo in the normal and injured central nervous system in the rat, rabbit, and human. *J Neurosci.* 1995;15(1 Pt 1):152–164.
51. De Keyser J, Wilczak N, Leta R, Streetland C. Astrocytes in multiple sclerosis lack beta-2 adrenergic receptors. *Neurology.* 1999;53(8):1628–1633.
52. De Keyser J, Wilczak N, Walter JH, Zurbriggen A. Disappearance of beta2-adrenergic receptors on astrocytes in canine distemper encephalitis: possible implications for the pathogenesis of multiple sclerosis. *Neuroreport.* 2001;12(2):191–194.
53. De Keyser J, Zeinstra E, Wilczak N. Astrocytic beta2-adrenergic receptors and multiple sclerosis. *Neurobiol Dis.* 2004;15(2):331–339.
54. Venugopalan VV, Ghali Z, Senecal J, Reader TA, Descarries L. Catecholaminergic activation of G-protein coupling in rat spinal cord: further evidence for the existence of dopamine and noradrenaline receptors in spinal grey and white matter. *Brain Res.* 2006;1070(1):90–100.
55. Papay R, Gaivin R, Jha A, et al. Localization of the mouse alpha1A-adrenergic receptor (AR) in the brain: alpha1AAR is expressed in neurons, GABAergic interneurons, and NG2 oligodendrocyte progenitors. *J Comp Neurol.* 2006;497(2):209–222.
56. Papay R, Gaivin R, McCune DF, et al. Mouse alpha1B-adrenergic receptor is expressed in neurons and NG2 oligodendrocytes. *J Comp Neurol.* 2004;478(1):1–10.
57. Khorchid A, Cui Q, Molina-Holgado E, Almazan G. Developmental regulation of alpha 1A-adrenoceptor function in rat brain oligodendrocyte cultures. *Neuropharmacology.* 2002;42(5):685–696.
58. Cohen RI, Almazan G. Norepinephrine-stimulated PI hydrolysis in oligodendrocytes is mediated by alpha 1A-adrenoceptors. *Neuroreport.* 1993;4(9):1115–1118.
59. Kastritsis CH, McCarthy KD. Oligodendroglial lineage cells express neuroligand receptors. *Glia.* 1993;8(2):106–113.
60. Ventimiglia R, Greene MI, Geller HM. Localization of beta-adrenergic receptors on differentiated cells of the central nervous system in culture. *Proc Natl Acad Sci USA.* 1987;84(14):5073–5077.
61. Ghiani CA, Eisen AM, Yuan X, DePinho RA, McBain CJ, Gallo V. Neurotransmitter receptor activation triggers p27(Kip1)and p21(CIP1) accumulation and G1 cell cycle arrest in oligodendrocyte progenitors. *Development.* 1999;126(5):1077–1090.
62. Larson VA, Zhang Y, Bergles DE. Electrophysiological properties of NG2 cells: matching physiological studies with gene expression profiles. *Brain Res.* 2015;1638(Pt B):138–160.
63. Mori K, Ozaki E, Zhang B, et al. Effects of norepinephrine on rat cultured microglial cells that express alpha1, alpha2, beta1 and beta2 adrenergic receptors. *Neuropharmacology.* 2002;43(6):1026–1034.
64. Dello Russo C, Boullerne AI, Gavrilyuk V, Feinstein DL. Inhibition of microglial inflammatory responses by norepinephrine: effects on nitric oxide and interleukin-1beta production. *J Neuroinflammation.* 2004;1(1):9.
65. Tanaka KF, Kashima H, Suzuki H, Ono K, Sawada M. Existence of functional beta1- and beta2-adrenergic receptors on microglia. *J Neurosci Res.* 2002;70(2):232–237.
66. Markus T, Hansson SR, Cronberg T, Cilio C, Wieloch T, Ley D. beta-Adrenoceptor activation depresses brain inflammation and is neuroprotective in lipopolysaccharide-induced sensitization to oxygen-glucose deprivation in organotypic hippocampal slices. *J Neuroinflammation.* 2010;7:94.

67. Gyoneva S, Traynelis SF. Norepinephrine modulates the motility of resting and activated microglia via different adrenergic receptors. *J Biol Chem*. 2013;288(21):15291–15302.
68. Kriegler S, Chiu SY. Calcium signaling of glial cells along mammalian axons. *J Neurosci*. 1993;13(10):4229–4245.
69. Bernstein M, Lyons SA, Moller T, Kettenmann H. Receptor-mediated calcium signalling in glial cells from mouse corpus callosum slices. *J Neurosci Res*. 1996;46(2):152–163.
70. Ding F, O'Donnell J, Thrane AS, et al. alpha1-Adrenergic receptors mediate coordinated $Ca2^{+}$ signaling of cortical astrocytes in awake, behaving mice. *Cell Calcium*. 2013;54(6):387–394.
71. Paukert M, Agarwal A, Cha J, Doze VA, Kang JU, Bergles DE. Norepinephrine controls astroglial responsiveness to local circuit activity. *Neuron*. 2014;82(6):1263–1270.
72. Constantinou S, Fern R. Conduction block and glial injury induced in developing central white matter by glycine, GABA, noradrenalin, or nicotine, studied in isolated neonatal rat optic nerve. *Glia*. 2009;57(11):1168–1177.
73. Honmou O, Young W. Norepinephrine modulates excitability of neonatal rat optic nerves through calcium-mediated mechanisms. *Neuroscience*. 1995;65(1):241–251.
74. Alix JJ, Dolphin AC, Fern R. Vesicular apparatus, including functional calcium channels, are present in developing rodent optic nerve axons and are required for normal node of Ranvier formation. *J Physiol*. 2008;586(Pt 17):4069–4089.
75. Muyderman H, Sinclair J, Jardemark K, Hansson E, Nilsson M. Activation of beta-adrenoceptors opens calcium-activated potassium channels in astroglial cells. *Neurochem Int*. 2001;38(3):269–276.
76. Roy ML, Sontheimer H. Beta-adrenergic modulation of glial inwardly rectifying potassium channels. *J Neurochem*. 1995;64(4):1576–1584.
77. Hertz L, Xu J, Song D, et al. Astrocytic glycogenolysis: mechanisms and functions. *Metab Brain Dis*. 2015;30(1):317–333.
78. Parpura V, Scemes E, Spray DC. Mechanisms of glutamate release from astrocytes: gap junction "hemichannels", purinergic receptors and exocytotic release. *Neurochem Int*. 2004;45(2–3):259–264.
79. Pankratov Y, Lalo U. Role for astroglial alpha1-adrenoreceptors in gliotransmission and control of synaptic plasticity in the neocortex. *Front Cell Neurosci*. 2015;9:230.
80. Paris A, Mantz J, Tonner PH, Hein L, Brede M, Gressens P. The effects of dexmedetomidine on perinatal excitotoxic brain injury are mediated by the alpha2A-adrenoceptor subtype. *Anesth Analg*. 2006;102(2):456–461.
81. Butt AM, Kalsi A. Inwardly rectifying potassium channels (Kir) in central nervous system glia: a special role for Kir4.1 in glial functions. *J Cell Mol Med*. 2006;10(1):33–44.
82. Rinholm JE, Hamilton NB, Kessaris N, Richardson WD, Bergersen LH, Attwell D. Regulation of oligodendrocyte development and myelination by glucose and lactate. *J Neurosci*. 2011;31(2):538–548.
83. Lee Y, Morrison BM, Li Y, et al. Oligodendroglia metabolically support axons and contribute to neurodegeneration. *Nature*. 2012;487(7408):443–448.
84. Bekar LK, He W, Nedergaard M. Locus coeruleus alpha-adrenergic-mediated activation of cortical astrocytes in vivo. *Cereb Cortex*. 2008;18(12):2789–2795.
85. Griffith R, Sutin J. Reactive astrocyte formation in vivo is regulated by noradrenergic axons. *J Comp Neurol*. 1996;371(3):362–375.

86. Sutin J, Griffith R. Beta-adrenergic receptor blockade suppresses glial scar formation. *Exp Neurol.* 1993;120(2):214–222.
87. Rivera A, Vanzuli I, Arellano JJ, Butt A. Decreased regenerative capacity of oligodendrocyte progenitor cells (NG2-Glia) in the ageing brain: a vicious cycle of synaptic dysfunction, myelin loss and neuronal disruption? *Curr Alzheimer Res.* 2016;13(4):413–418.
88. Hughes EG, Kang SH, Fukaya M, Bergles DE. Oligodendrocyte progenitors balance growth with self-repulsion to achieve homeostasis in the adult brain. *Nat Neurosci.* 2013;16(6):668–676.
89. Simonini MV, Polak PE, Sharp A, McGuire S, Galea E, Feinstein DL. Increasing CNS noradrenaline reduces EAE severity. *J Neuroimmune Pharmacol.* 2010;5(2):252–259.
90. Sherwin C, Fern R. Acute lipopolysaccharide-mediated injury in neonatal white matter glia: role of TNF-alpha, IL-1beta, and calcium. *J Immunol.* 2005;175(1):155–161.
91. Butt AM, Jenkins HG. Morphological changes in oligodendrocytes in the intact mouse optic nerve following intravitreal injection of tumour necrosis factor. *J Neuroimmunol.* 1994;51(1):27–33.
92. Pappas AC, Koide M, Wellman GC. Purinergic signaling triggers endfoot high-amplitude $Ca2^{+}$ signals and causes inversion of neurovascular coupling after subarachnoid hemorrhage. *J Cereb Blood Flow Metab.* 2016;36(11):1901–1912.
93. Mulligan SJ, MacVicar BA. Calcium transients in astrocyte endfeet cause cerebrovascular constrictions. *Nature.* 2004;431(7005):195–199.
94. Filosa JA, Bonev AD, Straub SV, et al. Local potassium signaling couples neuronal activity to vasodilation in the brain. *Nat Neurosci.*. 2006;9(11):1397–1403.
95. Pelligrino DA, Vetri F, Xu HL. Purinergic mechanisms in gliovascular coupling. *Semin Cell Dev Biol.* 2011;22(2):229–236.

Chapter 4

Role for Astroglial α_1-Adrenergic receptors in Glia-Neuron Communications and Aging-Related Metaplasticity in the Neocortex

Ulyana Lalo[1,✉] and Yuriy Pankratov[1,2,✉]

[1]University of Warwick, Coventry, United Kingdom, [2]Immanuel Kant Baltic Federal University, Kaliningrad, Russia

Chapter Outline

Introduction 82

Role for Astroglia in Brain Signaling and Metaplasticity 83

Astrocytic Ca^{2+} Signaling: Specific Role for Adrenergic Receptors 84

Adrenergic Receptors Induce the Release of Gliotransmitters From Neocortical Astrocytes 87

Age- and Environment-Related Alterations in Astroglial Adrenergic Signaling 89

Astroglial α_1-Adrenergic Receptors Modulate Synaptic Transmission and Plasticity in the Neocortex 91

Summary and Perspectives 99

Abbreviations 99

References 100

ABSTRACT

Communication between neuronal and glial cells is very important for mainlining brain function across a lifetime. Astroglia can receive and integrate signals from neurons and respond by modulating neuronal activity via release of gliotransmitters. Astrocytes therefore are strategically positioned to mediate the positive influence of enhanced mental and physical activity on the brain function. Still, age- and environment-related changes in astroglial signaling and their impact on brain function remain largely unexplored.

✉Correspondence address

E-mail: u.lalo@warwick.ac.uk; y.pankratov@warwick.ac.uk

Noradrenergic Signaling and Astroglia. DOI: http://dx.doi.org/10.1016/B978-0-12-805088-0.00004-9

© 2017 Elsevier Inc. All rights reserved.

Recently we showed the importance of noradrenaline-mediated signaling in the neocortical astrocytes. Noradrenaline did not evoke direct events in neurons but caused significant elevation in the cytosolic Ca^{2+} in astrocytes that in turn promoted vesicular release of gliotransmitters leading to enhancement of the long-term synaptic plasticity. We found out that adrenergic signaling in the neocortical astrocytes undergoes a considerable age-related decline, but enriched environment can ameliorate this and enhance the adrenergic receptor-triggered release of gliotransmitters.

Combined, our data suggest that adrenergic component of astrocyte-neuron communication is important for maintaining the balance between excitatory and inhibitory transmission in the neocortex. This interaction can significantly decline with aging and thus contribute to the age- and pathology-related cognitive impairment.

Keywords: Astrocyte; gliotransmitters; adrenergic receptors; AMPA receptors; GABA receptors; metaplasticity; synaptic plasticity

INTRODUCTION

Although maintaining brain function is very important for healthy aging, fundamental mechanisms of brain aging and longevity are not fully understood. Cognitive functions involve changes in the strength of synaptic connections between neurons, generally termed plasticity that is as an active process continuously occurring across a lifespan. Nowadays, age-related changes in the cognitive functions are viewed as consequence of alterations in the mechanisms rather than complete loss of synaptic plasticity.[1,2] These alterations are driven by variety of genetic and biochemical mechanisms that are yet to be fully studied.[2–5] Importantly, neural networks and synapses are remarkably responsive to environmental stimuli, physiological modifications, and experiences.[6]

In many regions of adult animal and human brain, particularly in the cortex, synapses can alter their biochemistry, structure, and function in response to an enriched environment (EE).[7,8] First described by Hebb,[9] the experimental paradigm of EE in animal models refers to an environment that provides greater possibilities for physical activity and physical and social stimulation than standard housing (SH) conditions.[10] EE is frequently used in animal models of experience-induced alteration of synaptic plasticity and is widely viewed as animal proxy of an active life-style in humans.[1–3,6,10] There are accumulating reports that EE and physical activity can, at least partially, rescue the long-term synaptic plasticity and memory in animal models of brain injury and neurodegenerative disorders.[2,3,11]

A dynamic regulation of synaptic plasticity, termed "metaplasticity," is regarded nowadays as very important for brain computational and cognitive mechanisms. Often metaplasticity is defined as "the plasticity of synaptic plasticity." Metaplasticity can occur when priming synaptic or cellular activity or inactivity leads to persistent change in the direction or degree of synaptic plasticity.[12] Astrocytes are gaining an increasing attention as a very important element of heterosynaptic metaplasticity.[13]

Significantly, the release of noradrenaline (NA) in the brain cortex occurs by "volume transmission," i.e., from varicosities of noradrenergic neurons into brain extracellular fluid rather than into a cleft of specialized synapses. Due to such remote mode of release, astrocytes might be the first cell target to receive and integrate diffuse noradrenergic input and pass the information to the local network. Indeed, adrenergic signaling has been reported to modulate the activity of astrocyte networks according to the behavioral state or sensory inputs.[14,15] Thus, astroglial adrenergic receptors (ARs) may be of particular importance for brain metaplasticity caused by environmental factors or aging.

This chapter will review the recent data demonstrating an important role for adrenergic astroglial signaling in aging- and environment-related brain metaplasticity.

Role for Astroglia in Brain Signaling and Metaplasticity

Results of last two decades suggest that neuronal networks cannot be considered as the sole substrate of higher brain function and neuronal-glial networks coordinate information processing in the brain.[13,16,17] A key element of this coordination is the ability of astrocytes to monitor the activity of neurons over large territory (often referred to as astrocytic "functional domains") by virtue of high-affinity receptors and provide a feedback mainly by modulation of synaptic strength or neuronal metabolism.[18–20] In comparison to interaction between neurons, astroglial modulation of synaptic transmission gains many peculiar features, such as slower timescale but longer spatial-scale and dependence on combined activity of numerous synapses.[16] These features can render a particular importance for astrocyte-neuron communication in the heterosynaptic metaplasticity.[13]

The central role in this communication belongs to the ability of astrocytes to release gliotransmitters in Ca^{2+}-dependent manner.[16,21] It has been widely reported[22,23] that release of glutamate and D-serine from astrocytes can modulate the activity of neuronal excitatory and inhibitory synapses,[23–25] long-term synaptic plasticity,[16,18,22,26,27] and vascular coupling.[20,28] Although detailed mechanisms of gliotransmission remain uncertain and are widely debated, importance of Ca^{2+}-dependent vesicular and nonvesicular pathways has been recently reported.[16,23,28,29] In the neocortex, Ca^{2+}-dependent exocytosis of adenosine triphosphate (ATP) and D-serine can be triggered by astrocytic PAR1 receptors, CB1 endocannabinoid receptors, and α_1-ARs[23,26,30] and facilitate induction of long-term potentiation (LTP) of synaptic transmission,[26,30] including in the aged animals.[22]

Our recent data show that glia-derived ATP can downregulate the GABAergic tonic and phasic inhibitory transmission.[23] At the same time, there is an evidence of the positive impact of ATP on activity and trafficking of AMPA receptors.[24,31] Hence, glia-derived ATP may be instrumental for maintenance of balance between excitation and inhibition and thereby affect the induction of synaptic plasticity. This may underlie another pathway of glial regulation of heterosynaptic metaplasticity, complementing the widely discussed cascade involving conversion of glia-derived ATP into adenosine and activation presynaptic A1 receptors.[13,16,32]

Astrocytic Ca^{2+} Signaling: Specific Role for Adrenergic Receptors

Elevation of cytosolic Ca^{2+} is the major way of glial activation and integration of information within the glial network.[16,17] Astrocytes express a variety of G_q protein-coupled neurotransmitter receptors that trigger inositol trisphosphate (IP_3)-mediated release of Ca^{2+} from endoplasmic reticulum (ER).[16,18,33] In addition, Ca^{2+}-permeable ligand-gated ion channels can contribute to astroglial signaling.[34,35] There is accumulating evidence that Ca^{2+} signaling in brain astrocytes can be initiated by (1) synaptic release of neurotransmitters[23,34,36,37]; (2) local autocrine release of gliotransmitters such as glutamate or ATP[16,38]; and (3) via diffuse volume-transmitted neuromodulators such as serotonin (5-hydroxytryptamine (5-HT)), acetylcholine (ACh), or NA.[14,15,33,39] The physiological roles of these pathways are intensively debated.[16,17,33] There is growing opinion that astrocytes are very important for a slower timescale but long-spatial scale modulation of synaptic transmission.[13,16,33] Astrocytic α_1-ARs have been recently highlighted as important participants in Ca^{2+} signaling in astrocytes, in contrast to 5-HT and ACh receptors.[14,39]

Our recent data reveal the specific features of adrenergic signals in neocortex that have been overlooked or not extensively studied so far.[30] First, astroglial α_1-ARs turned out to be very sensitive. Application of sub- and low-micromolar concentrations of NA to astrocytes of neocortical slices induced robust Ca^{2+} elevations both in their somata and branches and increased the amplitude and frequency of fast spontaneous Ca^{2+} transients (Fig. 4.1A). The EC_{50} for NA-activated Ca^{2+} response in neocortical astrocytes was about 370 nM (Fig. 4.1B), the EC_{50} for the specific α_1-AR agonist A61603 was about 19 nM.[30]

Second, astroglial α_1-ARs exhibited rapid desensitization. We observed that repetitive application of NA or specific α_1-AR agonist A61603 lead to significant reduction in the astroglial Ca^{2+}-response amplitude. After desensitization, astroglial NA signaling required rather long (10–15 min) time period to recover (Fig. 4.1C). Vulnerability of astroglial adrenergic signaling to desensitization suggests the existence of molecular mechanisms preventing their overstimulation, likely to avoid Ca^{2+}-overload. The desensitization of ARs upon prolonged exposure to an agonist was also

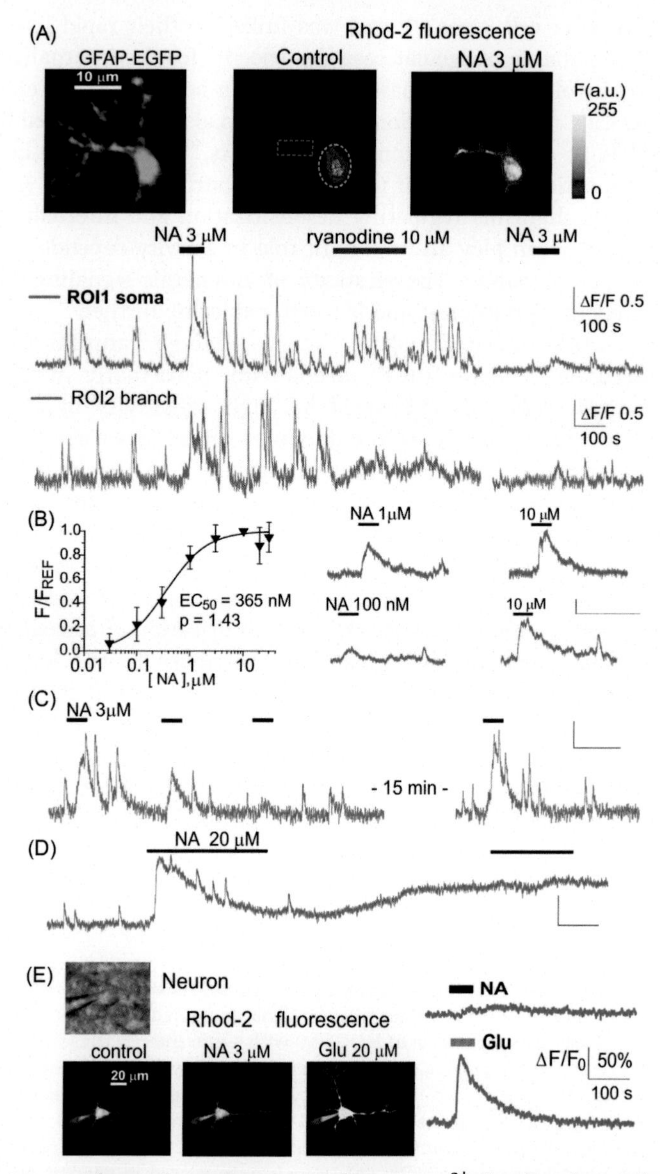

FIGURE 4.1 α_1-Adrenergic receptors contribute to Ca^{2+}-signaling in cortical astrocytes. (A) Representative multiphoton images of EGFP fluorescence and presudocolor images of Rhod-2 fluorescence recorded in the astrocytes from GFAP-EGFP (GFEC) mouse before (control) and after the application of noradrenaline (NA). Graphs below show the time course of Rhod-2 fluorescence averaged over regions indicated in fluorescence images. Note the significant NA-activated response in the astrocytic branch and inhibition of response to NA 10 min after application of ryanodine. (B) The concentration-dependence of net Ca^{2+}-transients evoked by NA in cortical astrocytes; each point shows mean $\pm$ SD for 4–5 cells.

(*Continued*)

observed in other cell types[40,41] and was linked to their rapid internalization.[40,42] The putative molecular cascades specific for the internalization of α_1-ARs may involve protein kinase C, which can be activated by elevations in cytosolic Ca^{2+}.[40] Internalization of α_1-ARs can also be influenced by their association with the cytoskeleton and lipid rafts,[40,43] which can be linked to other astrocytic receptors, for instance P2X1 purinoceptors.[44,45] Existence of feedback mechanisms regulating desensitization and internalization of astrocytic α_1-ARs can play an important role in activity-dependent plasticity of astroglial signaling. The plasticity of adrenergic signaling in astrocytes is surely of high interest and is worth exploring further.

Third, α_1-AR-mediated signals can undergo amplification by Ca^{2+}-induced Ca^{2+}-release (CICR) mechanism, particularly in the astrocytic branches. This was evidenced by high sensitivity of NA-evoked responses in cortical astrocytes to ryanodine—an inhibitor of Ca^{2+}-release from ER. Application of 10_μM ryanodine caused a transient initial augmentation of spontaneous Ca^{2+}-signaling in the astrocytes which was followed by irreversible inhibition of both spontaneous and NA-evoked Ca^{2+}-transients (Fig. 4.1A). Such behavior is typical for ryanodine modulation of CICR mechanism.[46] These results suggest the significant role of ryanodine receptor-mediated CICR in the amplification of astroglial Ca^{2+}-signaling. This amplification could also contribute to the long-lasting ("tonic") elevation of cytosolic Ca^{2+} in astrocytes evoked by prolonged application of high (10–30 μM) concentrations of NA (Fig. 4.1D).

Finally, cortical neurons did not exhibit any significant direct contribution of α_1-ARs into Ca^{2+} or synaptic signaling.[30] In our experiments in the neocortical pyramidal neurons, the amplitude of responses evoked by 3 μM NA was much smaller than the amplitude of glutamate-evoked response (Fig. 4.1E). We also did not observe any significant effect of

---

◀ Net response was evaluated as an integral Ca^{2+}-signal measured during 3 min after NA application and normalized to the integral Ca^{2+} signal measured during 3 min before NA application. (C) Repetitive application of NA (3 μM) with 5-min interval causes the desensitization of the response; (D) prolonged application of 20 μM NA leads to the elevated Ca^{2+} level and nonresponsiveness of neocortical astrocytes. Fluorescent signals shown in panels (B)–(D) were integrated over the cell somata; all scale bars are ΔF 0.5 and 200 s. (E) The pyramidal neurons of somatosensory cortex layer 2/3 of wild-type mice were loaded with Ca^{2+}-indicator Rhod-2 via patch-pipette. Ca^{2+}-signals were evoked in neurons by 60-s-long rapid bath application of 3 μM NA and 100 μM L-glutamate to cortical slices. The membrane holding potential during Ca^{2+}-measurements was −40 mV. *Right*, the representative gradient-contrast image and pseudocolor fluorescent images recorded at rest and at the peak of Ca^{2+}-responses to NA. *Left*, the representative Ca^{2+}-transients evoked in the neuron of dn-SNARE mouse by application of NA and glutamate. Note that application of NA did not evoke significant cytosolic Ca^{2+}-elevation, in contrast to L-glutamate. *Modified from Pankratov Y., Lalo U. Role for astroglial alpha1-adrenoreceptors in gliotransmission and control of synaptic plasticity in the neocortex.* Front Cell Neurosci. *2015;9:230.*

α_1-AR antagonist terazosin on the field excitatory synaptic potentials (fEPSPs) in the neocortical slices.[30] Furthermore, neither NA nor terazosin exhibited significant effects on synaptic currents in neocortical synapses when neurons were isolated from influence of astrocytes.[30]

It is worth noting that NA-evoked Ca^{2+}-signaling in the neocortical astrocytes was effectively blocked by specific α_1-AR antagonist terazosin.[30] Also, the effects of NA on astroglial Ca^{2+}-signaling and release of gliotransmitters were reproduced by the specific α_1-AR agonist A61603.[30] These results suggest a dominant role for α_1-subtype of ARs in the adrenergic signaling in neocortical astrocytes. Still, apart from α_1-, astrocytes can also express α_2- and β_1-ARs that may potentially influence the membrane trafficking and function of other astroglial receptors acting via G_i- and G_s-proteins.[14,47] The putative physiological role for astroglial α_2- and β_1-ARs is yet to be studied.

Combined, these data substantiate an important role for α_1-ARs in Ca^{2+}-signaling in neocortical astrocytes. Also, our results verify the suitability of application of NA in low micromolar dosage as an efficient, physiologically relevant, and specific method of activating astrocytes in the cortical slices. Our results on "used-dependent" augmentation of cytosolic Ca^{2+} by higher concentration of NA (Fig. 4.1D) could also have an important practical implication. Currently, some therapeutic strategies intended to enhance NA neurotransmission to slow down the progression of neurodegenerative diseases use a selective NA reuptake inhibition, and a treatment with NA pro-drug L-threo-3,4-dihydroxyphenylserine.[48] As our data suggest, such approach can have unwanted side-effects leading to long-lasting calcium overload in astrocytes.

Adrenergic Receptors Induce the Release of Gliotransmitters From Neocortical Astrocytes

Based on ability of α_1-ARs to activate astroglial Ca^{2+}-signaling, one could expect their involvement into triggering release of gliotransmitters. To verify this, we used a combination of approaches including recording the neuronal signals activated by gliotransmitters, direct measurements of gliotransmitters concentrations in the brain tissues, usage of transgenic mice with inducible astroglial expression of dominant-negative SNARE (soluble N-ethylmaleimide-sensitive factor attachment protein receptor) domain (dn-SNARE),[32] intracellular perfusion of astrocytes with Ca^{2+}-chelators, and inhibitors of SNARE proteins.

To monitor the concentration of gliotransmitters in the neocortical tissue, we used microelectrode biosensors to ATP, D-serine and glutamate, as described previously.[26,49] As NA can be oxidized by the sensors and thus generate a measurable sensor current, we used the specific α_1-AR agonist A61603. Activation of astroglial α_1-ARs by A61603 induced a

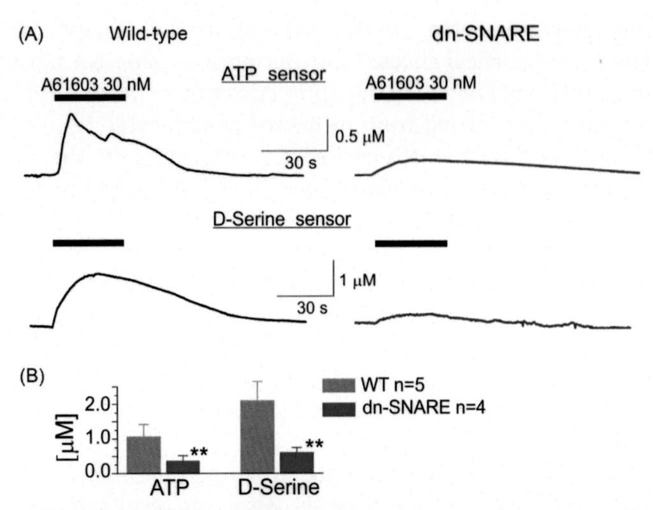

FIGURE 4.2 Adrenergic receptor-activated release of ATP and D-serine in the neocortical slices. (A) The representative responses of cortical slices of wild-type and dn-SNARE mice to the application of selective α_1-AR agonist A61603 (100 nM) were recorded using microelectrode sensors to ATP and D-serine placed in the layer II/III. The data are shown as an elevation relative to the resting concentration. (B) The pooled data on the peak magnitude of ATP- and D-serine transients evoked by application of A61603; data shown as mean ± SD for number of experiments indicated. Asterisks (**) indicate statistical significance of difference in the magnitude of ATP- and D-serine responses between wild-type and dn-SNARE mice, $P < 0.01$ (t-test). The significant reduction in the NA-evoked responses in the cortical slices from dn-SNARE mice strongly supports the vesicular mechanism of ATP and D-serine release from astrocytes. *First published in Pankratov Y., Lalo U. Role for astroglial alpha$_1$-adrenoreceptors in gliotransmission and control of synaptic plasticity in the neocortex.* Front Cell Neurosci. *2015;9:230.*

significant elevation of concentration of extracellular ATP and D-serine in the cortical tissues of wild-type mice (Fig. 4.2A). In comparison to the wild-type littermates, the amplitudes of ATP and D-serine transients were significantly reduced in the dn-SNARE expressing mice (Figs. 4.2A and B). This result suggested the significant role for vesicular mechanism in AR-activated astroglial release of ATP and D-serine. The incomplete inhibition of release of ATP and D-serine might be explained by partial expression of dn-SNARE in neocortical astrocytes of dn-SNARE mice.[32]

As an alternative approach to directly verify that astroglial α_1-ARs contribute to triggering Ca^{2+}-dependent exocytosis, we used ability of neocortical pyramidal neurons to respond to glia-derived ATP by virtue of expression of functional P2X receptors.[23] We recorded the whole-cell currents in neocortical pyramidal neurons at a membrane potential of −80 mV in the presence of 6,7-dinitroquinoxaline-2,3-dione (DNQX; 30 μM), D-2-Amino-5-phosphonopentanoic acid(D-APV; 30 μM), and picrotoxin (100 μM), to eliminate signals mediated correspondingly by the

AMPA, NMDA, and GABAA receptors. Similar to our previous experiments,[23,50] we observed residual nonglutamatergic miniature excitatory spontaneous synaptic currents (mEPSCs) mediated by P2X receptors (Fig. 4.3). In order to inhibit astroglial exocytosis, we perfused individual astrocytes with intracellular solution containing 3-nM tetanus neurotoxin (TeNTx) and 30 μM of the Ca^{2+}-indicator Calcium Green-2 (Fig. 4.3A) and recorded purinergic mEPSCs in a neuron lying in vicinity of perfused astrocyte. Perfusion of astrocytes with solution containing only Calcium Green-2 was used as a control.

The purinergic mEPSCs recorded under control conditions had an average amplitude of 8.4 ± 2.5 pA and an average decay time of 9.6 ± 2.6 ms. Application of NA (2 μM) caused a dramatic increase in the frequency of purinergic mEPSCs in pyramidal neurons. The burst of purinergic currents was accompanied by a decrease in the average amplitude to 6.7 ± 1.7 pA and an increase in the decay time to 13.3 ± 3.5 ms ($n = 7$). When astrocytes were perfused with TeNTx, activation of α_1-ARs caused much smaller burst of purinergic currents (Fig. 4.3C). Consistent with our previous reports,[23] purinergic currents in pyramidal neurons (Fig. 4.3D) exhibited a bimodal amplitude distribution with peaks at 5.8 ± 1.5 and 9.8 ± 2.4 pA ($n = 7$). The distributions of mEPSCs decay time in these neurons had peaks at 9.1 ± 1.1 and 15.2 ± 2.1 ms. Previously, we demonstrated that purinergic mEPSCs of smaller amplitude and slower kinetics originated from the vesicular release of ATP from astrocytes.[23,26] Application of NA dramatically increased the proportion of these smaller and slower currents (Fig. 4.3E). Perfusion of astrocytes with TeNTx selectively decreased the frequency of slower purinergic mEPSCs, both in control conditions and after application of NA (Fig. 4.3F). These results strongly suggest that smaller and slower purinergic mEPSCs originated directly from vesicular release of ATP from neighboring astrocytes. The NA-evoked bursts of purinergic mEPSCs were also significantly inhibited in the dn-SNARE mice, supporting their astroglial origin (Fig. 4.3F).

Age- and Environment-Related Alterations in Astroglial Adrenergic Signaling

Mechanisms of glia-neuron interaction, in particular release of gliotransmitters, gain a new importance in the context of brain aging. Despite accumulating evidence of the benefits of exercise and EE on neurogenesis and synaptic plasticity in aged brain,[2–4,11] their impact on astroglial signaling and release of gliotransmitters in old age remains elusive.

In our previous work, we showed that maturation and aging of the neocortex in mice over 1–24 months affected the density of purinergic and glutamatergic receptors and their contribution to the synaptically activated glial signaling.[23,37] Following the changes in the density of P2X

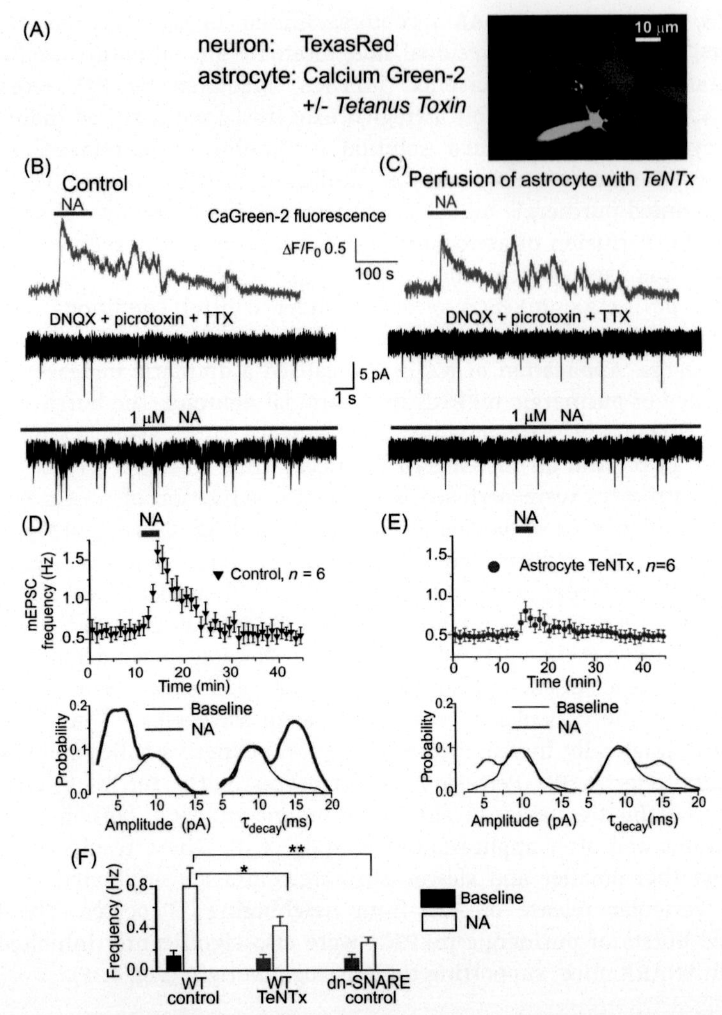

FIGURE 4.3 Astrocyte-derived ATP activates purinergic receptors in the neocortical neurons. (A) Whole-cell currents were recorded in the pyramidal neuron of layer 2/3 simultaneously with perfusion of astrocyte with intracellular solution containing either fluorescent dye Ca-Green-2 alone or Ca-Green-2 and 3 nM of tetanus neurotoxin (TeNTx) light chain. (B and D) Activation of astrocytic Ca^{2+}signaling elicited the burst of purinergic currents in the neuron. (B) The astrocytic Ca^{2+} transients (green) evoked by application of 1 μM NA and neuronal whole-cell currents (black) recorded in the presence of picrotoxin, 6,7-dinitroquinoxaline-2,3-dione (DNQX), and tetrodotoxin (TTX) at −80 mV. Spontaneous currents recorded in these conditions are mediated by the P2X receptors. Upper and lower trace show currents recorded correspondingly before and 1 min after NA application. Note the appearance of the large number of spontaneous currents after NA. (D) *Upper graph*: Each dot shows the average frequency of spontaneous purinergic currents recorded in 1 min time window in the pyramidal neurons; data are presented as mean ± SD for 6 neurons. *Lower graphs*

(*Continued*)

and NMDA receptors, the synaptically activated Ca^{2+}-signaling was maximal at 3–6 months and then declined steeply.[23] Consistent with age-related decline in astroglial Ca^{2+}-signaling, the exocytosis of ATP from neocortical astrocytes significantly decreased in the old age.[22] Hence, one might expect the astroglial adrenergic signaling to follow the similar trend of dramatic aging-related decline.

We explored the difference in the NA-elicited cytosolic Ca^{2+}-transients in the neocortical astrocytes of 3–6 months old (mature adult) and 14–15 months old (old) mice and compared them to Ca^{2+}-signals elicited by ATP and NMDA (Fig. 4.4). Rather surprisingly, the net Ca^{2+}-response to NA in cortical astrocytes showed only moderate (albeit statistically significant) decrease in the old mice (Fig. 4.4A). This behavior contrasted to the glutamate and ATP receptor-mediated Ca^{2+}-signaling (Fig. 4.4B). Importantly, the NA-mediated astroglial response was upregulated in mice kept in the EE (Fig. 4.4B).

To evaluate the age-related alterations in gliotransmission in the same age group of mice, we measured extracellular concentrations of ATP, D-serine and glutamate in the neocortical tissue with microelectrode biosensors and activated astrocytes with specific α_1-AR agonist A61603. Similarly to Ca^{2+}-signaling, release of all gliotransmitters significantly decreased in the old age but was strongly upregulated by the EE (Fig. 4.4C and D).

These results have two important implications. First, they strongly support the notion of substantial age- and environment-related alterations in the function of astrocytes. Second, out data suggest that relative contribution of ARs (as compared to ATP or glutamate receptors) into astroglial signaling and release of gliotransmitters can increase with aging.

Astroglial α_1-Adrenergic Receptors Modulate Synaptic Transmission and Plasticity in the Neocortex

Our data show that astroglial α_1-ARs can trigger exocytosis of gliotransmitters, in particular ATP and D-serine (Figs. 4.2–4.4). ATP and D-serine

---

◀ show the amplitude and decay time distributions for the purinergic currents recoded before (baseline) and 1–3 min after application of NA. Application of NA elicited the burst of purinergic currents which had slower kinetics and smaller-quantal size than currents recorded in the baseline conditions. (C and E) Perfusion of astrocyte with TeNTx inhibited the burst of purinergic currents verifying their origin from astrocytic exocytosis. (F) Diagram shows the frequency of slow purinergic spontaneous currents in the neurons of wild-type and dn-SNARE mice averaged within 3 min time window before (baseline) and after application of NA in control and during perfusion of astrocytes with TeNTx. Data are shown as mean ± SD for the 6 neurons. The statistical significance of difference from the wild-type control values was as indicated (*) $P < 0.05$ and (**) $P < 0.01$ (unpaired *t*-test). *First published in Pankratov Y., Lalo U. Role for astroglial alpha$_1$-adrenoreceptors in gliotransmission and control of synaptic plasticity in the neocortex.* Front Cell Neurosci. *2015;9:230.*

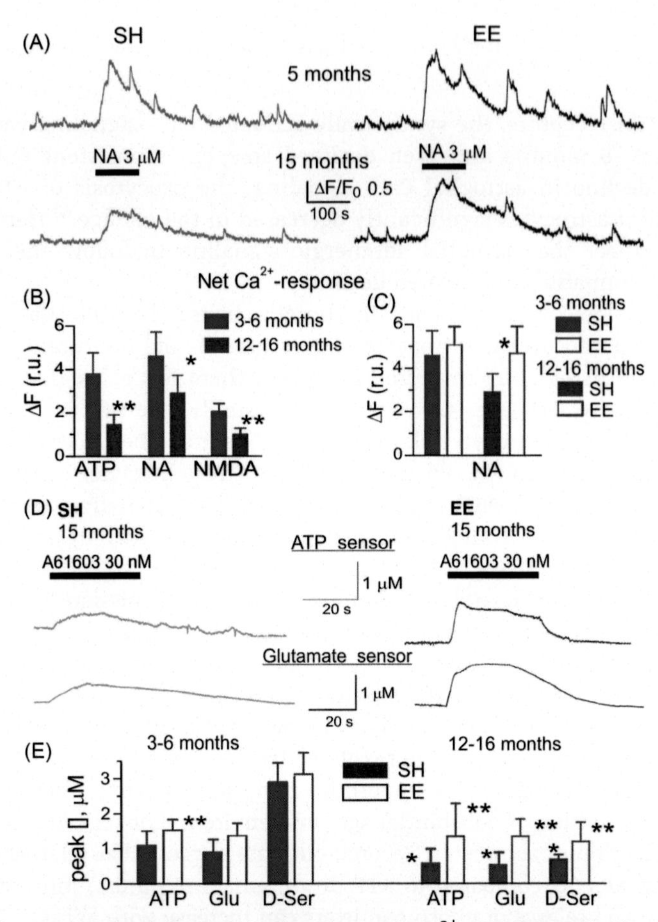

FIGURE 4.4 Age-related changes in the adrenergic Ca^{2+} signaling and release of gliotransmitters in the neocortex. Astroglial Ca^{2+}-signaling (A–C) and release of ATP, D-serine and glutamate (D and E) were evaluated in the neocortex of 3–6 month-old and 12–16 month-old mice as described previously in[22,26,30] and illustrated in Figs. 4.1 and 4.2. Wild-type (C57bl6) mice were kept either in standard housing (SH) or exposed to environmental enrichment (EE) as described in[52]. (A) The representative Ca^{2+}-responses to noradrenaline (NA) recorded in the layer 2/3 astrocytes of mice of different age and housing conditions. (B and C) The pooled data on the net response to the agonists of P2, NMDA, and α_1-ARs receptors recorded in astrocytes of mice of different age and environment group (C); data are shown as mean ± SD for 5 to 7 cells. Asterisks indicate statistical significance of the differences between two age groups (**) and between wild-type and APP/PS1tg mice (*); $P < 0.05$ as given by unpaired *t*-test. Note the significant age-related changes and increase in Ca^{2+}-signaling in the EE-mice of old age. (D and E) Glial release of ATP, glutamate, and D-serine in the neocortical slice was activated by application of 30 nM of selective α_1-AR agonist A61603 and detected using microelectrode biosensors. (D) The representative A61603-evoked responses of ATP- and Glu-sensors placed in the layer 2/3 of neocortical slices. (E) Pooled data on the magnitude of A61603-evoked transients recorded in the neocortex of mice of different age and environment groups; data are shown as mean ± SD for seven slices in the SH and six slices in the EE conditions. Asterisks indicate statistical significance of the differences between two age groups (**) and between wild-type and APP/PS1tg mice (*); $P < 0.05$ as given by unpaired *t*-test. Note the significant decrease in the concentrations of all transmitters in EE.

have been previously shown to modulate synaptic plasticity in the hippocampus[27,32] and the neocortex.[22,26] Although control of synaptic strength by glia-derived ATP and D-serine is gaining increasing attention nowadays,[16,17,51] the detailed molecular mechanisms underlying action of these gliotransmitters are yet to be explored. Although effects of D-serine on synaptic plasticity and development originate predominantly from coagonistic action on NMDA receptors,[32,51] ATP can act via different pathways, both pre- and postsynaptic. The presynaptic action of glia-derived ATP, which involves conversion to adenosine and modulation of release of neurotransmitters via A1 receptors, is widely acknowledged.[13,16,32] The postsynaptic mechanisms of ATP action, which potentially could strongly affect the synaptic strength, have been overlooked until recently.

However, recent data have shown the capability of P2X receptors, activated by ATP released from astrocytes, to modulate excitatory synapses in the hippocampal and magnocellular neurons[24,31] and GABAergic inhibition in the neocortical neurons.[23] In these works, the predominant role in the glia-driven purinergic modulation of synaptic strength was suggested for the extra-synaptic P2X receptors. Significantly, that purinergic modulation of AMPA receptors was evoked in both cases[24,31] via activation of astrocytes by NA. Hence, activation of ATP release via astroglial α_1-ARs could affect both excitatory and inhibitory synapses.

So far, it remains unknown whether activation of postsynaptic P2X receptors by glia-derived ATP can efficiently modulate the strength of excitatory synapses in the principal neurons of the neocortex. It is worth noting that ATP may exert both positive[24] and negative[31] effects on AMPA receptor-mediated signaling. In the magnocellular neurons, activation of extrasynaptic P2X receptors causes synaptic upscaling via PI3K-dependent insertion of AMPA receptors into dendritic spines,[26] whereas in the CA1 neurons P2X receptors triggered CamKII-dependent internalization of AMPA receptors and synaptic depression.[31] Potentially, the interference between these opposing mechanisms can undermine the net impact of glia-derived ATP on the strength of excitatory synapses. So, putative effect of ATP on the balance between excitation and inhibition cannot be predicted a priori and would, very likely, depend on the physiological context. The diversity and uncertainty in the mechanisms by which glia-derived ATP can regulate synaptic strength contributed to the current debate on physiological relevance of gliotransmission.[17]

To evaluate the effect of astrocytes on glutamatergic and GABAergic transmission in the neocortex, we monitored the miniature excitatory and inhibitory synaptic currents (mEPSCs and mIPSCs) in the neocortical pyramidal neurons before and after activation of astroglial α_1-ARs (Fig. 4.5). The AMPA receptors-mediated mEPSCs were recorded and analyzed as described previously[50] at membrane potential of -80 mV in presence of tetrodotoxin (TTX) (1 μM) and picrotoxin (100 μM); the GABAA

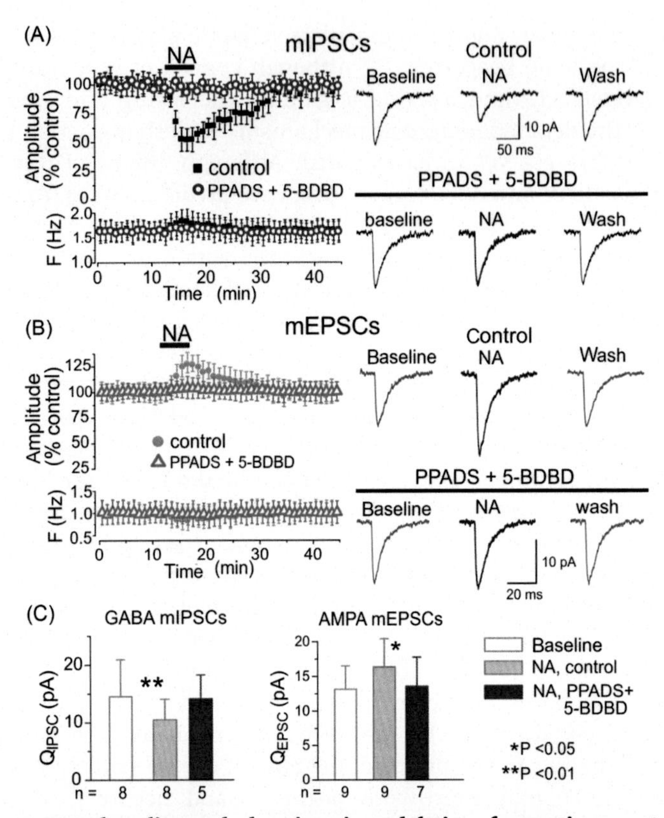

FIGURE 4.5 **Noradrenaline-evoked purinergic modulation of synaptic currents.** Release of gliotransmitters was elicited by application of noradrenaline (NA) as shown in Figs. 4.1–4.3. Spontaneous miniature inhibitory and excitatory synaptic currents were recorded in the somatosensory cortex layer 2/3 pyramidal neurons at −80 mV in the presence of 1 μM tetrodotoxin (TTX), 3_μM DPCPX (8-cyclopentyl-1,3-dipropylxanthine), and 50 μM DNQX (6,7-dinitroquinoxaline-2,3-dione; mIPSCs) or 100 μM picrotoxin (mEPSCs). Currents were detected and analyzed as described previously.[22,50] (A) *Left* panel, plots show the time course of mIPSCs recorded in the neurons of wild-type mice during application of NA (1 μM) in control and in presence of P2 purinoreceptors antagonists PPADS (pyridoxal-phosphate-6-azophenyl-2',4'-disulfonic acid; 10 μM) and 5-BDBD (5-(3-bromophenyl)-1,3-dihydro-2H-benzofuro[3,2-e]-1,4-diazepin-2-one; 5 μM). Data points represent the relative changes in the mean amplitude and frequency for mIPSCs recorded within 1-min time window; data are shown as mean ± SD for number of neurons indicated in (C). *Right* panel shows the average waveforms (25 mIPSCs each) of synaptic currents recorded before (baseline) and during application of NA and 15 min after its washout. (B) Time course of the amplitude and frequency and average waveforms of spontaneous mEPSCs recorded using similar protocol to (A) but under picrotoxin instead of DNQX. (C) The pooled data (mean ± SD) on the quantal amplitude (evaluated as a main peak at amplitude histogram) of mIPSCs and mEPSCs recorded before and during application of NA in control and in the presence of ATP receptors antagonists. Lack of changes in the frequency of spontaneous currents and clear changes in the quantal amplitude strongly support the postsynaptic locus of effects. Note that reduction in the inhibitory currents and increase of excitatory currents were strongly attenuated by antagonist of ATP receptors.

receptor-mediated mIPSCs were recorded at −80 mV in presence of TTX (1 μM) and DNQX (6,7-dinitroquinoxaline-2,3-dione; 50 μM). To dissect the role of ATP receptors, the mixture of P2 purinoreceptors inhibitors PPADS (pyridoxal-phosphate-6-azophenyl-2',4'-disulfonic acid; affects most of P2X and P2Y receptors, except P2X4) and 5-BDBD (5-(3-bromophenyl)-1,3-dihydro-2H-benzofuro[3,2-e]-1,4-diazepin-2-one; selectively inhibits P2X4 receptors) was used.

In the wild-type mice, the average baseline amplitude of mEPSCs was 12.4 ± 3.1 pA, the decay time 8.9 ± 2.2 ms. The mEPSCs amplitude increased by 22.7% ± 5.9% ($n = 7$, $P < 0.05$) after application of 3 μM NA (Fig. 4.5A). The effect was transient and the mEPSCs gradually decreased to the baseline size in 10–20 min after removal of NA. The increase in the amplitude was not accompanied by significant increase in the frequency of the events; this supports the postsynaptic locus of effect. The baseline amplitude of mIPSCs was 19.6 ± 5.8 pA, the decay time 23.9 ± 5.5 ms. In contrast to the excitatory currents, the amplitude of mIPSCs decreased by 34.3% ± 6.7% ($n = 7$) after activation of astrocytes via ARs (Fig. 4.5B). This result closely agrees with our previous observations, where decrease in GABAergic mIPSCs was caused by activation of astrocytes via PAR1 receptors.[23] Both upregulation of EPSCs and downregulation of IPSCs were abolished when NA was applied in presence of 10μM PPADS and 5μM 5-BDBD (Fig. 4.5A and B) and were significantly diminished in the dn-SNARE mice (Fig. 4.5C). So, both effects were most likely caused by vesicular release of ATP from astrocytes.

The glia-derived ATP-mediated downregulation of GABA receptors (Fig. 4.5B), in synergy with upregulation of AMPA receptors (Fig. 4.5A) and positive modulation of NMDA receptors by glia-derived D-serine (Fig. 4.2), could have a positive influence on the long-term synaptic plasticity in the neocortex. To test this hypothesis, we investigated the impact of glial α_1-ARs on LTP of the fEPSPs in layer II/III of somatosensory cortex of the wild-type and dn-SNARE mice of two age groups. The fEPSPs were evoked by the stimulation of neuronal afferents descending from layers IV-V; LTP was induced by 2 or 5 episodes of high-frequency theta-burst stimulation (TBS).[22,26] The magnitude of LTP induced by 5 episodes of TBS was around 160% in all 20 experiments (Fig. 4.5A). Application of α_1-AR antagonist terazosin inhibited the induction of LTP, suggesting the importance of glial adrenergic signaling that was supported by further experiments with weaker stimulation (Fig. 4.6B–D). Under control conditions, 2 TBS failed to induce the LTP (Fig. 4.6B). Application of 1 μM NA enabled the induction of LTP with weaker stimulation. Washout of NA 3 min after TBS did not affect the potentiation suggesting the importance of additional NA-mediated activation of astrocytes in the initial period of LTP induction.

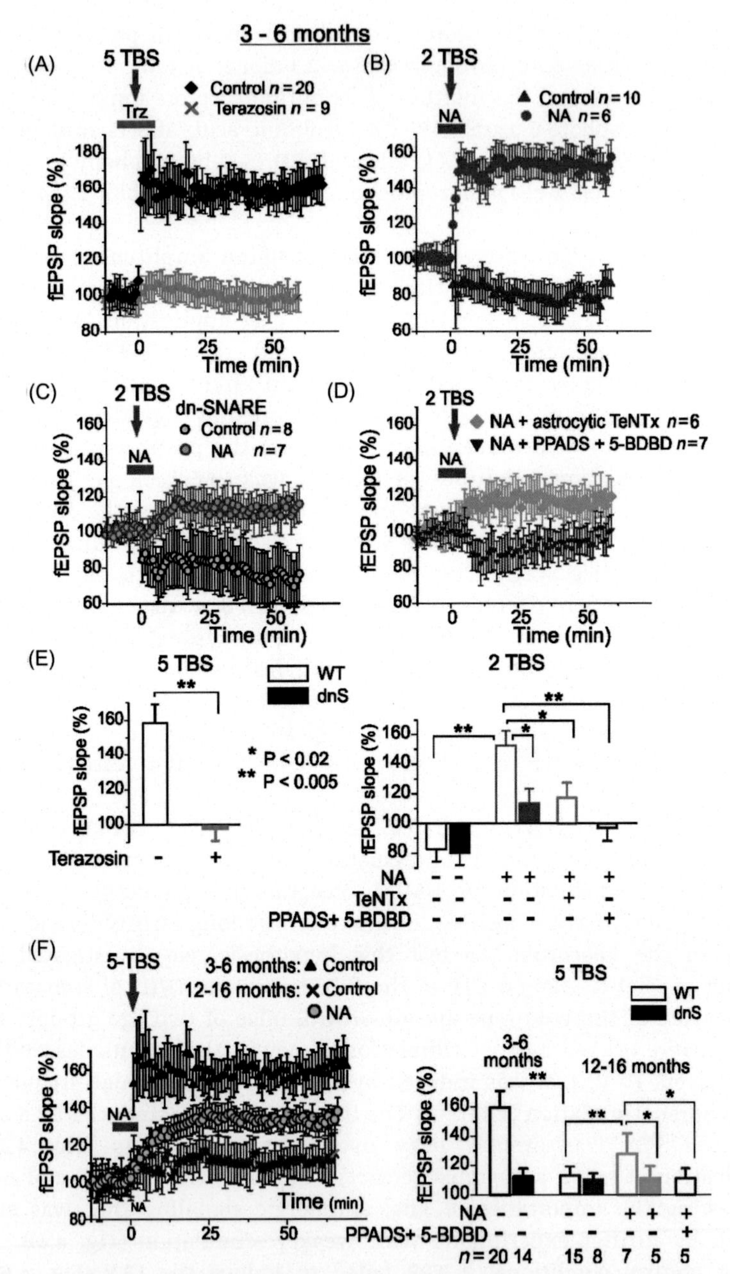

FIGURE 4.6 Impact of astroglial adrenergic signaling on synaptic plasticity in the neocortex. Long-term potentiation of field EPSPs (LTP) was induced in the layer 2/3 of somatosensory cortex of the wild-type and dn-SNARE mice by several theta-bursts of high-frequency stimulation (TBS). (A–E) Effects of modulation of adreneregic signaling and glial

(Continued)

Importantly, the α_1-AR-mediated facilitation of LTP was significantly reduced in the dn-SNARE mice (Fig. 4.5C and E). In addition, the magnitude of LTP induced by 2 TBS in presence of NA considerably decreased when fEPSPs were recorded in the vicinity of astrocytes perfused with TeNTx (similarly to the experiment with purinergic mEPSCs shown in Fig. 4.3). The facilitatory effect of NA was abolished by selective inhibition of P2 purinoreceptors with 10 μM PPADS and 5 μM 5-BDBD (Fig. 4.5D and E). This result demonstrates the importance of activation of ATP receptors for the facilitation of LTP. Combined, our data imply that astroglial α_1-ARs can trigger the exocytosis of gliotransmitters and thereby modulate synaptic plasticity in the neocortex.

One might suggest that aging-related decline in astroglial Ca^{2+}-signaling and release of gliotransmitters could contribute to age-related impairment of synaptic plasticity. This hypothesis has been confirmed by our data. Indeed, the magnitude of LTP, even induced by strong stimuli (5 or more TBS) was considerably lower in the neocortex of 12–16 months-old mice, as compared to 3–6 months old (Fig. 4.6F). Importantly, the LTP in the neocortex of old mice could be rescued by additional activation of astrocytic adrenergic signaling (Fig. 4.6F). Similarly to the facilitation of 2 TBS-induced LTP in the younger mice, the facilitation of LTP in the old mice was impaired in the dn-SNARE mice and was prevented by P2 receptor antagonists (Fig. 4.5H). These results are in a good agreement with our previous observations of facilitatory effects of astroglia-driven ATP on long-term synaptic potentiation in the neocortex[26,31] where glial exocytosis was triggered via PAR-1 and CB1 receptors. The data presented above highlight the role for astroglial ARs.

◀ exocytosis on LTP in the mature adult (3–6 month-old) mice. (A) The magnitude of LTP was significantly reduced by specific antagonist α_1-ARs terazosin (30 nM). (B) Delivering just 2 TBS did not induce LTP in the control but was able to induce LTP when NA (1 μM) was applied during induction as indicated in the graph. (C) Application of NA did not facilitate LTP induction in the dn-SNARE mice suggesting the importance of glial exocytosis for the action of NA. (D) Effect of NA was decreased by perfusion of astrocyte with TeNTx (similarly to the experiment shown in Fig. 4.4) or by inhibition of neuronal ATP receptors with PPADS (pyridoxalphosphate-6-azophenyl-2′,4′-disulfonic acid) and 5-BDBD (5-(3-bromophenyl)-1,3-dihydro-2H-benzofuro[3,2-e]-1,4-diazepin-2-one). (E) Pooled data on the magnitude of LTP evaluated as relative increase in the fEPSP slope at 60th min, averaged across 10 min time window. Each data point shows mean ± SD for the number of experiments indicated in (A–D). Asterisks (**) indicate statistical significance of difference in the LTP magnitude wild-type control values (unpaired *t*-test). (F) Comparison of the LTP induced by the stronger stimulus (5 TBS) in the neocortex of mature adult and the old (12–16 months) mice. The magnitude of LTP was significantly reduced in old mice in the control, as compared to LTP in younger age shown in the panel (A). The magnitude of LTP in the old mice significantly increased after activation of α_1-ARs in the wild-type but not in the dn-SNARE mice. *Modified from Pankratov Y., Lalo U. Role for astroglial alpha$_1$-adrenoreceptors in gliotransmission and control of synaptic plasticity in the neocortex.* Front Cell Neurosci. *2015;9:230.*

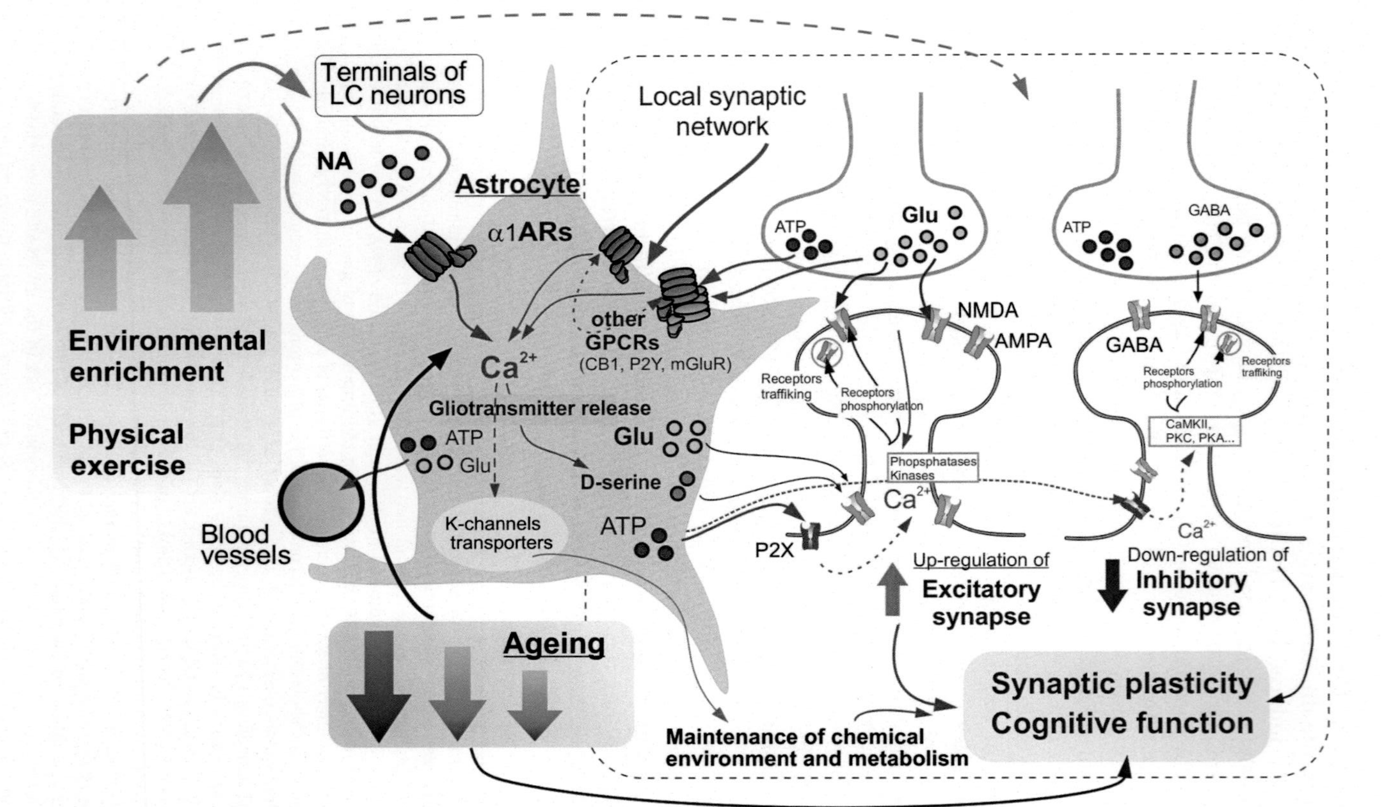

FIGURE 4.7 The putative mechanisms underlying the impact of astroglial adrenergic signaling on the age- and environment-related brain plasticity. Astrocytes act a "Brain Hub" receiving signals from neurons and regulating synaptic transmission, neuronal metabolism, and neurovascular coupling. Astroglial Ca^{2+} signaling is activated by release of glutamate and other transmitters from local synapses and noradrenergic "volume transmission" from *Locus Coeruleus* axons. Astrocytes upregulate excitatory and downregulate inhibitory synapse by releasing gliotransmitters thereby facilitating synaptic plasticity and cognitive functions. Age-related decline in astrocytic signaling compromises glial control of synaptic homeostasis. Environmental enrichment and exercise enhance adrenergic signaling in astrocytes and ameliorate the age-related decline in their function.

Summary and Perspectives

Results of the last two decades suggest that neuronal networks cannot be considered as the sole substrate of higher brain function and neuronal-glial networks coordinate information processing in the brain. A key element of this coordination is ability of astrocytes to integrate neuronal activity over large spatial domains and provide a feedback by modulation of synaptic strength. Highly sensitive α_1-ARs of cortical astrocytes are strategically positioned to receive and integrate diffuse adrenergic input from remote adrenergic neurons and modulate the plasticity of neuronal network. Combined, our data demonstrate the importance of astroglial α_1-ARs for glial modulation of excitatory and inhibitory synapses. The pivotal role in this modulation belongs to ability of ARs to elevate cytosolic Ca^{2+}-level and trigger the exocytosis of gliotransmitters. Based on our experimental and literature data, we suggest a working model of involvement of astroglial ARs in aging- and environment-related brain plasticity (Fig. 4.7). This model suggests that astroglial adrenergic signaling can be instrumental for mediating the positive influence of enriched environment and activity. Of course, the mechanisms, highlighted by this model, need to be explored further, in particular in relation to neurodegenerative disorders.

ABBREVIATIONS

ACh	acetylcholine
AR	adrenergic receptor
ATP	adenosine triphosphate
5-BDBD	5-(3-bromophenyl)-1,3-dihydro-2*H*-benzofuro[3,2-e]-1,4-diazepin-2-one
D-APV	D-2-amino-5-phosphonopentanoic acid
DNQX	6,7-dinitroquinoxaline-2,3-dione
DPCPX	8-cyclopentyl-1,3-dipropylxanthine
EE	enriched environment
fEPSP	field excitatory postsynaptic potential
LTP	long-term synaptic potentiation
mEPSC	miniature excitatory postsynaptic current
5-HT	5-hydroxytryptamine
mIPSC	miniature inhibitory postsynaptic current
NA	noradrenaline, norepinephrine
PAR1	protease-activated receptor 1
PPADS	pyridoxalphosphate-6-azophenyl-2′,4′-disulfonic acid
SH	standard housing
SNARE	soluble N-ethylmaleimide-sensitive factor attachment protein receptor
TBS	theta-burst stimulation
TeNTx	tetanus neurotoxin
TTX	tetrodotoxin

REFERENCES

1. Mishra J, de Villers-Sidani E, Merzenich M, Gazzaley A. Adaptive training diminishes distractibility in aging across species. *Neuron.* 2014;84(5):1091–1103.
2. Merzenich MM, Van Vleet TM, Nahum M. Brain plasticity-based therapeutics. *Front Hum Neurosci.* 2014;8:385.
3. Nithianantharajah J, Hannan AJ. The neurobiology of brain and cognitive reserve: mental and physical activity as modulators of brain disorders. *Prog Neurobiol.* 2009;89 (4):369–382.
4. Hillman CH, Erickson KI, Kramer AF. Be smart, exercise your heart: exercise effects on brain and cognition. *Nat Rev Neurosci.* 2008;9(1):58–65.
5. van Praag H. Exercise and the brain: something to chew on. *Trends Neurosci.* 2009;32 (5):283–290.
6. Pascual-Leone A, Amedi A, Fregni F, Merabet LB. The plastic human brain cortex. *Annu Rev Neurosci.* 2005;28:377–401.
7. Nithianantharajah J, Hannan AJ. Enriched environments, experience-dependent plasticity and disorders of the nervous system. *Nat Rev Neurosci.* 2006 Sep;7(9):697–709.
8. Ward NS, Cohen LG. Mechanisms underlying recovery of motor function after stroke. *Arch Neurol.* 2004;61:1844–1848.
9. Hebb D. The effects of early experience on problem-solving at maturity. *Am Psychol.* 1947;2:306–307.
10. Will B, Galani R, Kelche C, Rosenzweig MR. Recovery from brain injury in animals: relative efficacy of environmental enrichment, physical exercise or formal training. *Prog Neurobiol.* 2004;72:167–182.
11. Rodriguez JJ, Terzieva S, Olabarria M, Lanza RG, Verkhratsky A. Enriched environment and physical activity reverse astrogliodegeneration in the hippocampus of AD transgenic mice. *Cell Death Dis.* 2013;4:e678.
12. Abraham WC, Bear MF. Metaplasticity: the plasticity of synaptic plasticity. *Trends Neurosci.* 1996;19:126–130.
13. Hulme SR, Jones OD, Raymond CR, Sah P, Abraham WC. Mechanisms of heterosynaptic metaplasticity. *Philos Trans R Soc B.* 2014;369:20130148.
14. Ding F, O'Donnell J, Thrane AS, Zeppenfeld D, Kang H, Xie L, et al. Alpha1-adrenergic receptors mediate coordinated $Ca2^{+}$ signaling of cortical astrocytes in awake, behaving mice. *Cell Calcium.* 2013;54(6):387–394.
15. Paukert M, Agarwal A, Cha J, Doze VA, Kang JU, Bergles DE. Norepinephrine controls astroglial responsiveness to local circuit activity. *Neuron.* 2014;82(6):1263–1270.
16. Araque A, Carmignoto G, Haydon PG, Oliet SH, Robitaille R, Volterra A. Gliotransmitters travel in time and space. *Neuron.* 2014;81(4):728–739.
17. Bazargani N, Attwell D. Astrocyte calcium signaling: the third wave. *Nat Neurosci.* 2016;19(2):182–189.
18. Halassa MM, Haydon PG. Integrated brain circuits: astrocytic networks modulate neuronal activity and behavior. *Annu Rev Physiol.* 2010;72:335–355.
19. Giaume C, Koulakoff A, Roux L, Holcman D, Rouach N. Astroglial networks: a step further in neuroglial and gliovascular interactions. *Nat Rev Neurosci.* 2010;11 (2):87–99.
20. Attwell D, Buchan AM, Charpak S, Lauritzen M, Macvicar BA, Newman EA. Glial and neuronal control of brain blood flow. *Nature.* 2010;468(7321):232–243.

21. Halassa MM, Fellin T, Haydon PG. The tripartite synapse: roles for gliotransmission in health and disease. *Trends Mol Med.* 2007;13(2):54–63.
22. Lalo U, Rasooli-Nejad S, Pankratov Y. Exocytosis of gliotransmitters from cortical astrocytes: implications for synaptic plasticity and aging. *Biochem Soc Trans.* 2014;42: 1275–1281.
23. Lalo U, Palygin O, Rasooli-Nejad S, Andrew J, Haydon PG, Pankratov Y. Exocytosis of ATP from astrocytes modulates phasic and tonic inhibition in the neocortex. *PLoS Biol.* 2014;12(1):e1001747.
24. Gordon GR, Iremonger KJ, Kantevari S, Ellis-Davies GC, MacVicar BA, Bains JS. Astrocyte-mediated distributed plasticity at hypothalamic glutamate synapses. *Neuron.* 2009;64(3):391–403.
25. Panatier A, Theodosis DT, Mothet JP, Touquet B, Pollegioni L, Poulain DA, et al. Glia-derived D-serine controls NMDA receptor activity and synaptic memory. *Cell.* 2006;125:775–784.
26. Rasooli-Nejad S, Palygin O, Lalo U, Pankratov Y. Cannabinoid receptors contribute to astroglial Ca(2)(+)-signalling and control of synaptic plasticity in the neocortex. *Philos Trans R Soc Lond B Biol Sci.* 2014;369(1654):20140077.
27. Henneberger C, Papouin T, Oliet SH, Rusakov DA. Long-term potentiation depends on release of D-serine from astrocytes. *Nature.* 2010;463(7278):232–236.
28. Gourine AV, Kasymov V, Marina N, et al. Astrocytes control breathing through pH-dependent release of ATP. *Science.* 2010;329(5991):571–575.
29. Woo DH, Han KS, Shim JW, et al. TREK-1 and Best1 channels mediate fast and slow glutamate release in astrocytes upon GPCR activation. *Cell.* 2012;151 (1):25–40.
30. Pankratov Y, Lalo U. Role for astroglial alpha1-adrenoreceptors in gliotransmission and control of synaptic plasticity in the neocortex. *Front Cell Neurosci.* 2015;9:230.
31. Pougnet JT, Toulme E, Martinez A, Choquet D, Hosy E, Boue-Grabot E. ATP P2X receptors downregulate AMPA receptor trafficking and postsynaptic efficacy in hippocampal neurons. *Neuron.* 2014;83(2):417–430.
32. Pascual O, Casper KB, Kubera C, Zhang J, Revilla-Sanchez R, Sul JY, et al. Astrocytic purinergic signaling coordinates synaptic networks. *Science.* 2005;310:113–116.
33. Khakh BS, McCarthy KD. Astrocyte calcium signaling: from observations to functions and the challenges therein. *Cold Spring Harb Perspect Biol.* 2015;7(4): a020404.
34. Palygin O, Lalo U, Verkhratsky A, Pankratov Y. Ionotropic NMDA and P2X1/5 receptors mediate synaptically induced $Ca2^+$ signalling in cortical astrocytes. *Cell Calcium.* 2010;48:225–231.
35. Pankratov Y, Lalo U. Calcium permeability of ligand-gated $Ca2^+$ channels. *Eur J Pharmacol.* 2014;739:60–73.
36. Panatier A, Vallee J, Haber M, Murai KK, Lacaille JC, Robitaille R. Astrocytes are endogenous regulators of basal transmission at central synapses. *Cell.* 2011;146 (5):785–798.
37. Lalo U, Palygin O, North RA, Verkhratsky A, Pankratov Y. Age-dependent remodelling of ionotropic signalling in cortical astroglia. *Aging Cell.* 2011;10(3):392–402.
38. Figueiredo M, Lane S, Stout Jr. RF, Liu B, Parpura V, Teschemacher AG, et al. Comparative analysis of optogenetic actuators in cultured astrocytes. *Cell Calcium.* 2014;56:208–214.

39. Schipke CG, Heidemann A, Skupin A, Peters O, Falcke M, Kettenmann H. Temperature and nitric oxide control spontaneous calcium transients in astrocytes. *Cell Calcium.* 2008;43(3):285–295.
40. Akinaga J, Lima V, Kiguti LR, et al. Differential phosphorylation, desensitization, and internalization of alpha1A-adrenoceptors activated by norepinephrine and oxymetazoline. *Mol Pharmacol.* 2013;83(4):870–881.
41. Jiang T, Yu JT, Tan MS, Zhu XC, Tan L. Beta-arrestins as potential therapeutic targets for Alzheimer's disease. *Mol Neurobiol.* 2013;48:812–818.
42. Mohan ML, Vasudevan NT, Gupta MK, Martelli EE, Naga Prasad SV. G-protein coupled receptor resensitization-appreciating the balancing act of receptor function. *Curr Mol Pharmacol.* 2012;May 30.
43. Morris DP, Lei B, Wu YX, Michelotti GA, Schwinn DA. The alpha1a-adrenergic receptor occupies membrane rafts with its G protein effectors but internalizes via clathrin-coated pits. *J Biol Chem.* 2008;283(5):2973–2985.
44. Allsopp RC, Lalo U, Evans RJ. Lipid raft association and cholesterol sensitivity of P2X1-4 receptors for ATP: chimeras and point mutants identify intracellular amino-terminal residues involved in lipid regulation of P2X1 receptors. *J Biol Chem.* 2010;285 (43):32770–32777.
45. Lalo U, Roberts JA, Evans RJ. Identification of human P2X1 receptor-interacting proteins reveals a role of the cytoskeleton in receptor regulation. *J Biol Chem.* 2011;286 (35):30591–30599.
46. Zucchi R, Ronca-Testoni S. The sarcoplasmic reticulum Ca^{2+} channel/ryanodine receptor: modulation by endogenous effectors, drugs and disease states. *Pharmacol Rev.* 1997;49(1):2–37.
47. Hertz L, Lovatt D, Goldman SA, Nedergaard M. Adrenoceptors in brain: cellular gene expression and effects on astrocytic metabolism and $[Ca(2^{+})]i$. *Neurochem Int.* 2010;57 (4):411–420.
48. Espay AJ, LeWitt PA, Kaufmann H. Norepinephrine deficiency in Parkinson's disease: the case for noradrenergic enhancement. *Mov Disord.* 2014;29(14):1710–1719.
49. Frenguelli BG, Wigmore G, Llaudet E, Dale N. Temporal and mechanistic dissociation of ATP and adenosine release during ischaemia in the mammalian hippocampus. *J Neurochem.* 2007;101:1400–1413.
50. Pankratov Y, Lalo U, Verkhratsky A, North RA. Quantal release of ATP in mouse cortex. *J Gen Physiol.* 2007;129(3):257–265.
51. Sultan S, Li L, Moss J, Petrelli F, Casse F, Gebara E, et al. Synaptic integration of adult-born hippocampal neurons is locally controlled by astrocytes. *Neuron.* 2015;88 (5):957–972.
52. Correa SA, Hunter CJ, Palygin O, Wauters SC, Martin KJ, McKenzie C, et al. MSK1 regulates homeostatic and experience-dependent synaptic plasticity. *J Neurosci.* 2012;32 (38):13039–13051.

Chapter 5

Adrenergic Ca^{2+} and cAMP Excitability: Effects on Glucose Availability and Cell Morphology in Astrocytes

Robert Zorec[1,2,✉], **Marko Kreft**[1,2] **and Nina Vardjan**[1,2,✉]
[1]*University of Ljubljana, Ljubljana, Slovenia,* [2]*Celica Biomedical, Ljubljana, Slovenia*

Chapter Outline

Introduction 104
Adrenergic Modulation of Cytosolic Ca^{2+} and cAMP Excitability in Cultured Astrocytes 105
- Adrenergic Activation Triggers Phasic Ca^{2+} and Tonic cAMP/PKA Responses in Cultured Astrocytes 106
- Simultaneous Activation of α- and β-ARs Potentiates Ca^{2+} and cAMP/PKA Responses in Astrocytes 108
- Characteristics of Adrenergic Ca^{2+} Signaling in Astrocytes In Situ and In Vivo 109

Adrenergic Excitability and Availability of Glycogen-Derived Cytosolic Glucose in Astrocytes 110
Adrenergic Excitability and Astrocyte Morphologic Plasticity 113
- β-Adrenergic Activation and Stellation of Astrocytes 114
- Adrenergic Activation and Astrocyte Morphology In Vivo: Prevention of CNS Cellular Edema 117

Conclusions 118
Abbreviations 119
References 120

ABSTRACT

A number of recent studies revealed that astrocytes are a target of the locus coeruleus (LC)-nerve terminals that supply noradrenaline (NA) throughout the central nervous system (CNS). LC activation triggers a global CNS excitation, associated with arousal, which affects many processes including metabolism and memory formation. Although electrically silent, astrocytes can respond to NA via G-protein coupled membrane receptors with increased levels in cytosolic second

✉Correspondence address
E-mail: robert.zorec@mf.uni-lj.si; nina.vardjan@mf.uni-lj.si

Noradrenergic Signaling and Astroglia. DOI: http://dx.doi.org/10.1016/B978-0-12-805088-0.00005-0
© 2017 Elsevier Inc. All rights reserved.

messengers Ca^{2+} and cyclic adenosine monophosphate (cAMP), termed cytoplasmic excitability. Here we overview the temporal characteristics of the adrenergic Ca^{2+} and cAMP excitability and the mechanisms that synergize these two pathways to generate the optimal cellular and global CNS response to LC activation. Moreover, we also consider how adrenergic excitability in astrocytes regulates glucose availability, cellular morphology and prevents cytotoxic brain edema. Understanding the dynamics and molecular mechanisms underlying astrocyte noradrenergic signaling processes is of key importance to develop new therapies for neurological diseases.

Keywords: Astrocytes; locus coeruleus; noradrenergic neurons; noradrenaline; adrenergic receptors; Ca^{2+}; cAMP; cytosolic excitability; metabolism; morphology

INTRODUCTION

The human neocortex represents over 80% of brain mass and consists of neurons and nonneuronal cells.[1] However, nonneuronal cells outnumber neurons and among them are also astrocytes, highly branched glial cells,[2–4] with many brain functions. They not only play important homeostatic housekeeping roles in the central nervous system (CNS) but are also involved in the higher brain functions. Although electrically silent, they actively participate in the information processing in the brain, underlying learning, and memory formation.[5] Astrocytic fine processes ensheath synapses, limiting the synaptic cleft in which bidirectional communication between astrocytes and neurons takes place. Through their surface, G-protein coupled receptors (GPCRs), astrocytes respond to neurotransmitters with "cytosolic excitability," represented by rises in intracellular levels of second messengers Ca^{2+} and/or cyclic adenosine monophosphate (cAMP), that triggers the release of their own signaling molecules, which in turn affect neuronal signaling.[6]

Astrocytes may interact also with neurons that form a spatially diffuse nonsynaptic system, targeting large volumes of the CNS (i.e., volume transmission[7]), including the noradrenergic neurons.[8] Noradrenergic neurons originate in the locus coeruleus (LC), a brainstem nucleus, the primary source of the CNS noradrenaline (NA; or norepinephrine). Upon LC activation, catecholamine NA is released from the numerous varicosities on the noradrenergic axons that project bilaterally from the LC into almost all brain areas and into the spinal cord. It is estimated that an individual rat LC neuron possesses more than 1.2 million varicosities, thus possibly affecting a large area of the cortex.[9] When released, NA binds to the plasma membrane G-protein coupled α- and β-adrenergic receptors (ARs) present on all cells in the CNS, including astrocytes, and activates the intracellular Ca^{2+} and cAMP signaling pathways. This triggers a global excitatory response in the CNS, manifesting in augmented alertness, arousal, and attention. Among other functions, LC/noradrenergic system also regulates the sleep/wake cycle, the clearance of waste from the interstitial fluid (glymphatic system), memory formation, it affects behavior, CNS energy metabolism, and neuroinflammation.[10,11] Such a system is not exclusive for mammalian brain but appears to be present in insects as well.[12]

Although the noradrenergic modulation of target neurons has been addressed in the past, it is becoming clear that astrocytes are also a key substrate of this system. Understanding the molecular mechanisms underlying the astrocyte noradrenergic signaling system is of extreme importance as any malfunction may affect normal brain processes, as discussed previously.[13] With the development of genetically encoded fluorescence resonance energy transfer (FRET) nanosensors for measuring intracellular second messenger (Ca^{2+}, cAMP) and metabolite (glucose) levels in living cells, we have recently gained new insights on the temporal dynamics of the adrenergic signaling and AR-mediated metabolite pathways and morphologic plasticity in astrocytes. In this chapter, we first address how the activation of α- and β-ARs affects intracellular Ca^{2+} and cAMP excitability in single astrocytes. In particular, we focus into the temporal characteristics of the two aforementioned second messenger signals and their cross-talk pathways in vitro, in situ, and in vivo in intact brain. Then we discuss the effect of noradrenergic signaling on the availability of glycogen-derived cytosolic-free glucose in time in single living astrocytes. Finally, we consider how adrenergic excitability affects astrocyte morphology within minutes, and how this prevents the formation of hypotonicity- and neurotrauma-induced cellular/cytotoxic edema in the CNS.

ADRENERGIC MODULATION OF CYTOSOLIC CA^{2+} AND cAMP EXCITABILITY IN CULTURED ASTROCYTES

Astrocytes may respond to catecholamines, such as NA or adrenaline (ADR), by activation of surface α- or β-ARs and intracellular Ca^{2+}- and cAMP-mediated signaling pathways.[14,15] α-ARs (α_1 and α_2) and β-ARs (β_1, β_2, and β_3) are coupled to different G-protein types. Coupling of α_1-ARs to the G_q-protein linked to phospholipase C (PLC) signaling pathway causes the formation of inositol 1,4,5-trisphosphate (IP_3). Binding of IP_3 to IP_3 receptors/Ca^{2+} channels (IP_3Rs) in the membrane of the endoplasmic reticulum (ER) releases Ca^{2+} from the ER and increases the intracellular concentration of Ca^{2+} ($[Ca^{2+}]_i$).[13,16] On the other hand, α_2-ARs are generally coupled to G_i-proteins that inhibit adenylate cyclase (AC), an enzyme that catalyzes the conversion of adenosine triphosphate (ATP) to cAMP, leading to a reduction in the intracellular concentration of cAMP ($[cAMP]_i$). In contrast, activation of G_s-proteins coupled to β-ARs potentiates AC activity and leads to increased $[cAMP]_i$ and the activation of cAMP effectors such as protein kinase A (PKA), hyperpolarizing cyclic nucleotide-gated (HCN) channels, exchange factor directly activated by cAMP (Epac), and Popeye domain–containing (Popdc) proteins.[13]

AR-induced Ca^{2+} and cAMP signaling pathways mediate various downstream cellular processes. Some of these processes occur relatively rapidly upon AR stimulation, such as astrocyte glycogenolysis and glycolysis,[17,18] gliotransmitter uptake[19] and release,[6,20] gap-junction permeability,[21] and membrane transport,[13] but the others are relatively slow and may include gene expression, such as astrocyte antigen-presenting function[22] and long-term morphologic plasticity.[23,24] The molecular mechanisms underlying these processes, such as the temporal characteristics of the AR-induced Ca^{2+} and cAMP signaling pathways, are of extreme importance for understanding the CNS function upon LC/noradrenergic activation in normal and pathologic conditions.[13] Impairment of the LC/noradrenergic system is present in Alzheimer's disease (AD), Parkinson's disease, which is associated with demise of LC neurons,[25,26] and in multiple sclerosis, where a disappearance of astrocytic β_2-ARs in white matter has been observed[22] as well as LC degeneration.[27]

As NA can activate both types of receptors on the same cell simultaneously, the NA-triggered downstream processes in astrocytes could be mediated via either α- or β-AR signaling pathways or most likely a combination of both. By simultaneous activation of α- and β-ARs on the same cell, astrocytes could optimize their key support functions during periods of increased demand, as observed in other cell types.[28–30] In this section, the current knowledge on temporal characteristics of AR-mediated Ca^{2+} and cAMP signaling in astrocytes and the interaction between the two signaling pathways is discussed, separately for cultured astrocytes and astrocytes in situ and in vivo.

Adrenergic Activation Triggers Phasic Ca^{2+} and Tonic cAMP/PKA Responses in Cultured Astrocytes

Recent results obtained on single cultured cortical rat astrocytes using real-time confocal microscopy and the intracellular Ca^{2+} and cAMP indicators, Fluo4-AM and Epac1-camps/AKAR2, respectively, have shown for the first time that temporal profiles of the α_1-AR and β-AR secondary messenger pathways are distinct (Fig. 5.1).[31]

It is well known that the activation of cultured astrocytes with the α-/β-AR (ADR, NA) and selective α_1-AR agonists [isoproterenol (ISO)] elicits oscillations in $[Ca^{2+}]_i$, within 10 s upon agonist application, which then decline back to resting level.[16,31–34] These Ca^{2+} oscillations likely occur through the activation of α_1-AR/G_q-proteins and the IP_3 signaling.[34–36] On the other hand, the selective β-AR activation with ISO that is coupled to G_s-proteins triggers no or only sparse $[Ca^{2+}]_i$

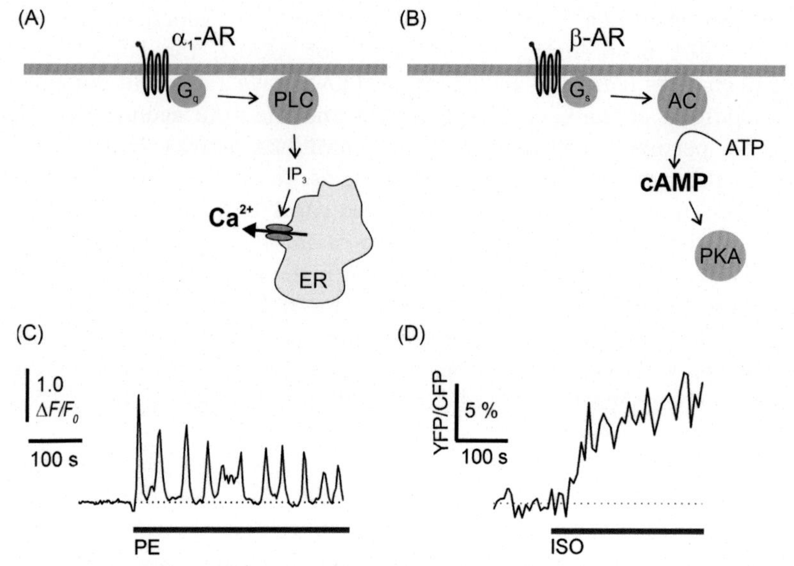

FIGURE 5.1 Temporal differences in α1-adrenergic and β-adrenergic signaling in rat primary astrocytes. (A and B) Schematic representation of α_1-AR (A) and β-AR (B) signaling pathways. (C) Representative fluorescence intensity changes in $\Delta F/F_0$ in astrocytes loaded with green fluorescent Ca^{2+} indicator, Fluo4-AM and stimulated with α_1-AR agonist phenylephrine (PE; 10 μM). The increases in $\Delta F/F_0$ signal after PE stimulation reflect increases in $[Ca^{2+}]_i$. (D) Representative time course of the FRET-based cAMP/PKA activity reporter AKAR2 emission ratio after stimulation with β-AR agonist isoprenaline (ISO) in cultured astrocyte. The monophasic exponential increase in the FRET (fluorescence resonance energy transfer) signal (represented by the YFP/CFP ratio) after ISO stimulation reflects an increase in $[cAMP]_i$ and subsequent PKA activation in cultured astrocyte. Note that activation of astrocytic α_1-ARs coupled to G_q-proteins leads to phasic oscillations in $[Ca^{2+}]_i$ (C), whereas the activation of β-ARs coupled to G_s-proteins leads to tonic increase in cAMP-dependent PKA activity without oscillations at the steady state (D). The maximal Ca^{2+} response occurs faster than cAMP/PKA response. *Modified with permission from Horvat A., Zorec R., Vardjan N. Adrenergic stimulation of single rat astrocytes results in distinct temporal changes in intracellular $Ca^{(2+)}$ and cAMP-dependent PKA responses. Cell Calcium. 2016.*

oscillations.[31,37] α_1-AR-induced $[Ca^{2+}]_i$ oscillations in astrocytes may be explained by cyclical release of Ca^{2+} from the ER via IP_3Rs. This has been reported in vascular smooth muscle cells, where each α_1-AR-induced Ca^{2+} wave depends on an initial sarcoplasmic reticulum Ca^{2+} release that is followed by refilling of the sarcoplasmic reticulum Ca^{2+} store through SERCA (sarcoplasmic–endoplasmic reticulum Ca^{2+}-ATPase). Ca^{2+} oscillations could also be supported by plasma membrane Ca^{2+} channels and exchangers.[38] Moreover, the Ca^{2+} shuttling between the ER and mitochondria could represent an essential component of the Ca^{2+} oscillation mechanism and may have a pacemaker role in the generation of Ca^{2+} oscillations.[39]

In contrast to Ca^{2+} signals, in cells expressing the genetically encoded cAMP FRET nanosensors, Epac1-camps[40] or AKAR2, the latter reports cAMP changes indirectly via activity of PKA[41], have a different dynamics. The addition of the α-/β-AR agonist NA and the β-AR agonist ISO triggered a persistent exponential rise in cAMP/PKA activity that probably occurs through the activation of G_s-proteins. In contrast to α_1-AR-mediated Ca^{2+} oscillations, NA- or ISO-induced cAMP/PKA activity was devoid of oscillations.[23,31,42] The initial increase in cAMP (measured with Epac1-camps) was threefold faster compared with the increase in PKA activity (measured with AKAR2) [time constant for ISO was $\sim$15 s (cAMP) vs. $\sim$50 s (PKA)], indicating that the cAMP increase precedes PKA activation, consistent with PKA being a downstream effector of cAMP. The α_1-AR agonist phenylephrine (PE) did not affect cAMP/PKA activity,[23,31] as observed in biochemical studies.[46]

Thus, astrocytes in vitro respond to AR activation with phasic Ca^{2+} and tonic cAMP/PKA responses (Fig. 5.1). The Ca^{2+} response is $>$4-fold faster than the cAMP response and $>$10-fold faster than the PKA response.[23] This indicates that AR-induced Ca^{2+} mobilizations in astrocytes affect downstream cellular processes within a 4-to-10-fold faster time-domain than AR-induced cAMP/PKA responses.

Simultaneous Activation of α- and β-ARs Potentiates Ca^{2+} and cAMP/PKA Responses in Astrocytes

Concomitant activation of distinct GPCRs can result in the cross-talk between the signaling pathways in various cell types.[43,44] It has been shown that such an interaction also exists between α- and β-AR pathways in cultured astrocytes.[31,45] Real-time confocal microscopy of individual living astrocytes, labeled with fluorescent Ca^{2+} (Fluo4 AM) and cAMP/PKA sensors (AKAR2), has shown that nonselective α-/β-AR activation (NA/ADR) produces a $\sim$2-fold larger increase in Ca^{2+} response and $\sim$3-fold larger increase in cAMP/PKA activity than selective α_1-AR (PE) and β-AR (ISO) activation, respectively. Moreover, the Ca^{2+} response to selective α_1/G_q-coupled ARs was 1.5-fold higher in cells with preactivated β/G_s-coupled ARs than in untreated cells. When a biochemical approach was used to evaluate multiple cells and the changes in intracellular cAMP levels, activation of both β/G_s- and α_1/G_q-coupled ARs in astrocytes also triggered a potentiation of cAMP response.[45] Thus, simultaneous activation of α- and β-ARs in cultured astrocytes potentiates the Ca^{2+} and cAMP/PKA responses. Potentiation of cAMP response can be observed also in astrocytes upon the simultaneous activation of β-AR and serotonin ($5\text{-}HT_2$)[46] or group I metabotropic glutamate receptors.[47] Consistent with this, the simultaneous activation of β/G_s and purinergic/G_q-coupled receptor was shown to potentiate the Ca^{2+} response in astrocytes.[48]

Moreover, selective α_1- and β-AR activation induced Ca^{2+} and cAMP/PKA responses in only 70%–80% of cultured astrocytes, but in astrocytes stimulated with nonselective α-/β-AR agonists (NA/ADR), the responsiveness of cells increased to 100%. Consistent with this, when β-ARs were preactivated, the responsiveness of astrocytes to α_1-AR-induced Ca^{2+} elevations increased from 83% to 100%. β-AR activation also increased the number and frequency of α_1-AR-induced Ca^{2+} events up to twofold and vice versa.[31] These results clearly indicate that the simultaneous activation of different types of GPCR can lead to the cross-talk and potentiation of Ca^{2+} and cAMP signaling pathways in astrocytes and may increase responsiveness of cells to the stimuli, which may affect the downstream cellular processes as well.

Ca^{2+} and cAMP signaling may interact to generate the optimal final response in cells.[30] The interaction between cAMP/PKA pathway and Ca^{2+} levels may occur via regulating Ca^{2+} transport pathways, involving ER Ca^{2+} channels, IP_3Rs and ryanodine receptors (RyRs), membrane Na^+/Ca^{2+} exchangers, Ca^{2+} ATPase pumps, voltage-dependent Ca^{2+} channels, and TRP (transient receptor potential) channels. The cAMP/PKA pathway may also mediate Ca^{2+} clearance through SERCA activity and may modulate the activity of various PLC isoforms, leading to changes in IP_3 production. Significantly, PLC isoforms are sensitive to Gβ/γ-protein subunits, which may affect the activity of IP_3Rs.[29,43,44] It is possible that at least some of these mechanisms participate in the cross-talk between the AR-induced Ca^{2+} and cAMP/PKA responses in astrocytes.

In addition to the acute effects of cAMP on Ca^{2+} response, activation of cAMP/PKA pathway can also upregulate the long-term gene expression and synthesis of key proteins involved in Ca^{2+} signaling.[29] Furthermore, activation of the cAMP/PKA pathway induces astrocytes to attain stellate morphology,[23] which is known to be because of cytoskeleton rearrangements and can lead to spatial reorganization of the plasma membrane and the ER.[49] This may affect functional coupling of key Ca^{2+} signaling elements, in particular ER Ca^{2+} release channels (RyR and IP_3R) and plasma membrane Ca^{2+} entry, leading to an enhanced GPCR-induced Ca^{2+} response.[49] Moreover, the observed cAMP-mediated potentiation of Ca^{2+} oscillations may affect a number of cellular processes in astrocytes because the regulation of gene transcription in cells was reported to critically depend on the Ca^{2+} oscillation frequency.[50,51]

Characteristics of Adrenergic Ca^{2+} Signaling in Astrocytes In Situ and In Vivo

AR-mediated intracellular signaling pathways in astrocytes in situ and in vivo and their functional role in the CNS are less studied. NA and α_1-AR agonist PE triggers transient $[Ca^{2+}]_i$ increases in Calcium

Orange-labeled astrocyte networks in situ within the mouse hippocampal brain slice.[52] This was confirmed also in astrocytes in vivo in intact brain preparations of adult living, but anesthetized mice, where topical application of either α_1-AR or β-AR agonists, methoxamine and ISO, respectively, as well as direct LC activation (stimulation electrodes) triggered transient Ca^{2+} oscillations in Fluo-4/AM-labeled astrocytes.[53] Furthermore, in awake, behaving mice during LC activation [triggered by sensory (whisker) stimulation, air-puff startle response or locomotion] NA released from noradrenergic neuronal projections causes in multiple brain regions simultaneous and coordinated α_1-AR-mediated Ca^{2+} mobilizations in rhod2 AM/*Glt-1*-eGFP labeled astrocytes[54] and in astrocytes expressing genetically encoded Ca^{2+} indicator GCaMP3.[55] These oscillations were blocked with α_1-AR antagonist but not with β-AR antagonists.[54] Astrocyte stimulation was widespread, likely affecting astrocyte networks, as Ca^{2+} excitability can propagate from an excited astrocyte to its neighboring unstimulated astrocytes in the form of intercellular Ca^{2+} waves that can travel 10–20 μm/s[56] and are achieved by diffusion of IP_3 or Ca^{2+} through gap junctions.[57] It has been suggested that gap-junction permeability may also be affected by the noradrenergic system; it can be reduced by α_1-AR activation and increased by β-AR activation as observed in cultured striatal astrocytes. Thus activation of different AR signaling pathways may have opposite effects on the permeability of gap junctions.[21] The studies on awake mice have positioned astrocytes as the primary target of noradrenergic neurons and have shown that NA enhances the sensitivity of astrocytes to local circuit activity, as pairing of NA release with for instance light stimulation markedly enhanced astrocyte Ca^{2+} signaling in visual cortex.[55]

However, the spatiotemporal characteristics of β-AR-mediated intracellular cAMP signaling in astrocytes in vivo and whether cAMP excitability can be propagated between neighboring astrocytes through gap junctions still needs to be determined. The ability to stably express genetically encoded cAMP nanosensors selectively in astrocytes in vivo using conditional, for example, Epac1-camps mice will help us to find the answers to these questions in the future.

ADRENERGIC EXCITABILITY AND AVAILABILITY OF GLYCOGEN-DERIVED CYTOSOLIC GLUCOSE IN ASTROCYTES

Glycogen in the brain represents the source of energy that can extend the neuronal function for 20 min or longer. In the adult brain glycogen is found predominantly in astrocytes, where it is stored throughout the cell body, including astrocytic processes. The level of glycogen is regulated by several hormones/neurotransmitters, including NA that controls both glycogen breakdown (in the short term within minutes after exposure) and synthesis (in the long term following several hours of exposure and involves gene expression).[58,59] By providing energy substrates such as glucose and lactate to neurons, glycogen has been shown to play a fundamental role in supporting learning and memory.[60–62]

NA increases glycogenolysis in cultured astrocytes, mainly by a β-AR/cAMP signaling pathway that activates cAMP-depended PKA. PKA phosphorylates glycogen phosphorylase, an enzyme that catalyzes the rate-limiting step in glycogenolysis in animals and increases its activity. In accordance with in vitro studies, the involvement of β-AR in glycogenolysis was confirmed also in mice[63] and chicken brain slices.[64] Besides β-AR, α_2-ARs also contribute to the enhanced glycogenolysis as complete inhibition of NA-induced glycogenolysis was achieved only in the presence of both β- and α_2-AR antagonists.[65] α_2-ARs may via $G_{i/o}$ protein β/γ-subunit enhance the activity of PKC and Ca^{2+} release from internal stores, which may permit glycogenolysis under certain conditions.[66] Recent real-time measurement of cytosolic glucose in astrocytes expressing a FRET-based glucose nanosensor has shown that astrocytes respond to ADR/NA stimulation with an exponential rise in cytosolic glucose concentration from 0.3 to 0.5 mM with the time constant of ~155 s. The initial glucose increase rate was 1.6 μM/s.[17] Astrocytes preincubated with the inhibitor of the glycogen phosphorylase DAB (1,4-dideoxy-1,4-imino-d-arabinitol) displayed a 70% reduction in the NA-induced glucose increase (Fig. 5.2).[17] This indicates that the majority of glucose increase upon NA stimulation can be attributed to glycogen degradation and not to the glucose uptake from the extracellular space, although glucose uptake upon β-AR activation may occur as pointed out in astrocytes in β-AR knockout mice.[67] After uptake, the locally increased intracellular glucose concentration in astrocytes can be rapidly spread throughout the cytosol with an apparent diffusion coefficient (D_{app}) of $2.38 \pm 0.41 \times 10^{-10}$ m^2/s (at 22–24 °C).[68]

It has been reported that NA activation also increases glycogenesis (glycogen synthesis) specifically via α_2-AR coupled to inhibitory $G_{i/o}$ protein α-subunit, which inhibits AC activity and cAMP production, thus promoting glycogenesis. In chick astrocytes, α_2-AR-induced increase in glycogenesis was shown to depend on the activation of $G_{i/o}$/phosphoinositide 3-kinase (PI3K) pathway.[69] When compared to the rapid glycogen degradation, glycogenesis is much slower and appears in astrocytes several hours after exposure to NA.[59] The process likely appears during neuronal activity in the resting state and is necessary for the maintenance of the astrocytic glycogen shunt and normal brain functions as the inhibition of α_2-AR stimulation of glycogen synthesis has been associated with the cognitive impairment.[70]

Many studies indicate that during periods of high demand, such as hypoglycemia and intense neuronal activity, glycogen in astrocytes is initially broken down to glucose, which is however not directly

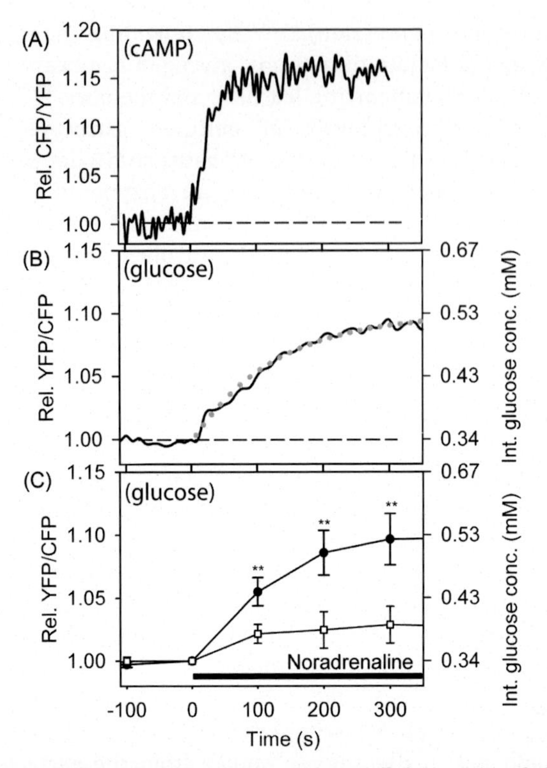

FIGURE 5.2 **Noradrenaline activation increases intracellular cAMP and glucose concentration in rat primary astrocytes.** (A) Real-time FRET (fluorescence resonance energy transfer) measurements of cAMP in primary rat astrocytes transfected with cAMP Epac1-camps nanosensor after 1-μM noradrenaline stimulation [black bar in (C)]. (B) Real-time FRET measurements of intracellular glucose in astrocytes transfected with glucose FLII12PGLU-700μΔ6 nanosensor after 200-μM noradrenaline stimulation [black bar in panel (C)]. (C) Mean FRET response of intracellular glucose recorded in astrocytes upon noradrenaline stimulation in control cells (filled circles; $n = 8$ cells) and in cells pretreated with DAB, a glycogen phosphorylase inhibitor (empty squares; $n = 11$ cells). Mean FRET ratio after noradrenaline stimulation in cell pretreated with DAB was significantly lower than in untreated cells (Student's t test for paired data, $^{**}P < 0.01$). FRET ratio in glucose measurements stimulation increased after noradrenaline stimulation with a time constant of 115.9 ± 8.2 s [gray dotted line represents the exponential fit to the curve in (B)]. Note the faster increase in intracellular cAMP than glucose concentration. *Modified with permission from Prebil M., Vardjan N., Jensen J., Zorec R., Kreft M. Dynamic monitoring of cytosolic glucose in single astrocytes.* Glia. *2011;59(6):903–913.*

transported to neurons, but preferentially metabolized to lactate in the process of aerobic glycolysis. FRET-based nanosensor for lactate has been developed recently,[71,72] and future studies are needed to evaluate the temporal characteristics of NA-induced lactate production in astrocytes in vitro and in vivo.

ADRENERGIC EXCITABILITY AND ASTROCYTE MORPHOLOGIC PLASTICITY

Astrocytes are morphologically very dynamic, plastic cells that rapidly change shape in response to alterations in their environment and their physiological status. Unlike neurons, protoplasmic astrocytes in the brain cortex have an extremely branched morphology. Individual mouse protoplasmic astrocytes have soma diameters of ~7 μm with 5–10 major processes, which further branch into fine astrocytic protrusions to give astrocytes a sponge-like appearance that extends ~40 μm from the soma.[73] Human protoplasmic astrocytes are even larger with 27-times greater volume than their rodent counterparts; their processes span 100–200 μm from the soma.[74] This higher span in human astrocytes may contribute to enhance plasticity and learning as was shown in human glial chimeric mouse[75]. Both rodent and human protoplasmic astrocytes occupy nonoverlapping domains throughout the synaptic neuropil, which are at least in normal physiological conditions[76] not invaded by the neighboring astrocytes.[2,3,73,74] Within its domain, an individual protoplasmic astrocyte ensheaths different synapses (see the tripartite synapse concept[77]), contacts neuronal cell bodies and the dendrites of different neurons, as well as blood vessels and other brain cells such as microglia.

Despite a "rigid" domain organization, the morphology of an individual astrocyte within its space domain is dynamic. Under physiologic conditions, such as reproduction, sensory stimulation, memory formation, learning, and sleep/wake cycle, they display a structural plasticity, particularly at the level of their fine protrusions, termed perisynaptic astrocytic processes, which normally ensheath synapses. These processes can undergo morphological changes within minutes which have the potential of modification of the geometry and diffusion properties of the extracellular space in relation with neuronal synapses.[78] It is estimated that an individual human protoplasmic astrocyte can associate via its perisynaptic processes with roughly 270 thousand to 2 million synapses.[74] Thus, changes in morphology of an astrocyte within its domain can affect many neurons. Moreover, this information on morphological plasticity could be translated to neighboring astrocytes as astrocytes are able to communicate and coordinate activity within and around their domain also through the intercellular and autocellular gap junctions.[79]

Many new data support the view that changes in astrocytes morphology are important for normal brain function. For instance, memory formation thought to occur only at the level of structural remodeling of synaptic connections[80] has been shown to involve also remodeling of astroglial protrusions.[4,81] Moreover, recent in vivo studies on mouse intact brain suggest that morphologic plasticity of astrocytes (shrinking during sleep and swelling during wakefulness) controls the flow of cerebrospinal fluid through the glymphatic system and seems to function as a waste clearance mechanism in the brain.[11,82] Astrocytes can also undergo dramatic volume changes. As they abundantly express water channels on their plasma membrane, they are the only cells in brain that can rapidly swell as a result of ionic dysregulation (e.g., increase in extracellular potassium[83]) during brain pathologies, such as ischemia, trauma, and epilepsy.[84] This causes cellular (cytotoxic) brain edema and reduces the volume of brain extracellular space. Furthermore, reactive tissue astrocytes in neurotrauma, ischemia, inflammation, or neurodegeneration undergo hypertrophy of cellular processes.[85] These data clearly show that astrocyte morphologic plasticity is of great importance for normal brain function, but the molecular and cellular mechanisms that control astrocyte restructuring still remains poorly understood.

β-Adrenergic Activation and Stellation of Astrocytes

Astrocyte morphology could be influenced by many factors, including NA, an arousing neuromodulator released from noradrenergic neurons that raises and coordinates the function of multiple cell types to optimize CNS performance.[10] A remarkable potential for morphologic plasticity of astrocytes upon AR activation was initially detected using culture studies. These revealed that although astrocytes in culture have flattened polygonal morphology, they may transform on activation of the β-AR/cAMP pathway via a series of cytoskeletal reorganizations, which include the restructuring of actin filaments, microtubules, and intermediate filaments,[86,87] into stellate, process-bearing cells.[88–93] The process does not appear to involve changes in membrane area, the cell membrane only retracts and the cell body rounds up with the formation of thin branching processes.[94]

Recent studies on cultured astrocytes expressing Epac1-camps FRET-based cAMP nanosensor have shown that astrocyte attain stellate morphology within 30 min to 1 h on β-AR activation associated with the rise in cAMP (Fig. 5.3A).[23,89,91] Intracellular cAMP levels together with changes in cell morphology were evaluated during the first 10 min of real-time recordings and then 30 min to 1 h after the recording. In the majority of cells, small changes in morphology with the concomitant rise in cAMP were detected immediately (within seconds) after β-AR stimulation. These early changes were not addressed in previous studies and suggest that astrocytes have the potential to rapidly change their morphology on β-AR activation. Within 1 h, up to 20% reduction in the cell cross-sectional area was measured, suggesting the repositioning of cytoplasm to the perinuclear region due to restructuring of the cytoskeleton.[91] At the same time, the cell perimeter increased by 30%–50% and short protrusions on the plasma membrane with some elongated processes appeared. The treatment of astrocytes with forskolin (FSK), an AC

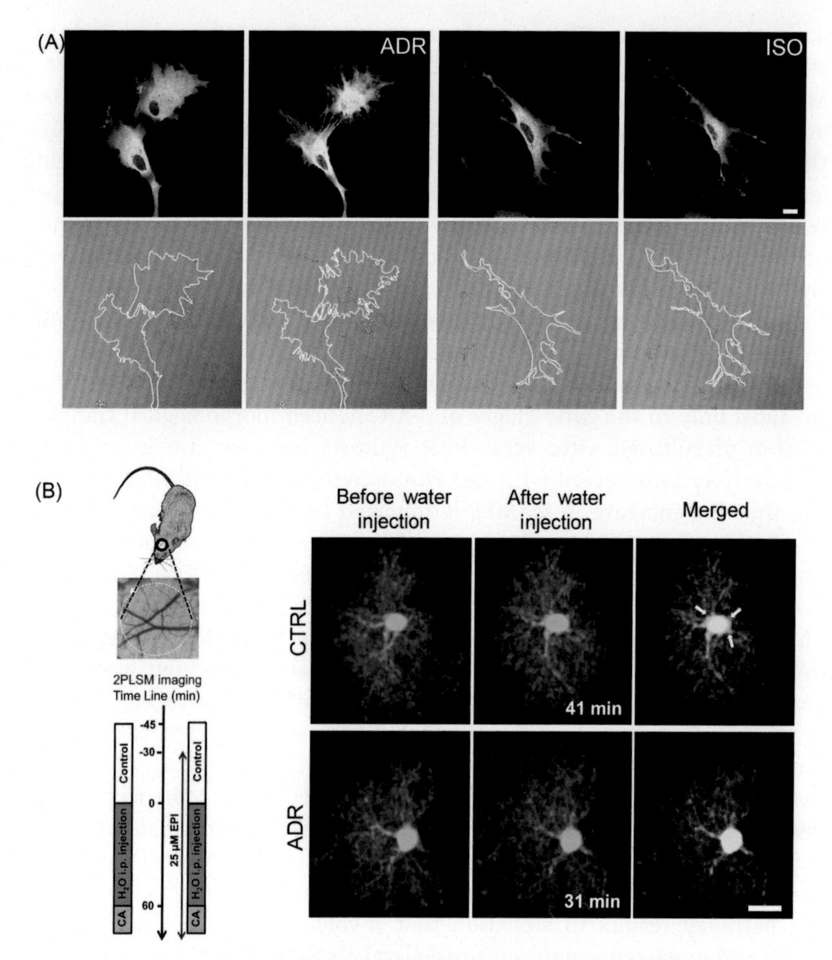

FIGURE 5.3 **Effects of adrenergic receptor activation on astrocyte morphology.** (A) AR activation induces rapid changes in astrocyte morphology. Representative images of Epac1-camps transfected astrocytes (above) and their corresponding DIC images (below) before (left) and within 30 min to 1 h after (right) the addition of a nonselective α-/β-AR agonist adrenaline (ADR) and β-AR agonist isoprenaline (ISO). The perimeter of individual cells expressing Epac1-camps was traced using LSM510 META software, which also outlines the cross-sectional area of the cell. Scale bar: 20 μm. (B) ADR treatment attenuates swelling of cortical astrocytes in vivo after intraperitoneal injection (*i.p.*) of water in mice. [B (left)] General experimental schematics and time line. CCD camera image reveals cortical vasculature directly below craniotomy over mouse somatosensory cortex in which 2PLSM (two-photon laser scanning microscopy) imaging of astrocytes was performed according to the experimental time line. In mice untreated with ADR (left time line), 2PLSM images were acquired before and during 1 h after *i.p.* distilled water injection (150 mL/kg) and then during cardiac arrest (CA). ADR (25 μM) was directly applied for 30 min to the open

(*Continued*)

activator, and dbcAMP, a membrane permeable cAMP analog, induced similar morphologic changes to those observed with β-AR agonists.[91] Moreover, pretreatment of astrocytes with propranolol (Pro), a β-AR antagonist, prevented NA induced astrocyte stellation, thus excluding the involvement of α-ARs in stellation. As the FRET signal reporting $[cAMP]_i$ and changes in astrocyte morphology have shown a bell-shaped relationship, it was concluded that in cultured astrocytes maximal morphological changes on β-AR activation are limited to an optimal and relatively narrow range of $[cAMP]_i$.[23]

Biochemical studies have shown that β-AR activation induces only transient increases in cAMP, which shortly after reaching a peak (in ~5 min) decay toward the initial levels,[91] suggesting that cAMP is involved only in the early stages of β-AR-induced morphological transformation of cultured astrocytes. β-AR agonists increase also intracellular PKA activity as observed with the PKA activity reporter AKAR2,[31] indicating that the increase in $[cAMP]_i$ is followed by the activation of PKA. It was reported that β-AR stimulation affects the activity of PKA only for up to 2 h; however, the morphological changes induced by continuous β-AR activation progress over time and reach their maximum after 48 h. This indicates that downstream effectors of the β-AR/cAMP pathway are involved at the later stages of astrocyte stellation,[89] which also affect gene expression. Delayed and sustained activation of mitogen-activated protein kinase p-ERK (phosphorylated extracellular signal-regulated kinase) has been proposed to be critical for the later stages of β-AR-induced morphological transformation of astrocytes.[89] Moreover, it has been suggested that astrocyte stellation also depends on small GTP-ases, such as RhoA, Rac1, and Cdc42.[94,95] Tonic activation of RhoA pathway maintains the polygonal shape of cultured astrocytes, but inactivation of the pathway results in stellation that involves the loss of actin stress fibers and a rearrangement of cytoskeletal elements.[94]

◀ craniotomy and then replaced with agarose containing EPI for the duration of the experiment (right time line). 2PLSM imaging was performed before, during *i.p.* water injection, and after CA. [B (right)] Paired maximal intensity projection 2PLSM images showing an astrocyte (green) before (left) and after (center) injection of water in untreated control mice (CTRL) and in ADR-pretreated mice (ADR). (right) Merged images; arrows point to green areas illustrating swelling. Note the swelling of an astrocyte after *i.p.* injection of water in control untreated mice, as revealed by the overlay image. An astrocyte in mice pretreated with ADR does not appear swollen after *i.p.* injection of water. Scale bar: 10 μm. *(A) Modified with permission from Vardjan N., Kreft M., Zorec R. Dynamics of β-adrenergic/cAMP signaling and morphological changes in cultured astrocytes.* Glia. *2014. (B) Modified with permission from Vardjan N., Horvat A., Anderson J.E., et al. Adrenergic activation attenuates astrocyte swelling induced by hypotonicity and neurotrauma.* Glia. *2016.*

Stellation of cultured astrocytes might be irrelevant in the brain tissue. However, local restructuring of the cytoskeleton may also coordinate surface expansions/retractions of astrocytic processes in vivo. Such restructuring may be supported by mechanisms analogous to those controlling stellation in cultured astrocytes and could be of importance in in vivo astrocyte physiology and pathology.

Adrenergic Activation and Astrocyte Morphology In Vivo: Prevention of CNS Cellular Edema

Many new evidences implicate the role of noradrenergic signaling in morphological plasticity of astrocytes also in vivo. For instance, blockade of β-AR prevents NA-induced stellation of cultured astrocytes, but it also suppresses reactive gliosis in vivo,[96,97] indicating the involvement of noradrenergic stimulation and cAMP signaling in the transformation of resting into reactive astrocytes in vivo. Furthermore, β-AR/cAMP-triggered stellation of cultured astrocytes can be blocked by neurotransmitter glutamate[92,93] and in embryonic astrocytes by ATP.[98] This indicates that in the CNS, glial morphology changes induced by NA are dynamically regulated by the release of other neurotransmitters from surrounding cells as well. Recently, it has been proposed that in early stages of AD,[99] Parkinson's disease,[25,26] and in multiple sclerosis,[27] degeneration of LC leads to the loss of cortical noradrenergic innervation and a reduction of NA. This may affect β-AR/cAMP signaling in the brain and possibly astrocyte morphology. Consistent with this astrocyte atrophy was reported in some brain areas in the early stages of AD model mice.[100]

Astrocytes have been strongly implicated in the development of CNS edema[101,102] as they are the only CNS cells that undergo rapid changes in volume[84,101,103] caused by ionic dysregulation.[104] Recent studies have revealed that hypotonicity-induced swelling of cultured cortical astrocytes is reduced by ADR and a selective β-AR agonist ISO, indicating the involvement of the β-AR/cAMP pathway.[105] This was confirmed also in vivo in real-time imaging of fluorescently labeled astrocytes in mouse neocortex by two-photon microscopy. Topical application of ADR (25 μM) to the rat neocortex prevented the astrocyte swelling, i.e., the typical 45% increase in astrocyte size seen upon 1 h after intraperitoneal injection of water (Fig. 5.3B).[84] ADR also reduces astrocyte swelling in a pathologic setting[105] in a rat model of spinal-cord injury.[106] ADR administration into the epicenter of the lesion starting 5 min after contusion injury reduced the overall cell body size of injury-affected astrocytes in the gray matter. This reduction in astrocytic edema appeared to be beneficial, as swelling of neurons was reduced as well. Thus adrenergic agonists robustly reduce cellular edema in neural cells, which represents potential new targets for the treatment of cellular edema in the CNS.[105]

The mechanism underlying cAMP-mediated reduction in astrocyte swelling is poorly understood. Exposure of astrocytes to hypotonicity elicits repetitive transient increases in $[Ca^{2+}]_i$,[103,107] which may contribute to astrocyte swelling. After pretreatment with ADR or ISO for 5 min, astrocytes responded to hypotonicity with a single transient increase in $[Ca^{2+}]_i$ that is much briefer than in untreated cells. This finding implies that ADR-mediated reduction of hypotonicity-induced swelling of cortical astrocytes in vivo is associated with a reduction in $[Ca^{2+}]_i$ that is linked to a β-AR-induced increase in $[cAMP]_i$. Hypotonicity-induced increases in $[Ca^{2+}]_i$ in astrocytes may be associated with Ca^{2+} influx resulting from stretching of mechanosensitive plasma membrane channels such as TRPV-4[108] or from activation of L-type Ca^{2+} channels.[107] Moreover, Ca^{2+} release from intracellular stores[109] upon autocrine activation of astrocyte surface purinergic P_2 receptors by hypotonicity-released ATP[101] and subsequent activation of IP_3 receptors may be involved. How increased $[Ca^{2+}]_i$ controls volume changes in astrocytes is unclear. Downregulation of TRPV-4 function may be mediated by cAMP-depended PKA activity,[110] but it is possible that cAMP-dependent mechanisms also target other proteins that contribute to Ca^{2+} homeostasis, such as L-type Ca^{2+} channels[111] and IP_3 receptors[112] as well as processes that govern the phosphorylation[83] and the translocation of water channels (e.g., AQP4) to the plasma membrane[113], and the rearrangement of cytoskeletal elements involved in morphological restructuring of astrocytes.[23]

Further studies are needed to elucidate the mechanisms and potential role of noradrenergic signaling on astrocyte morphology in vivo, in normal and pathologic conditions of the CNS, in particular during edema and reactive gliosis, when astrocytes undergo morphologic changes such as swelling and hypertrophy, respectively.

CONCLUSIONS

Recent data on astrocyte NA-mediated cytosolic excitability imply that astrocytes are the key players in governing the noradrenergic effects in the CNS, such as states of attention, memory formation, and brain metabolism. It seems that the activation of both α- and β-ARs is needed to achieve optimal Ca^{2+} and cAMP/PKA responses in astrocytes. Such synergism might be necessary to generate the optimal downstream cellular and global CNS responses to LC activation. Any malfunction in the LC status and/or the astrocytic response to LC activation may affect normal brain physiology. For instance, astrocytes in multiple sclerosis lack β-ARs. As these are involved in the control of astrocyte antigen presenting function, reduced β-ARs likely contribute to the pathogenesis of multiple sclerosis by augmenting neuroinflammation. Moreover, during multiple sclerosis and AD, noradrenergic neurons in the LC begin to deteriorate, which has been associated with the astrocyte shape changes (cell atrophy). The latter may affect the structural interactions between astrocytes and synapses, which are of extreme importance in synapse restructuring during memory formation. Thus, understanding the temporal and molecular mechanisms underlying the noradrenergic signaling and its downstream cellular processes in astrocytes in normal and diseased brain is of extreme importance. Recent results on noradrenergic Ca^{2+} signals in astrocytes in vivo during states of attention imply that NA enhances the sensitivity of local circuit activity via astrocytes.[12] However, future studies are needed to gain information also for cAMP signals. The expression of fluorescent cAMP/PKA nanosensors under astrocyte-restricted promoters in vivo would help us to understand how β-AR/cAMP-dependent signaling mechanisms operate in connection with other residential brain cells upon LC activation.

ABBREVIATIONS

5-HT2	serotonin
AC	adenylate cyclase
AD	Alzheimer's disease
ADR	adrenaline or epinephrine
ARs	adrenergic receptors
ATP	adenosine triphosphate
cAMP	cyclic adenosine monophosphate
CNS	central nervous system
dbcAMP	dybutyryl cAMP
Epac	exchange factor directly activated by cAMP
ER	endoplasmic reticulum
FRET	fluorescence resonance energy transfer
GPCR	G-protein coupled receptors
HCN	hyperpolarizing cyclic nucleotide-gated channels
IP_3	inositol 1,4,5-trisphosphate
IP_3R	inositol 1,4,5-trisphosphate activated Ca^{2+} release channels
ISO	isoproterenol or isoprenaline
LC	locus coeruleus
NA	noradrenaline or norepinephrine
PDEs	phosphodiesterases
PE	phenylephrine
PK	protein kinase
PLC	phospholipase C
Popdc	popeye domain—containing proteins
Pro	propranolol
RyR	ryanodine-activated Ca^{2+} release channels

SERCA sarcoplasmic-endoplasmic reticulum Ca^{2+}-ATPase
TRP transient receptor potential channels

REFERENCES

1. Azevedo FA, Carvalho LR, Grinberg LT, et al. Equal numbers of neuronal and nonneuronal cells make the human brain an isometrically scaled-up primate brain. *J Comp Neurol.* 2009;513(5):532–541.
2. Oberheim NA, Goldman SA, Nedergaard M. Heterogeneity of astrocytic form and function. *Methods Mol Biol.* 2012;814:23–45.
3. Bushong EA, Martone ME, Jones YZ, Ellisman MH. Protoplasmic astrocytes in CA1 stratum radiatum occupy separate anatomical domains. *J Neurosci.* 2002;22(1):183–192.
4. Heller JP, Rusakov DA. Morphological plasticity of astroglia: understanding synaptic microenvironment. *Glia.* 2015;63(12):2133–2151.
5. Rusakov DA, Zheng K, Henneberger C. Astrocytes as regulators of synaptic function: a quest for the Ca^{2+} master key. *Neuroscientist.* 2011;17(5):513–523.
6. Vardjan N, Zorec R. Excitable astrocytes: $Ca^{(2+)}$- and cAMP-regulated exocytosis. *Neurochem Res.* 2015;40(12):2414–3414.
7. Hirase H, Iwai Y, Takata N, Shinohara Y, Mishima T. Volume transmission signalling via astrocytes. *Philos Trans R Soc Lond B Biol Sci.* 2014;369(1654):20130604.
8. Vizi ES, Fekete A, Karoly R, Mike A. Non-synaptic receptors and transporters involved in brain functions and targets of drug treatment. *Br J Pharmacol.* 2010;160 (4):785–809.
9. Audet MA, Doucet G, Oleskevich S, Descarries L. Quantified regional and laminar distribution of the noradrenaline innervation in the anterior half of the adult rat cerebral cortex. *J Comp Neurol.* 1988;274(3):307–318.
10. O'Donnell J, Zeppenfeld D, McConnell E, Pena S, Nedergaard M. Norepinephrine: a neuromodulator that boosts the function of multiple cell types to optimize CNS performance. *Neurochem Res.* 2012;37(11):2496–2512.
11. Xie L, Kang H, Xu Q, et al. Sleep drives metabolite clearance from the adult brain. *Science.* 2013;342(6156):373–377.
12. Ma Z, Stork T, Bergles DE, Freeman MR. Neuromodulators signal through astrocytes to alter neural circuit activity and behaviour. *Nature.* 2016;539(7629):428–432.
13. Hertz L, Chen Y, Gibbs ME, Zang P, Peng L. Astrocytic adrenoceptors: a major drug target in neurological and psychiatric disorders? *Curr Drug Targets CNS Neurol Disord.* 2004;3(3):239–267.
14. Perrais D, Kleppe I, Taraska J, Almers W. Recapture after exocytosis causes differential retention of protein in granules of bovine chromaffin cells. *J Physiol.* 2004;560 (Pt 2):413–428.
15. Rahamimoff R, Fernandez J. Pre- and postfusion regulation of transmitter release. *Neuron.* 1997;18(1):17–27.
16. Salm AK, McCarthy KD. Norepinephrine-evoked calcium transients in cultured cerebral type 1 astroglia. *Glia.* 1990;3(6):529–538.
17. Prebil M, Vardjan N, Jensen J, Zorec R, Kreft M. Dynamic monitoring of cytosolic glucose in single astrocytes. *Glia.* 2011;59(6):903–913.
18. Rupnik M, Kreft M, Sikdar S, et al. Rapid regulated dense-core vesicle exocytosis requires the CAPS protein. *Proc Natl Acad Sci USA.* 2000;97(10):5627–5632.

19. Kreft M, Milisav I, Potokar M, Zorec R. Automated high through-put colocalization analysis of multichannel confocal images. *Comput Methods Programs Biomed.* 2004;74(1):63–67.
20. Hartmann J, Lindau M. A novel $Ca^{(2+)}$-dependent step in exocytosis subsequent to vesicle fusion. *FEBS Lett.* 1995;363(3):217–220.
21. Giaume C, Marin P, Cordier J, Glowinski J, Premont J. Adrenergic regulation of intercellular communications between cultured striatal astrocytes from the mouse. *Proc Natl Acad Sci USA.* 1991;88(13):5577–5581.
22. De Keyser J, Wilczak N, Leta R, Streetland C. Astrocytes in multiple sclerosis lack beta-2 adrenergic receptors. *Neurology.* 1999;53(8):1628–1633.
23. Vardjan N, Kreft M, Zorec R. Dynamics of β-adrenergic/cAMP signaling and morphological changes in cultured astrocytes. *Glia.* 2014;62(4):566–579.
24. Sankaranarayanan S, Ryan T. Real-time measurements of vesicle-SNARE recycling in synapses of the central nervous system. *Nat Cell Biol.* 2000;2(4):197–204.
25. Wilson RS, Nag S, Boyle PA, et al. Neural reserve, neuronal density in the locus ceruleus, and cognitive decline. *Neurology.* 2013;80(13):1202–1208.
26. Feinstein DL, Kalinin S, Braun D. Causes, consequences, and cures for neuroinflammation mediated via the locus coeruleus: noradrenergic signaling system. *J Neurochem.* 2016;139(Suppl 2):154–178.
27. Polak PE, Kalinin S, Feinstein DL. Locus coeruleus damage and noradrenaline reductions in multiple sclerosis and experimental autoimmune encephalomyelitis. *Brain.* 2011;134(Pt 3):665–677.
28. Antoni FA. Interactions between intracellular free Ca^{2+} and cyclic AMP in neuroendocrine cells. *Cell Calcium.* 2012;51(3-4):260–266.
29. Bruce JI, Straub SV, Yule DI. Crosstalk between cAMP and Ca^{2+} signaling in non-excitable cells. *Cell Calcium.* 2003;34(6):431–444.
30. Ahuja M, Jha A, Maléth J, Park S, Muallem S. cAMP and Ca^{2+} signaling in secretory epithelia: crosstalk and synergism. *Cell Calcium.* 2014;55(6):385–393.
31. Horvat A, Zorec R, Vardjan N. Adrenergic stimulation of single rat astrocytes results in distinct temporal changes in intracellular $Ca^{(2+)}$ and cAMP-dependent PKA responses. *Cell Calcium.* 2016;59(4):156–163.
32. Sweeney S, Broadie K, Keane J, Niemann H, O'Kane C. Targeted expression of tetanus toxin light chain in Drosophila specifically eliminates synaptic transmission and causes behavioral defects. *Neuron.* 1995;14(2):341–351.
33. Washbourne P, Thompson P, Carta M, et al. Genetic ablation of the t-SNARE SNAP-25 distinguishes mechanisms of neuroexocytosis. *Nat Neurosci.* 2002;5(1):19–26.
34. Jahn R, Scheller R. SNAREs—engines for membrane fusion. *Nat Rev Mol Cell Biol.* 2006;7(9):631–643.
35. Chow R, von Rüden L, Neher E. Delay in vesicle fusion revealed by electrochemical monitoring of single secretory events in adrenal chromaffin cells. *Nature.* 1992;356(6364):60–63.
36. Deák F, Schoch S, Liu X, Südhof T, Kavalali E. Synaptobrevin is essential for fast synaptic-vesicle endocytosis. *Nat Cell Biol.* 2004;6(11):1102–1108.
37. Stenovec M, Poberaj I, Kreft M, Zorec R. Concentration-dependent staining of lactotroph vesicles by FM 4-64. *Biophys J.* 2005;88(4):2607–2613.
38. Lee CH, Poburko D, Sahota P, Sandhu J, Ruehlmann DO, van Breemen C. The mechanism of phenylephrine-mediated $[Ca^{(2+)}](i)$ oscillations underlying tonic contraction in the rabbit inferior vena cava. *J Physiol.* 2001;534(Pt 3):641–650.

39. Ishii K, Hirose K, Iino M. Ca^{2+} shuttling between endoplasmic reticulum and mitochondria underlying Ca^{2+} oscillations. *EMBO Rep.* 2006;7(4):390–396.
40. Nikolaev VO, Bünemann M, Hein L, Hannawacker A, Lohse MJ. Novel single chain cAMP sensors for receptor-induced signal propagation. *J Biol Chem.* 2004;279(36): 37215–37218.
41. Zhang J, Hupfeld CJ, Taylor SS, Olefsky JM, Tsien RY. Insulin disrupts beta-adrenergic signalling to protein kinase A in adipocytes. *Nature.* 2005;437(7058):569–573.
42. Pelkmans L, Zerial M. Kinase-regulated quantal assemblies and kiss-and-run recycling of caveolae. *Nature.* 2005;436(7047):128–133.
43. Hua S, Raciborska D, Trimble W, Charlton M. Different VAMP/synaptobrevin complexes for spontaneous and evoked transmitter release at the crayfish neuromuscular junction. *J Neurophysiol.* 1998;80(6):3233–3246.
44. Geppert M, Goda Y, Hammer R, et al. Synaptotagmin I: a major Ca2 + sensor for transmitter release at a central synapse. *Cell.* 1994;79(4):717–727.
45. Lollike K, Borregaard N, Lindau M. The exocytotic fusion pore of small granules has a conductance similar to an ion channel. *J Cell Biol.* 1995;129(1):99–104.
46. Hansson E, Simonsson P, Alling C. Interactions between cyclic AMP and inositol phosphate transduction systems in astrocytes in primary culture. *Neuropharmacology.* 1990;29(6):591–598.
47. Balázs R, Miller S, Chun Y, O'Toole J, Cotman CW. Metabotropic glutamate receptor agonists potentiate cyclic AMP formation induced by forskolin or beta-adrenergic receptor activation in cerebral cortical astrocytes in culture. *J Neurochem.* 1998;70 (6):2446–2458.
48. Jiménez AI, Castro E, Mirabet M, Franco R, Delicado EG, Miras-Portugal MT. Potentiation of ATP calcium responses by A2B receptor stimulation and other signals coupled to Gs proteins in type-1 cerebellar astrocytes. *Glia.* 1999;26(2):119–128.
49. Archer D, Graham M, Burgoyne R. Complexin regulates the closure of the fusion pore during regulated vesicle exocytosis. *J Biol Chem.* 2002;277(21):18249–18252.
50. Wang C, Grishanin R, Earles C, et al. Synaptotagmin modulation of fusion pore kinetics in regulated exocytosis of dense-core vesicles. *Science.* 2001;294(5544):1111–1115.
51. Han X, Wang C, Bai J, Chapman E, Jackson M. Transmembrane segments of syntaxin line the fusion pore of Ca^{2+}-triggered exocytosis. *Science.* 2004;304(5668): 289–292.
52. Duffy S, MacVicar BA. Adrenergic calcium signaling in astrocyte networks within the hippocampal slice. *J Neurosci.* 1995;15(8):5535–5550.
53. Bekar LK, He W, Nedergaard M. Locus coeruleus alpha-adrenergic-mediated activation of cortical astrocytes in vivo. *Cereb Cortex.* 2008;18(12):2789–2795.
54. Ding F, O'Donnell J, Thrane AS, et al. α1-Adrenergic receptors mediate coordinated Ca^{2+} signaling of cortical astrocytes in awake, behaving mice. *Cell Calcium.* 2013;54(6):387–394.
55. Paukert M, Agarwal A, Cha J, Doze VA, Kang JU, Bergles DE. Norepinephrine controls astroglial responsiveness to local circuit activity. *Neuron.* 2014;82(6):1263–1270.
56. Leybaert L, Sanderson MJ. Intercellular $Ca^{(2+)}$ waves: mechanisms and function. *Physiol Rev.* 2012;92(3):1359–1392.
57. Scemes E, Suadicani SO, Spray DC. Intercellular communication in spinal cord astrocytes: fine tuning between gap junctions and P2 nucleotide receptors in calcium wave propagation. *J Neurosci.* 2000;20(4):1435–1445.
58. Brown AM, Ransom BR. Astrocyte glycogen and brain energy metabolism. *Glia.* 2007;55 (12):1263–1271.

59. Sorg O, Magistretti PJ. Vasoactive intestinal peptide and noradrenaline exert long-term control on glycogen levels in astrocytes: blockade by protein synthesis inhibition. *J Neurosci.* 1992;12(12):4923–4931.
60. Gibbs ME. Role of glycogenolysis in memory and learning: regulation by noradrenaline, serotonin and ATP. *Front Integr Neurosci.* 2015;9:70.
61. Gibbs ME, Bowser DN. Astrocytic adrenoceptors and learning: alpha1-adrenoceptors. *Neurochem Int.* 2010;57(4):404–410.
62. Gibbs ME, Hutchinson DS, Summers RJ. Role of beta-adrenoceptors in memory consolidation: beta3-adrenoceptors act on glucose uptake and beta2-adrenoceptors on glycogenolysis. *Neuropsychopharmacology.* 2008;33(10):2384–2397.
63. Quach TT, Duchemin AM, Rose C, Schwartz JC. [3H]glycogenolysis in brain slices mediated by beta-adrenoceptors: comparison of physiological response and [3H]dihydroalprenolol binding parameters. *Neuropharmacology.* 1988;27(6):629–635.
64. Gibbs ME, Hutchinson D, Hertz L. Astrocytic involvement in learning and memory consolidation. *Neurosci Biobehav Rev.* 2008;32(5):927–944.
65. Subbarao KV, Hertz L. Effect of adrenergic agonists on glycogenolysis in primary cultures of astrocytes. *Brain Res.* 1990;536(1–2):220–226.
66. Hertz L, Peng L, Dienel GA. Energy metabolism in astrocytes: high rate of oxidative metabolism and spatiotemporal dependence on glycolysis/glycogenolysis. *J Cereb Blood Flow Metab.* 2007;27(2):219–249.
67. Catus SL, Gibbs ME, Sato M, Summers RJ, Hutchinson DS. Role of β-adrenoceptors in glucose uptake in astrocytes using β-adrenoceptor knockout mice. *Br J Pharmacol.* 2011;162(8):1700–1715.
68. Kreft M, Lukšič M, Zorec TM, Prebil M, Zorec R. Diffusion of D-glucose measured in the cytosol of a single astrocyte. *Cell Mol Life Sci.* 2013;70(8):1483–1492.
69. Hutchinson DS, Catus SL, Merlin J, Summers RJ, Gibbs ME. α2-Adrenoceptors activate noradrenaline-mediated glycogen turnover in chick astrocytes. *J Neurochem.* 2011;117(5):915–926.
70. Hertz L, Gibbs ME. What learning in day-old chickens can teach a neurochemist: focus on astrocyte metabolism. *J Neurochem.* 2009;109(Suppl 1):10–16.
71. San Martín A, Ceballo S, Ruminot I, Lerchundi R, Frommer WB, Barros LF. A genetically encoded FRET lactate sensor and its use to detect the Warburg effect in single cancer cells. *PLoS ONE.* 2013;8(2):e57712.
72. Barros LF, San Martín A, Sotelo-Hitschfeld T, et al. Small is fast: astrocytic glucose and lactate metabolism at cellular resolution. *Front Cell Neurosci.* 2013;7:27.
73. Ogata K, Kosaka T. Structural and quantitative analysis of astrocytes in the mouse hippocampus. *Neuroscience.* 2002;113(1):221–233.
74. Oberheim NA, Wang X, Goldman S, Nedergaard M. Astrocytic complexity distinguishes the human brain. *Trends Neurosci.* 2006;29(10):547–553.
75. Han X, Chen M, Wang F, et al. Forebrain engraftment by human glial progenitor cells enhances synaptic plasticity and learning in adult mice. *Cell Stem Cell.* 2013;12(3):342–353.
76. Oberheim NA, Tian GF, Han X, et al. Loss of astrocytic domain organization in the epileptic brain. *J Neurosci.* 2008;28(13):3264–3276.
77. Araque A, Parpura V, Sanzgiri RP, Haydon PG. Tripartite synapses: glia, the unacknowledged partner. *Trends Neurosci.* 1999;22(5):208–215.
78. Theodosis DT, Poulain DA, Oliet SH. Activity-dependent structural and functional plasticity of astrocyte-neuron interactions. *Physiol Rev.* 2008;88(3):983–1008.

79. Giaume C, Koulakoff A, Roux L, Holcman D, Rouach N. Astroglial networks: a step further in neuroglial and gliovascular interactions. *Nat Rev Neurosci.* 2010;11(2):87–99.
80. Chklovskii DB, Mel BW, Svoboda K. Cortical rewiring and information storage. *Nature.* 2004;431(7010):782–788.
81. Zorec R, Horvat A, Vardjan N, Verkhratsky A. Memory formation shaped by astroglia. *Front Integr Neurosci.* 2015;9:56.
82. Iliff JJ, Wang M, Liao Y, et al. A paravascular pathway facilitates CSF flow through the brain parenchyma and the clearance of interstitial solutes, including amyloid β. *Sci Transl Med.* 2012;4(147):147ra111.
83. Song Y, Gunnarson E. Potassium dependent regulation of astrocyte water permeability is mediated by cAMP signaling. *PLoS ONE.* 2012;7(4):e34936.
84. Risher WC, Andrew RD, Kirov SA. Real-time passive volume responses of astrocytes to acute osmotic and ischemic stress in cortical slices and in vivo revealed by two-photon microscopy. *Glia.* 2009;57(2):207–221.
85. Wilhelmsson U, Bushong EA, Price DL, et al. Redefining the concept of reactive astrocytes as cells that remain within their unique domains upon reaction to injury. *Proc Natl Acad Sci USA.* 2006;103(46):17513–17518.
86. Goldman JE, Abramson B. Cyclic AMP-induced shape changes of astrocytes are accompanied by rapid depolymerization of actin. *Brain Res.* 1990;528(2):189–196.
87. Safavi-Abbasi S, Wolff JR, Missler M. Rapid morphological changes in astrocytes are accompanied by redistribution but not by quantitative changes of cytoskeletal proteins. *Glia.* 2001;36(1):102–115.
88. Bicknell RJ, Luckman SM, Inenaga K, Mason WT, Hatton GI. Beta-adrenergic and opioid receptors on pituicytes cultured from adult rat neurohypophysis: regulation of cell morphology. *Brain Res Bull.* 1989;22(2):379–388.
89. Gharami K, Das S. Delayed but sustained induction of mitogen-activated protein kinase activity is associated with beta-adrenergic receptor-mediated morphological differentiation of astrocytes. *J Neurochem.* 2004;88(1):12–22.
90. Hatton GI, Luckman SM, Bicknell RJ. Adrenalin activation of beta 2-adrenoceptors stimulates morphological changes in astrocytes (pituicytes) cultured from adult rat neurohypophyses. *Brain Res Bull.* 1991;26(5):765–769.
91. Shain W, Forman DS, Madelian V, Turner JN. Morphology of astroglial cells is controlled by beta-adrenergic receptors. *J Cell Biol.* 1987;105(5):2307–2314.
92. Won CL, Oh YS. cAMP-induced stellation in primary astrocyte cultures with regional heterogeneity. *Brain Res.* 2000;887(2):250–258.
93. Shao Y, Enkvist MO, McCarthy KD. Glutamate blocks astroglial stellation: effect of glutamate uptake and volume changes. *Glia.* 1994;11(1):1–10.
94. Rodnight RB, Gottfried C. Morphological plasticity of rodent astroglia. *J Neurochem.* 2013;124(3):263–275.
95. Racchetti G, D'Alessandro R, Meldolesi J. Astrocyte stellation, a process dependent on Rac1 is sustained by the regulated exocytosis of enlargeosomes. *Glia.* 2012;60(3):465–475.
96. Griffith R, Sutin J. Reactive astrocyte formation in vivo is regulated by noradrenergic axons. *J Comp Neurol.* 1996;371(3):362–375.
97. Sutin J, Griffith R. Beta-adrenergic receptor blockade suppresses glial scar formation. *Exp Neurol.* 1993;120(2):214–222.
98. Abe K, Saito H. Effect of ATP on astrocyte stellation is switched from suppressive to stimulatory during development. *Brain Res.* 1999;850(1–2):150–157.

99. Chalermpalanupap T, Kinkead B, Hu WT, et al. Targeting norepinephrine in mild cognitive impairment and Alzheimer's disease. *Alzheimers Res Ther.* 2013;5(2):21.
100. Olabarria M, Noristani HN, Verkhratsky A, Rodríguez JJ. Concomitant astroglial atrophy and astrogliosis in a triple transgenic animal model of Alzheimer's disease. *Glia.* 2010;58(7):831–838.
101. Thrane AS, Rappold PM, Fujita T, et al. Critical role of aquaporin-4 (AQP4) in astrocytic Ca^{2+} signaling events elicited by cerebral edema. *Proc Natl Acad Sci USA.* 2011;108(2): 846–851.
102. Manley GT, Fujimura M, Ma T, et al. Aquaporin-4 deletion in mice reduces brain edema after acute water intoxication and ischemic stroke. *Nat Med.* 2000;6(2):159–163.
103. Pangrsic T, Potokar M, Haydon P, Zorec R, Kreft M. Astrocyte swelling leads to membrane unfolding, not membrane insertion. *J Neurochem.* 2006;99(2):514–523.
104. Seifert G, Schilling K, Steinhäuser C. Astrocyte dysfunction in neurological disorders: a molecular perspective. *Nat Rev Neurosci.* 2006;7(3):194–206.
105. Vardjan N, Horvat A, Anderson JE, et al. Adrenergic activation attenuates astrocyte swelling induced by hypotonicity and neurotrauma. *Glia.* 2016;64(6):1034–1049.
106. Teng YD, Choi H, Onario RC, et al. Minocycline inhibits contusion-triggered mitochondrial cytochrome c release and mitigates functional deficits after spinal cord injury. *Proc Natl Acad Sci USA.* 2004;101(9):3071–3076.
107. O'Connor ER, Kimelberg HK. Role of calcium in astrocyte volume regulation and in the release of ions and amino acids. *J Neurosci.* 1993;13(6):2638–2650.
108. Conner MT, Conner AC, Bland CE, et al. Rapid aquaporin translocation regulates cellular water flow: mechanism of hypotonicity-induced subcellular localization of aquaporin 1 water channel. *J Biol Chem.* 2012;287(14):11516–11525.
109. Fischer R, Schliess F, Häussinger D. Characterization of the hypo-osmolarity-induced Ca^{2+} response in cultured rat astrocytes. *Glia.* 1997;20(1):51–58.
110. Tong X, Shigetomi E, Looger LL, Khakh BS. Genetically encoded calcium indicators and astrocyte calcium microdomains. *Neuroscientist.* 2013;19(3):274–291.
111. Gao T, Yatani A, Dell'Acqua ML, et al. cAMP-dependent regulation of cardiac L-type Ca^{2+} channels requires membrane targeting of PKA and phosphorylation of channel subunits. *Neuron.* 1997;19(1):185–196.
112. Shigetomi E, Tong X, Kwan KY, Corey DP, Khakh BS. TRPA1 channels regulate astrocyte resting calcium and inhibitory synapse efficacy through GAT-3. *Nat Neurosci.* 2012;15(1):70–80.
113. Potokar M, Stenovec M, Jorgačevski J, et al. Regulation of AQP4 surface expression via vesicle mobility in astrocytes. *Glia.* 2013;61(6):917–928.

Chapter 6

Adrenergic Receptors on Astrocytes Modulate Gap Junctions

Eliana Scemes[✉], **Randy F. Stout, Jr,**[✉] **and David C. Spray**[✉]
Albert Einstein College of Medicine, Bronx, NY, United States

Chapter Outline

Gap Junction Subtypes in Glia and Their Consensus Sites of Phosphorylation by Adrenergic Receptor–Mediated Processes 128
Direct Effects of Adrenergic Receptors on Gap Junctions 130
Gap Junction Formation and Degradation 133
Indirect Effects of Adrenergic Signaling on Coupling Within the Astrocyte Network 136
Calcium Signaling 136
Diffusion of cAMP 137
Diffusion of Metabolites 138
Conclusions 140
Abbreviations 140
References 141

ABSTRACT
Adrenergic receptors in astrocytes have both direct and indirect impact on intercellular communication in the nervous system. Direct mechanisms include activation of second messenger pathways that produce posttranslational modifications of the gap junction proteins [connexins (Cxs)] or activate transcription factors or other processes such as miRNAs that dictate or modify Cx gene expression or protein translation. These direct effects change the strength of intercellular communication through modifying open probability of the gap junction channels, altering balance between delivery to the membrane and degradation or through regulation of transcript abundance. Indirect actions with potentially profound implications for glial and neuronal network activity include alterations in the intracellular concentration of metabolites, energy-supplying molecules and second messengers, thereby changing the diffusional driving force responsible for intercellular spread.

[✉]Correspondence address
E-mail: eliana.scemes@einstein.yu.edu; randy.stout@einstein.yu.edu; david.spray@einstein.yu.edu

Noradrenergic Signaling and Astroglia. DOI: http://dx.doi.org/10.1016/B978-0-12-805088-0.00006-2
© 2017 Elsevier Inc. All rights reserved.

Although direct effects target the gap junction channels themselves to affect gating, assembly, and turnover, indirect action to liberate intracellular molecules may also produce rapid dynamic changes in spread of these factors throughout the coupled networks.

Keywords: Connexin; Cx43, Cx32; Cx47; Cx30; glia; panglial syncytium; oligodendrocyte; cAMP; protein kinases

GAP JUNCTION SUBTYPES IN GLIA AND THEIR CONSENSUS SITES OF PHOSPHORYLATION BY ADRENERGIC RECEPTOR–MEDIATED PROCESSES

The membrane-spanning core proteins of gap junction channels are the connexin (Cx) family in vertebrates and the topologically similar but unrelated innexin family in invertebrates. For the Cxs, two nomenclature schemes were initially proposed: GJ followed by Greek letters (α through δ) indicating homologous gene families, followed by numbers indicating the order of discovery, *vs* Cx (Cx)N, where N indicates the molecular weight in kDa of the protein predicted from the cDNA. As a compromise that has further confused the field, genes use the former terminology and proteins the latter. At least seven Cx types are expressed in the brain parenchyma. Astrocytes express Cx *GJB2*, *GJB6*, and *GJA1* genes in man, encoding Cx26, Cx30, and Cx43, respectively. Oligodendrocytes express Cx29 (which is likely nonfunctional), Cx32, and Cx47, encoded by *GJC3*, *GJB1*, and *GJC2* genes. Although still controversial, microglia have been reported to express Cx36[1], Cx32,[2] and Cx43.[3] Neurons express predominantly Cx36 (*GJA9*).

Cxs possess four membrane-spanning domains with intracellular amino- and carboxyl-termini and two extracellular loops (see Fig. 6.1 for membrane topologies of the glial Cxs). A hexamer of Cxs forms a connexon or hemichannel in each cell, and hemichannels pair with apposing hemichannels through irreversible high affinity interactions of extracellular loops. Sequences of Cxs differ most in cytoplasmic domains. In these domains amino acid residues are located that may be posttranslationally modified through phosphorylation, nitrosylation, acetylation, ubiquination, SUMOylation, palmitoylation, but there is no glycosylation (reviewed in Ref. [4]). Most important from the standpoint of this review are Cx-specific serine, threonine, and tyrosine residues that can be phosphorylated by specific protein kinases (PKs) and dephosphorylated by phosphatases.

The first Cx for which phosphorylation was demonstrated and phosphorylation sites determined was Cx32.[5] For those experiments, Cx32 was isolated from rat liver, a tissue in which gap junctions are particularly abundant, occupying as much as 5% of the surface membrane. Among glia, Cx32 is found in myelinating Schwann cells and in oligodendrocytes.

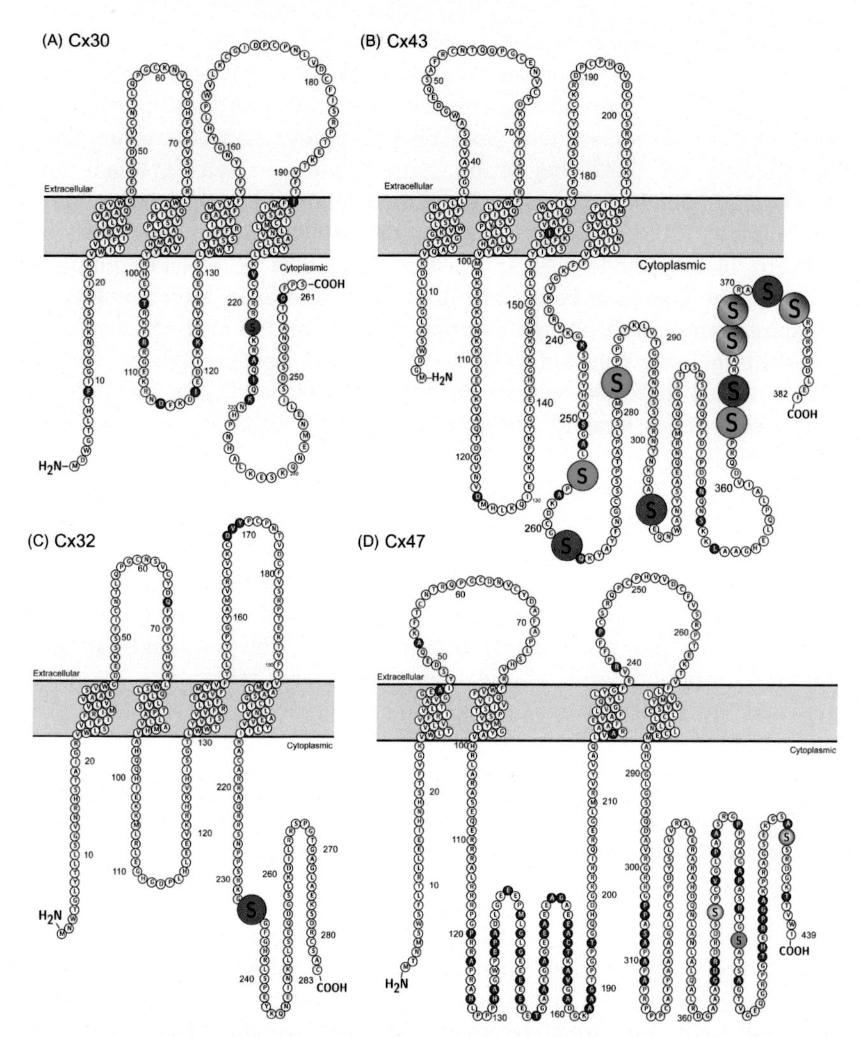

FIGURE 6.1 Human glial connexins with residues marked that are posttranslationally modified by kinases targeted by adrenergic signaling. (A and B) Astrocyte connexins Cx30 (*Gjb6*), and Cx43 (*Gja1*). (C and D) Oligodendrocyte connexins that form gap junctions, Cx32 (*Gjb1*), and Cx47 (*Gjc2*, a.k.a. *Gja12*). Residues in these human sequences that differ from the mouse orthologs are darkened. Experimentally validated phosphorylation sites are indicated by large colored residue circles in Cx43 and Cx32 as cited in the text. Phosphorylation sites for Cx30 and Cx47 have not been reported. The kinase-specific prediction program Scansite3[66] was used to identify potential phosphorylation sites (only protein kinase A (PKA) and protein kinase C (PKC) sites of potential cytoplasmic, C-terminal serine residues are shown). Predicted but not proven phosphorylation sites for PKA are shown in green and for PKC are shown in red. Connexin sequences were plotted by the online visualization tool Protter.[67]

Initial studies demonstrated that serine and possibly threonine residues were phosphorylated by protein kinase C (PKC),[5] and it was later shown that PKC and Ca^{2+}/calmodulin-dependent PK II (CaMKII) also phosphorylated Cx32.[6] Most subsequent work on phosphorylation of gap junctions has focused on Cx43, beginning with studies in which Musil and Goodenough[7] showed that phosphorylation retards electrophoretic mobility of Cx43. The multiple ~39–45 kDa bands readily detectable on Western blots provide a simple assay for extent of phosphorylation,[7] although the degree of band shift does not indicate the stoichiometry of phosphorylation, and phosphorylation of some amino acid residues does not shift mobility (see Ref. [8]). The Musil and Goodenough study also demonstrated that Cx43 within gap junction plaques is less mobile, and thus likely more heavily phosphorylated than Cx43 in nonjunctional domains.

Among the consequences of phosphorylation and dephosphorylation of Cx43 within gap junctions is the altered affinity of Cx43 for binding partners, which is discussed in more detail below. Cx phosphorylation also plays fundamental roles in gap junction regulation, affecting channel gating, gap junction assembly, and turnover, as well as Cx biogenesis. As described below in more detail, these changes in gap junction function and abundance profoundly affect the panglial network, dynamically expanding, shrinking, or contorting the "glia connectome."

DIRECT EFFECTS OF ADRENERGIC RECEPTORS ON GAP JUNCTIONS

Astrocytes express both α- and β-adrenergic receptors (ARs), primarily the α_1 and β_2 subtypes. As illustrated in Fig. 6.2, binding of appropriate ligands to β_2-AR results in activation of adenylate cyclase (AC) and stimulation of cyclic adenosine monophosphate (cAMP) and PKA. Binding to α_1-ARs primarily activates PKC. Although signal transduction pathways overlap, with considerable crosstalk, we focus on these two kinases in this review.

Of the five functional Cxs in astrocytes and oligodendrocytes (Cx26, Cx30, and Cx43 in astrocytes, Cx32 and Cx47 in oligodendrocytes), detailed information on phosphorylation sites is only available for Cx32 and Cx43 (Fig. 6.1B and C). However, consensus phosphorylation sites can be predicted from amino acid sequences with reasonable precision. With regard to the two targets of adrenergic stimulation, PKC and cAMP-dependent PKA, we illustrate the consensus phosphorylation sites on mouse and human isoforms of each glial Cx in Fig. 6.1.

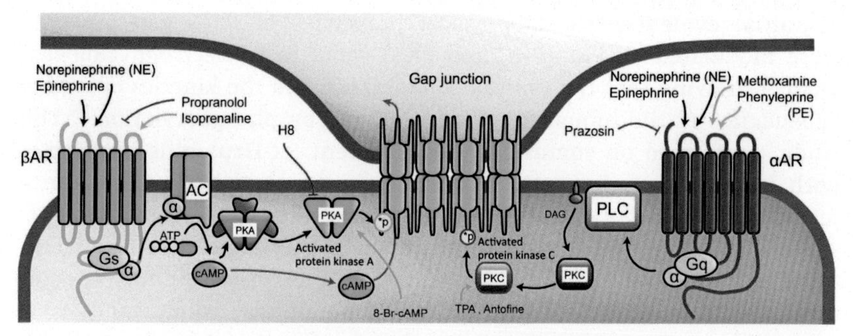

FIGURE 6.2 Adrenergic signaling pathways control phosphorylation of astrocyte connexins. A gap junction connecting two cells is shown in blue with the carboxyl-terminal tails extending into the cytoplasm. Pharmacological agonists and inhibitors mentioned in this text are listed. The pathway components on which agonists/antagonists act are indicated by arrows/lines. β-AR signaling (green, left-side pathway) is mediated through adenylate cyclase (AC), cyclic adenosine monophosphate (cAMP), and protein kinase A (PKA), shown with brown repressor subunit, which dissociates from the active PKA catalytic subunits when cAMP is bound and PKA is activated. α_1-AR signaling (red, right-side pathway) signals through $G_{q\text{-}\alpha}$ activation of protein kinase C (PLC), which produces DAG along with IP_3, DAG activates PKC, which can phosphorylate specific residues in the carboxyl-terminus of Cx43 (and other connexins) to regulate channel function and gap junction endocytosis as discussed in text.

Early reports of second messenger regulation of gap junction channels showed that cAMP enhanced gap junctional communication between hepatocytes and between cardiac myocytes, including the demonstration that β-AR stimulation caused similar changes.[5,9–11] These functional changes were shown to be associated with levels of Cx phosphorylation in several cases.[5,7,12,13] In the late 1980s, it became apparent that astrocytes express Cx43,[14,15] and several studies demonstrated that adrenergic stimulation affected functional coupling. For example, using scrape loading to evaluate dye coupling, Giaume et al.[16] reported effects of both β- and α-AR signaling in striatal astrocytes; activation of AR by prolonged noradrenaline (NA) application decreased intercellular spread of the dye Lucifer Yellow (LY) by 50%, whereas the selective β-AR agonist isoproterenol resulted in 16% increase in dye coupling among astrocytes. Based on the effects of a selective α-AR agonist and an antagonist, the authors suggested that the uncoupling effect of NA was mediated by α_1-AR because methoxamine mimicked the uncoupling effect of NA, and the α_1-AR antagonist prazosin blocked the NA action. Furthermore, the strong correlation obtained between the extent of AR second messenger levels [phospholipase (PL) C activation and accumulation of cAMP] and the degree of dye coupling suggested that astrocyte gap junctional communication was regulated by distinct pathways such as PLC and PLA_2 pathways and cAMP-dependent mechanisms.[16,17]

Initial evidence that biophysical properties of Cx43 gap junctions were modulated by PKs (PKC and PKA) came from electrophysiological studies showing that the unitary conductance and the kinetics of voltage dependence of the human Cx43 were altered by phosphorylation.[18] This study, performed on communication deficient SK-Hep1 cells transfected with human Cx43, indicated that exposure to phosphorylating agents, tetradecanoylphorbol acetate (TPA), a phorbol ester that stimulates PKC, and 8-Br-cAMP, which activates PKA, reduced the single channel conductance, whereas the PK inhibitor staurosporine increased unitary conductance.[18] Similar effects of PKC activation on unitary conductance were reported by Habo Jongsma's group, who also found unitary conductance decreased by membrane permeable cyclic guanosine monophosphate (cGMP).[19] Dye coupling was also reduced by these agents, indicating an overall reduction in coupling strength, whereas PKA stimulation had no effect.[20] The effects of cGMP were restricted to the rat Cx43 isoform and are not seen in human Cx43, likely reflecting a difference in amino acid sequence between these species.[21]

The effect of β-AR stimulation via PKA activation on Cx43 gap junctions is controversial. As mentioned above, Moreno et al.[18] reported a decrease in Cx43 unitary conductance induced by 8-Br-cAMP, and Kwak and Jongsma[20] did not detect a change. In contrast, enhanced coupling between cardiomyocytes (which express mainly Cx43) was found after stimulation with the β-AR agonist isoproterenol,[9,22] similar to that reported in astrocytes by Christian Giaume's group,[23] and leptomenigeal cells also increased junctional conductance when treated with 8-Br-cAMP.[24] These somewhat different results may relate to species differences (human Cx43 has four PKA phosphorylation sites, whereas rat Cx43 and mouse Cx43 have three sites), to the conformational state of the C-terminal of Cx43, which has been proposed to be a poorer substrate for PKA in transfected than in native cells,[25] and/or to the functional state of the β-AR/AC/cAMP/PKA pathway[26] reviewed in Ref. [27].

It was in 1994 that Konietzko and Mueller using acute hippocampal slices showed that astrocytes are extensively dye-coupled to each other in situ and that phorbol ester treatment greatly reduced the spread of dye LY through the astrocytic network.[28] These authors suggested that gap junctional communication could be modulated by neurotransmitters acting through second messengers, which would allow for the subcompartmentalization of astrocytic networks in a dynamic and neuronal activity dependent manner. Venance et al.[29] used Ca^{2+} imaging and the application of various transmitter molecules to determine the heterogeneity of astrocytes in slice preparations. They found that all cells responded to endothelin-1 (ET-1), whereas responses to the muscarinic agonist carbachol and the α_1-AR agonist methoxamine were heterogeneous. Responses to ET-1 were blocked by certain gap junction inhibitors (heptanol, halothane, octanol), but not others (e.g., 18βGA); 18βGA reduced the number of cells responding to carbachol and methoxamine, indicating that astrocytes may effectively share muscarinic receptor and AR.

Li et al.[8] showed that Cx43 binding to other proteins [β-actin, extracellular signal-regulated kinases 1/2 (ERK1/2) and mitogen-activated PK phosphatase 1 (MKP-1)] was altered in cultured astrocytes exposed to chemical ischemia/hypoxia and that this was accompanied by changes in Cx43 phosphorylation and binding of c-Src (a tyrosine PK named for its similarity to a sarcoma viral protein). These results identified new binding partners for Cx43 and demonstrated that interactions between Cx43 and PKs and phosphatases were dynamically altered as a consequence of Cx43 phosphorylation. Subsequent studies by the groups of Lampe and Sorgen[30] have elegantly shown conformational changes of the Cx43 carboxyl-terminal region induced by phosphorylation of a serine residue in this region (S365) that blocks PKC phosphorylation and uncoupling in response to ischemia in heart, further revealing the impact of phosphorylation on Cx43-protein interactions.

Gap Junction Formation and Degradation

The synthesis and transport of Cx43 follows the secretory pathway for integral membrane proteins (see Fig. 6.3). Cx43 lipophilic domains are integrated, during translation at the ribosome, into the membrane of the ER where its tertiary structure is formed. The immature Cx43 undergoes several phosphorylation steps to become the mature Cx43; these are essential steps for proper oligomerization of Cxs into connexons and for proper composition and function of a complete gap junction channel. Once a complete gap junction channel is assembled, it cannot be disassembled into two hemichannels; instead, one of the cells incorporates the whole gap junction channel by invagination of the junctional cell membrane, forming what are called "annular gap junctions", for review see Ref. [31]. Cx43 has a short half-life (90−180 min) compared to other integral membrane proteins, and it is degraded via lysosomal and proteosomal pathways. Modifiability of Cxs turnover through protein phosphorylation was demonstrated in studies showing that treatment of rat hepatocytes with cAMP prolonged coupling among these cells after dissociation into cell clusters by decreasing rate of Cx32 degradation.[32] That phosphorylation of Cx43 also plays a role in gap junction degradation was first shown by Lampe,[12] and it is now known that PKC plays an important role in Cx43 ubiquitination, endocytosis, and degradation (review in Ref. [33]). Using the cardiac myoblast cell line H9c6, Popolo et al.[34] showed that PKC activation accelerated degradation of Cx43 when using adenosine plus α_1-AR stimulation. The increased Cx43

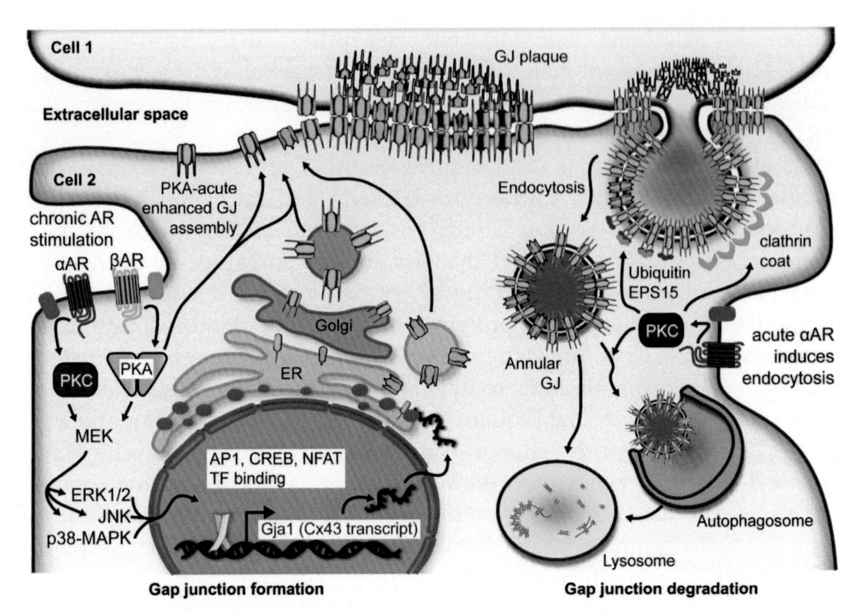

FIGURE 6.3 **Gap junction assembly and degradation are controlled by adrenergic signaling pathways.** Connexin Cx43 (blue) and Cx30 (green) subunits are thought to be assembled into connexons (hexameric hemichannels) in separate compartments. Vesicles containing connexons are trafficked to the plasma membrane where they are inserted. Connexons then move laterally in the membrane and dock with connexons in the membrane of an adjacent cell (cell 1 in this cartoon). Gap junction channels cluster as they dock to form gap junction plaques. Upon phosphorylation (by protein kinase C (PKC) and other kinases, not shown), ubiquitin, EPS15, clathrin, and other endocytic protein machinery are recruited to the gap junction (GJ) plaque and initiate endocytosis of a portion (or the entire structure) of the gap junction plaque. This results in formation of a double-membrane annular gap junction. Content of annular gap junctions undergoes fusion with autophagosomes or lysosomes and degradation. Pathways (arrows) on the left-side show chronic PKC and protein kinase A (PKA) activation by α- and β-adrenergic receptors (ARs) that lead to increased Cx43 expression, trafficking, and/or gap junction plaque formation. On the left, the effect of acute α-AR stimulation leads to PKC-mediated endocytosis and degradation by autophagy.

phosphorylation was shown to be necessary for Cx43 ubiquitination and then degradation via the proteasome system[34], which was hypothesized as underlying new mechanism for the antiadrenergic actions of adenosine on cardiomyocytes.

The involvement of PKC in Cx43 degradation in astrocytes was recently shown by Huang et al.[35] who found that an anticancer alkaloid (antofine, derived from a milkweed plant) caused dose-dependent inhibition of scrape loading dye coupling and decreased Cx43-protein level. Antofine was shown to induce endocytosis of surface gap junctions into the cytoplasm, where Cx43 was found to be colocalized with the early endosome marker EEA1. Inhibition of lysosomes or proteasomes by their respective specific inhibitors, NH_4Cl or MG132, partially inhibited the antofine-induced decrease in Cx43protein levels. The specific PKCβ inhibitor 3-(1-(3-imidazol-1-ylpropyl)-1*H*-indol-3-yl)-4-anilino-1*H*-pyrrole-2,5-dione prevented antofine-induced endocytosis of gap junctions and downregulation of Cx43.[35]

In contrast to the acute uncoupling effects of α_1-AR stimulation on Cx43, chronic stimulation is reported to enhance Cx43 expression via PKC, p38, ERK1/2, JNK, c-fos, and AP1 (reviewed in Ref. [27]). Twenty four hours of continuous infusion of phenylephrine in vivo resulted in increased expression of Cx43 in the heart of rats.[36] Similar effects were also observed in cultured rat cardiomyocytes, where both Cx43 mRNA and protein levels were found elevated.[36] Blockade of PKC but not of PKA prevented the increase in electrical coupling, which was attributed to blockade of upregulated protein expression.[36] Using specific inhibitors of several kinases, this group found that phenylephrine-induced increased expression of Cx43 was inhibited by blocking MEK, p38-MAP, ERK1/2, and JNK.[37] The nuclear translocation of transcription factors AP1 and c-fos is downstream of these kinases, with AP1 being among the most important transcription factors controlling Cx43 following α_1-AR stimulation.[38,39] Several other groups have indicated that among the α_1-AR subtypes, Cx43 expression is regulated by α_{1D}-ARs and not the α_{1A}-ARs involved in the regulation of other proteins.[40–42]

In cardiomyocytes, β-AR agonists activate PKA and the MAPK cascade and the calcineurin pathway, with chronic treatment, leading to the translocation to the nucleus of the transcription factors AP1, CREB, and NFAT. This ultimately leads to upregulation of Cx43 mRNA and protein levels paralleled by Cx43 phosphorylation (reviewed in Ref. [27]). Activation of MAPKs/PKA and the parallel increase in Cx43 expression and phosphorylation were shown to be inhibited by propranolol (β-AR antagonist) and H8 (PKA inhibitor) in isoprenaline-treated cardiomyocytes for 24 h.[36,43] In addition, propranolol, H8, p38, and MEK1 inhibitors were shown to block isoprenaline-induced translocation of AP1 and CREB,[43] two transcription factors that have been shown to bind to Cx43 promoter.[44–46]

Gap junction channels span two membranes, and are believed to form primarily by hemichannel accrual from the perimeter and linkage across extracellular space. It was initially believed that their turnover involved "unzipping" the membranes, reversing the formation process. However, numerous studies have now concluded that the pairing of hemichannels is virtually irreversible, so that gap junctions are endocytosed by one cell or the other as double-walled vesicles or "annular" gap junctions and then degraded through autophagosome and lysosomal pathways (see Refs. [47–50]). This raises the interesting question of what factors determine into which of the cells the annular gap junctions are transported.

As summarized in a recent review,[51] clathrin recruitment to the gap junction plaque has been shown to be facilitated by MAPK- and PKC-dependent phosphorylation/dephosphorylation of sites in the carboxyl-terminus of Cx43. These phosphorylation sites include at least three serine residues (S279, S282, and S368), and the phosphorylation of these sites is now believed to bias the retrieval of the annular gap junction vesicles into that member of the cell pair.

INDIRECT EFFECTS OF ADRENERGIC SIGNALING ON COUPLING WITHIN THE ASTROCYTE NETWORK

Spread of signals throughout the panglial syncytium can be regulated by changing coupling strength through gating, posttranslational or transcriptional mechanisms, as summarized in the preceding section. However, spread of signals through gap junctions is diffusional, with flux being driven by concentration gradients between donor and recipient cells. There are several ways in which adrenergic stimulation may alter concentration gradients in addition to the elevation of second messengers caused by binding of the adrenergic agonists to their receptors. Such effects include alterations in glucose abundance through altered glycogenolysis or glycogen synthesis, altered concentrations of intracellular ATP, Ca^{2+}, and IP_3 by α-ARs, and altered concentrations of amino acids through altered activity of transporters.

Calcium Signaling

That modulation of astrocyte gap junctional communication by second messengers has functional significance was first shown by Enkvist and McCarthy in 1992.[52] These authors showed that such modulation impacted on the transmission of intercellular Ca^{2+} waves among astrocytes. By combining the use of intracellularly injected dye LY and optical imaging of mechanically induced intracellular Ca^{2+} spread in cultured astrocytes, they showed that activation by PKC with a phorbol ester (PMA) or a synthetic diacylglycerol inhibited spread of LY and of Ca^{2+} waves. Similar uncoupling was obtained following activation of purinergic P2Y receptors with 2-methythiol-ATP, which elevates cytosolic IP_3 levels.[52]

That adrenergic signaling affected Ca^{2+} communication among astrocytes was revealed by Muyderman et al.,[53] who showed that the α_1-AR agonist phenylephrine decreased Ca^{2+} wave spread in mixed astrocyte/neuron cocultures, whereas the β-AR agonist isoproterenol increased this form of signaling. Based on experiments separately manipulating the two second messengers generated by phenylephrine stimulation by use of the PLC inhibitor U-73122 and a PKC-activating phorbol ester, these authors proposed that PLC played a critical role in the initial phase of Ca^{2+} wave spread, whereas PKC inhibited the transmission at a later phase.[53]

These studies and several others (reviewed in Ref. [54]) illustrate that Ca^{2+} signal transmission among astrocytes depends on the (sub)-type of membrane receptors, and on the degree of gap junction coupling, which can be differentially modulated by PKC and PKA.[54] Thus, the effective volume of the intracellular compartment provided by the gap junction channels to sustain long distance Ca^{2+} signal transmission will depend not only on the amount of Ca^{2+}-mobilizing second messengers (e.g., IP_3) generated by membrane receptors but also on the gap junction phosphorylating kinases that are activated by these G-protein coupled receptors.

Diffusion of cAMP

Lawrence et al.[55] elegantly showed that cAMP passed between cells through gap junctions in coculture experiments using cell types with distinct mechanisms of cAMP activation [β-AR agonist activation in cardiac myocytes, follicle-stimulating hormone (FSH) in ovarian granulosa cells].[55] Generation of cAMP by agents specifically acting on one cell was shown to indirectly affect the other. Thus, FSH applied to ovarian granulosa cells slowed spontaneous beat rate of adjacent cardiac myocytes, whereas β-AR stimulation of cardiac myocytes caused morphological changes typical of FSH treatment in adjacent granulosa cells. Numerous subsequent studies have supported such exchange, including the activation of nucleotide-gated chloride channels in one cell by cAMP introduced into the other.[56]

Although experiments have suggested that cAMP is permeable through gap junction hemichannels, the evidence is somewhat inconclusive. Using a reconstituted vesicle preparation in which Cx32 or Cx26 was incorporated, Bevans et al.[57] reported that both cAMP and cGMP were permeable through Cx32, whereas permeability was reduced when Cx32 and Cx26 were coexpressed, indicating lower permeability for Cx26 channels.[57] However, in a more recent study, Valiunas[58] used a cyclic nucleotide-modulated channel from sea urchin sperm (SpIH) as a cAMP reporter, enabling determination of cAMP permeability of Cx43 and Cx26 hemichannels expressed in HeLa cells.[58] SpIH-derived currents were induced by extracellular solution containing high K^+ to depolarize the membrane and cAMP and were inhibited by high extracellular Ca^{2+} and the gap junction blocker carbenoxolone. They concluded that both Cx43 and Cx26 hemichannels provided the pathway for influx of extracellular cAMP, a conclusion that is in contrast with the low cAMP permeability shown in the Bevans et al.[57] experiments.

Overall, these studies demonstrated that gap junctions composed of the astrocyte Cx26 and Cx43 are permeable to cAMP, although permeability of this second messenger through Cx26 channels is likely lower than for channels formed of Cx43. Of note, however, is that in the Valiunas[58] study, the properties of the Cx hemichannels included the activation of uptake of dyes by elevated extracellular [K^+] and the block by high extracellular [Ca^{2+}] or carbenoxolone. Such properties are striking characteristics of channels formed by pannexin1 (Panx1), a member of gap junction family of vertebrate proteins that do not form intercellular channels. Whether astrocyte hemichannels formed of Cxs are permeable to cAMP, and the possible relevance of such permeability, remains to be conclusively determined.

Diffusion of Metabolites

The metabolites glucose and lactate are readily permeable through gap junction channels. Astrocyte gap junctions provide an intercellular route for delivery of these molecules from the blood–brain barrier to oligodendrocytes and to neurons (see Refs. [59,60]). As pointed out in elegant studies by Rouach et al.,[60] because gap junction-mediated diffusion depends on driving force, delivery of metabolites can be biased toward metabolically active regions.[60] By comparing diffusion of inert gap junction permeant molecules (LY, Texas red, and biocytin) to that of the fluorescent glucose derivative, [2-[*N*-(7-nitrobenz-2-oxa-1,3-diazol-4-yl) amino]-2-deoxyglucose (2-NBDG)], they determined that glucose delivery among astrocytes was gap junction-mediated and could be directed toward highly active neuronal populations. In a commentary on that study and those illustrating similar impact of glucose diffusion among astrocytes by Dienel's group,[61,62] we linked this supply distribution to that of a "power grid" (see Fig. 6.4) to emphasize the potential impact of metabolite sharing among coupled cells to provide localized delivery.[63] A caveat of interpretation of the study by Rouach et al.[60] is that they utilized a deoxyglucose derivative, which like glucose is freely gap junction permeable, although glucose itself is rapidly converted into glucose-6-phosphate (G-6P) by cytoplasmic hexokinase, G-6P is the most abundant intracellular glucose form.[60] However, although measurements of isotope distribution following injection of tritiated G-6P were initially reported as providing evidence for gap junction permeability to glucose monophosphates,[64] later studies using several techniques indicate that its junctional permeability is 70%–90% lower than that of glucose itself.[60,65] Nevertheless, G-6P is to some extent gap junction permeable, and lactate is as permeable as glucose, and it follows that local variation in glucose or lactate levels would be expected to strongly impact metabolic spread throughout the panglial syncytium.

(A) Without adrenergic stimulation, smaller network spread

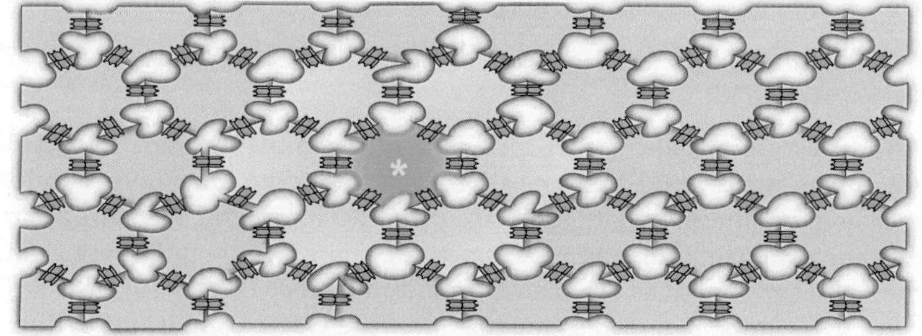

(B) Increased metabolite production in one astrocyte leads to expanded spread

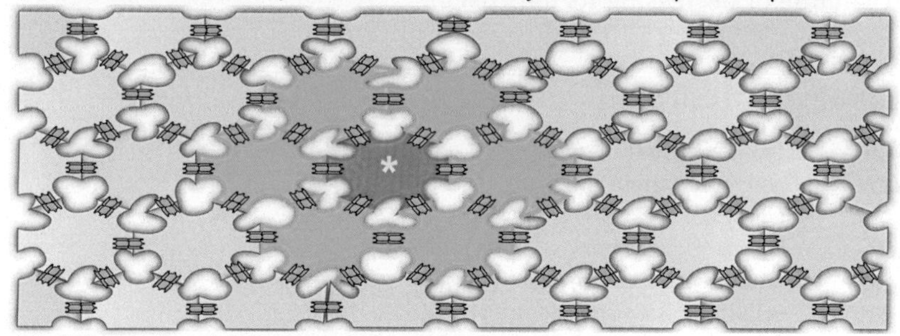

(C) Anisotropic metabolite spread due to local neuronal activity

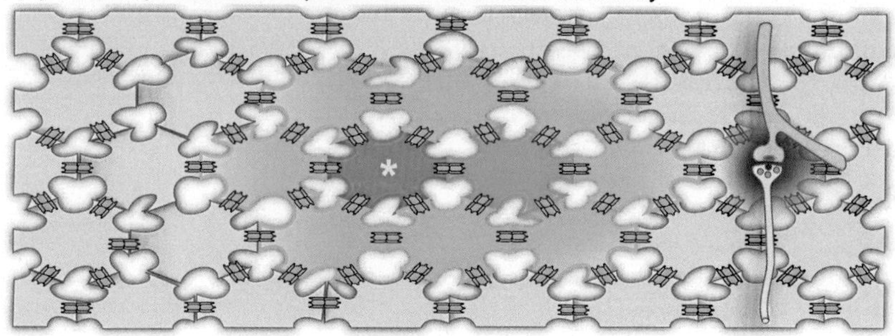

(D) Profiles of glucose across astrocyte syncytium

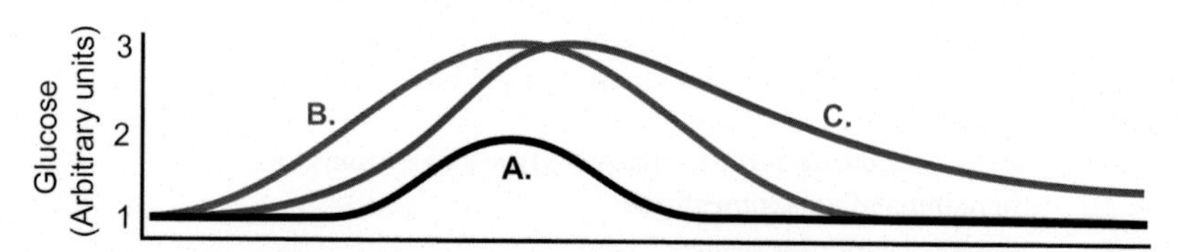

FIGURE 6.4 **Metabolite production and usage stimulated by adrenergic receptor activation leads to differential distribution of signaling molecules and metabolites through the astrocyte network.** (A and B) AR activation of one astrocyte (labeled with yellow asterisk) increases signaling molecule and/or metabolite production in astrocyte 1, with higher activation leading to more extensive spread through the astrocyte gap junction network. (C) Focal stimulation [yellow/orange synapse, right side in (C)] of metabolic demand may produce an anisotropic spread of metabolites through gap junctions toward the astrocyte at the location of increased activity due to a concentration gradient (indirect effect) or altering the network structure by localized control of gap junction channel activity (direct affects) as described by Rouach et al.[60] (D) Metabolite profiles across the astrocyte network are reshaped from panels (A–C) depending on AR-stimulated glucose/lactate production and/or local energy demand.

CONCLUSIONS

Astrocytes are a major target for adrenergic stimulation. The impact of AR on gap junctional communication depends on the AR subtypes and is mediated by phosphorylation of Cxs, with PKC in general decreasing conductance and coupling and PKA increasing coupling strength. Because in the CNS, gap junctions are mainly found in astrocytes and oligodrendrocytes, NA can be regarded as modulator of the entire panglial syncytium, expanding or decreasing its dimension.

ABBREVIATIONS

18βGA 18β-glycyrrhetinic acid
AP1 activator protein 1
AC adenylate cyclase
AR adrenergic receptor
CREB cAMP response element binding
c-Src cellular proto-oncogene receptor kinase
Cx connexin
cAMP cyclic adenosine monophosphate
cGMP cyclic guanosine monophosphate
DAG diacylglycerol
EEA1 early endosome antigene 1
ET-1 endothelin-1
ERK extracellular signal-regulated kinases
FSH follicle-stimulating hormone
G-6P glucose-6-phosphate
SK-Hep1 human endothelial adenocarcinoma-derived cell line
IP3 inositol trisphosphate
JNK Jun N-terminal kinase
LY Lucifer Yellow
MAPK mitogen-activated protein kinase
MKP mitogen-activated protein kinase phosphatase
MEK mitogen-activated kinase
2-NBDG 2-[*N*-(7-nitrobenz-2-oxa-1,3-diazol-4-yl)amino]-2-deoxyglucose
NA noradrenaline/norepinephrine
NFAT nuclear factor of activated T-cells
p38-MAP p38-mitogen-activated protein kinase
PMA phorbol myristate acetate
PL phospholipase
PK protein kinase
SpIH cyclic nucleotide-modulated channel from sea urchin sperm
SUMO small ubiquitin-like modifier
TPA tetradecanoylphorbol acetate

REFERENCES

1. Dobrenis K, Chang HY, Pina-Benabou MH, et al. Human and mouse microglia express connexin36, and functional gap junctions are formed between rodent microglia and neurons. *J Neurosci Res.* 2005;82(3):306–315.
2. Takeuchi H, Jin S, Wang J, et al. Tumor necrosis factor-alpha induces neurotoxicity via glutamate release from hemichannels of activated microglia in an autocrine manner. *J Biol Chem.* 2006;281(30):21362–21368.
3. Eugenin EA, Eckardt D, Theis M, Willecke K, Bennett MV, Saez JC. Microglia at brain stab wounds express connexin 43 and in vitro form functional gap junctions after treatment with interferon-gamma and tumor necrosis factor-alpha. *Proc Natl Acad Sci USA.* 2001;98(7):4190–4195.
4. Johnstone SR, Billaud M, Lohman AW, Taddeo EP, Isakson BE. Posttranslational modifications in connexins and pannexins. *J Membr Biol.* 2012;245(5–6):319–332.
5. Saez JC, Spray DC, Nairn AC, Hertzberg E, Greengard P, Bennett MV. cAMP increases junctional conductance and stimulates phosphorylation of the 27-kDa principal gap junction polypeptide. *Proc Natl Acad Sci USA.* 1986;83(8):2473–2477.
6. Saez JC, Nairn AC, Czernik AJ, et al. Phosphorylation of connexin 32, a hepatocyte gap-junction protein, by cAMP-dependent protein kinase, protein kinase C and Ca^{2+}/calmodulin-dependent protein kinase II. *Eur J Biochem/FEBS.* 1990;192(2):263–273.
7. Musil LS, Goodenough DA. Biochemical analysis of connexin43 intracellular transport, phosphorylation, and assembly into gap junctional plaques. *J Cell Biol.* 1991;115(5):1357–1374.
8. Li W, Hertzberg EL, Spray DC. Regulation of connexin43-protein binding in astrocytes in response to chemical ischemia/hypoxia. *J Biol Chem.* 2005;280(9):7941–7948.
9. Burt JM, Spray DC. Inotropic agents modulate gap junctional conductance between cardiac myocytes. *Am J Physiol.* 1988;254(6 Pt 2):H1206–H1210.
10. De Mello WC. Interaction of cyclic AMP and Ca^{2+} in the control of electrical coupling in heart fibers. *Biochim Biophys Acta.* 1986;888(1):91–99.
11. Saez JC, Nairn AC, Czernik AJ, Fishman GI, Spray DC, Hertzberg EL. Phosphorylation of connexin43 and the regulation of neonatal rat cardiac myocyte gap junctions. *J Mol Cell Cardiol.* 1997;29(8):2131–2145.
12. Lampe PD. Analyzing phorbol ester effects on gap junctional communication: a dramatic inhibition of assembly. *J Cell Biol.* 1994;127(6 Pt 2):1895–1905.
13. Lampe PD, Lau AF. Regulation of gap junctions by phosphorylation of connexins. *Arch Biochem Biophys.* 2000;384(2):205–215.
14. Dermietzel R, Hertberg EL, Kessler JA, Spray DC. Gap junctions between cultured astrocytes: immunocytochemical, molecular, and electrophysiological analysis. *J Neurosci.* 1991;11(5):1421–1432.
15. Dermietzel R, Traub O, Hwang TK, et al. Differential expression of three gap junction proteins in developing and mature brain tissues. *Proc Natl Acad Sci USA.* 1989;86(24):10148–10152.
16. Giaume C, Cordier J, Glowinski J. Endothelins inhibit junctional permeability in cultured mouse astrocytes. *Eur J Neurosci.* 1992;4(9):877–881.
17. Glowinski J, Marin P, Tence M, Stella N, Giaume C, Premont J. Glial receptors and their intervention in astrocyto-astrocytic and astrocyto-neuronal interactions. *Glia.* 1994;11(2):201–208.

18. Moreno AP, Fishman GI, Spray DC. Phosphorylation shifts unitary conductance and modifies voltage dependent kinetics of human connexin43 gap junction channels. *Biophys J.* 1992;62(1):51–53.
19. Takens-Kwak BR, Jongsma HJ. Cardiac gap junctions: three distinct single channel conductances and their modulation by phosphorylating treatments. *Pflugers Archiv.* 1992;422(2):198–200.
20. Kwak BR, Jongsma HJ. Regulation of cardiac gap junction channel permeability and conductance by several phosphorylating conditions. *Mol Cell Biochem.* 1996;157 (1–2):93–99.
21. Kwak BR, Saez JC, Wilders R, et al. Effects of cGMP-dependent phosphorylation on rat and human connexin43 gap junction channels. *Pflugers Archiv.* 1995;430(5):770–778.
22. De Mello WC. Further studies on the influence of cAMP-dependent protein kinase on junctional conductance in isolated heart cell pairs. *J Mol Cell Cardiol.* 1991;23(3):371–379.
23. Giaume C, Marin P, Cordier J, Glowinski J, Premont J. Adrenergic regulation of intercellular communications between cultured striatal astrocytes from the mouse. *Proc Natl Acad Sci USA.* 1991;88(13):5577–5581.
24. Spray DC, Moreno AP, Kessler JA, Dermietzel R. Characterization of gap junctions between cultured leptomeningeal cells. *Brain Res.* 1991;568(1–2):1–14.
25. TenBroek EM, Lampe PD, Solan JL, Reynhout JK, Johnson RG. Ser364 of connexin43 and the upregulation of gap junction assembly by cAMP. *The Journal of cell biology.* 2001;155(7):1307–1318.
26. De Mello WC. Impaired regulation of cell communication by beta-adrenergic receptor activation in the failing heart. *Hypertension.* 1996;27(2):265–268.
27. Salameh A, Dhein S. Adrenergic control of cardiac gap junction function and expression. *Naunyn Schmiedebergs Arch Pharmacol.* 2011;383(4):331–346.
28. Konietzko U, Muller CM. Astrocytic dye coupling in rat hippocampus: topography, developmental onset, and modulation by protein kinase C. *Hippocampus.* 1994;4 (3):297–306.
29. Venance L, Premont J, Glowinski J, Giaume C. Gap junctional communication and pharmacological heterogeneity in astrocytes cultured from the rat striatum. *J Physiol.* 1998;510(Pt 2):429–440.
30. Solan JL, Marquez-Rosado L, Sorgen PL, Thornton PJ, Gafken PR, Lampe PD. Phosphorylation at S365 is a gatekeeper event that changes the structure of Cx43 and prevents down-regulation by PKC. *J Cell Biol.* 2007;179(6):1301–1309.
31. Jordan K, Chodock R, Hand AR, Laird DW. The origin of annular junctions: a mechanism of gap junction internalization. *J Cell Sci.* 2001;114(Pt 4):763–773.
32. Saez JC, Gregory WA, Watanabe T, et al. cAMP delays disappearance of gap junctions between pairs of rat hepatocytes in primary culture. *Am J Physiol.* 1989;257(1 Pt 1):C1–C11.
33. Kjenseth A, Fykerud T, Rivedal E, Leithe E. Regulation of gap junction intercellular communication by the ubiquitin system. *Cell Signal.* 2010;22(9):1267–1273.
34. Popolo A, Morello S, Sorrentino R, Pinto A. Antiadrenergic effect of adenosine involves connexin 43 turn-over in H9c2 cells. *Eur J Pharmacol.* 2013;715(1–3):56–61.
35. Huang YF, Liao CK, Lin JC, Jow GM, Wang HS, Wu JC. Antofine-induced connexin43 gap junction disassembly in rat astrocytes involves protein kinase Cbeta. *Neurotoxicology.* 2013;35:169–179.
36. Salameh A, Frenzel C, Boldt A, et al. Subchronic alpha- and beta-adrenergic regulation of cardiac gap junction protein expression. *FASEB J.* 2006;20(2):365–367.

37. Salameh A, Krautblatter S, Baessler S, et al. Signal transduction and transcriptional control of cardiac connexin43 up-regulation after alpha 1-adrenoceptor stimulation. *J Pharmacol Exp Ther.* 2008;326(1):315–322.
38. De Leon JR, Buttrick PM, Fishman GI. Functional analysis of the connexin43 gene promoter in vivo and in vitro. *J Mol Cell Cardiol.* 1994;26(3):379–389.
39. Geimonen E, Jiang W, Ali M, Fishman GI, Garfield RE, Andersen J. Activation of protein kinase C in human uterine smooth muscle induces connexin-43 gene transcription through an AP-1 site in the promoter sequence. *J Biol Chem.* 1996;271(39):23667–23674.
40. Ponicke K, Schluter KD, Heinroth-Hoffmann I, et al. Noradrenaline-induced increase in protein synthesis in adult rat cardiomyocytes: involvement of only alpha1A-adrenoceptors. *Naunyn Schmiedebergs Arch Pharmacol.* 2001;364(5):444–453.
41. Zhang Y, Yan J, Chen K, et al. Different roles of alpha1-adrenoceptor subtypes in mediating cardiomyocyte protein synthesis in neonatal rats. *Clin Exp Pharmacol Physiol.* 2004;31(9):626–633.
42. Rojas Gomez DM, Schulte JS, Mohr FW, Dhein S. Alpha-1-adrenoceptor subtype selective regulation of connexin 43 expression in rat cardiomyocytes. *Naunyn Schmiedebergs Arch Pharmacol.* 2008;377(1):77–85.
43. Salameh A, Krautblatter S, Karl S, et al. The signal transduction cascade regulating the expression of the gap junction protein connexin43 by beta-adrenoceptors. *Br J Pharmacol.* 2009;158(1):198–208.
44. Echetebu CO, Ali M, Izban MG, MacKay L, Garfield RE. Localization of regulatory protein binding sites in the proximal region of human myometrial connexin 43 gene. *Mol Hum Reprod.* 1999;5(8):757–766.
45. Bailey J, Phillips RJ, Pollard AJ, Gilmore K, Robson SC, Europe-Finner GN. Characterization and functional analysis of cAMP response element modulator protein and activating transcription factor 2 (ATF2) isoforms in the human myometrium during pregnancy and labor: identification of a novel ATF2 species with potent transactivation properties. *J Clin Endocrinol Metab.* 2002;87(4):1717–1728.
46. Glover D, Little JB, Lavin MF, Gueven N. Low dose ionizing radiation-induced activation of connexin 43 expression. *Int J Radiat Biol.* 2003;79(12):955–964.
47. Laird DW. Connexin phosphorylation as a regulatory event linked to gap junction internalization and degradation. *Biochim Biophys Acta.* 2005;1711(2):172–182.
48. Laird DW. Life cycle of connexins in health and disease. *Biochem J.* 2006;394(Pt 3):527–543.
49. Falk MM, Kells RM, Berthoud VM. Degradation of connexins and gap junctions. *FEBS Lett.* 2014;588(8):1221–1229.
50. Bejarano E, Yuste A, Patel B, Stout Jr. RF, Spray DC, Cuervo AM. Connexins modulate autophagosome biogenesis. *Nat Cell Biol.* 2014;16(5):401–414.
51. Falk MM, Bell CL, Kells Andrews RM, Murray SA. Molecular mechanisms regulating formation, trafficking and processing of annular gap junctions. *BMC Cell Biol.* 2016;17 (Suppl):1–22.
52. Enkvist MO, McCarthy KD. Activation of protein kinase C blocks astroglial gap junction communication and inhibits the spread of calcium waves. *J Neurochem.* 1992;59 (2):519–526.
53. Muyderman H, Nilsson M, Blomstrand F, et al. Modulation of mechanically induced calcium waves in hippocampal astroglial cells. Inhibitory effects of alpha 1-adrenergic stimulation. *Brain Res.* 1998;793(1–2):127–135.

54. Scemes E, Giaume C. Astrocyte calcium waves: what they are and what they do. *Glia.* 2006;54(7):716–725.
55. Lawrence TS, Beers WH, Gilula NB. Transmission of hormonal stimulation by cell-to-cell communication. *Nature.* 1978;272(5653):501–506.
56. Qu Y, Dahl G. Function of the voltage gate of gap junction channels: selective exclusion of molecules. *Proc Natl Acad Sci USA.* 2002;99(2):697–702.
57. Bevans CG, Kordel M, Rhee SK, Harris AL. Isoform composition of connexin channels determines selectivity among second messengers and uncharged molecules. *J Biol Chem.* 1998;273(5):2808–2816.
58. Valiunas V. Cyclic nucleotide permeability through unopposed connexin hemichannels. *Front Pharmacol.* 2013;4:75.
59. Gandhi GK, Cruz NF, Ball KK, Theus SA, Dienel GA. Selective astrocytic gap junctional trafficking of molecules involved in the glycolytic pathway: impact on cellular brain imaging. *J Neurochem.* 2009;110(3):857–869.
60. Rouach N, Koulakoff A, Abudara V, Willecke K, Giaume C. Astroglial metabolic networks sustain hippocampal synaptic transmission. *Science.* 2008;322(5907):1551–1555.
61. Ball KK, Gandhi GK, Thrash J, Cruz NF, Dienel GA. Astrocytic connexin distributions and rapid, extensive dye transfer via gap junctions in the inferior colliculus: implications for [(14)C]glucose metabolite trafficking. *J Neurosci Res.* 2007;85(15):3267–3283.
62. Dienel GA, Cruz NF. Imaging brain activation: simple pictures of complex biology. *Ann N Y Acad Sci.* 2008;1147:139–170.
63. Stout Jr. RF, Spray DC, Parpura V. Astrocytic 'power-grid': delivery upon neuronal demand. *Cellscience.* 2009;5(3):34–43.
64. Tabernero A, Giaume C, Medina JM. Endothelin-1 regulates glucose utilization in cultured astrocytes by controlling intercellular communication through gap junctions. *Glia.* 1996;16(3):187–195.
65. Gandhi GK, Cruz NF, Ball KK, Dienel GA. Astrocytes are poised for lactate trafficking and release from activated brain and for supply of glucose to neurons. *J Neurochem.* 2009;111(2):522–536.
66. Obenauer JC, Cantley LC, Yaffe MB. Scansite 2.0: proteome-wide prediction of cell signaling interactions using short sequence motifs. *Nucleic Acids Res.* 2003;31(13):3635–3641.
67. Omasits U, Ahrens CH, Muller S, Wollscheid B. Protter: interactive protein feature visualization and integration with experimental proteomic data. *Bioinformatics.* 2014;30(6):884–886.

Chapter 7

Fluxes of Lactate Into, From, and Among Gap Junction-Coupled Astroglia and Their Interaction With Noradrenaline

Gerald A. Dienel[1,2,✉]

[1]University of Arkansas for Medical Sciences, Little Rock, AR, United States, [2]University of New Mexico, Albuquerque, NM, United States

Chapter Outline

Introduction 146
- Aerobic Glycolysis 146
- Lactate Release vs Lactate Shuttling-Oxidation 148
- Thematic Sequence 149

Lactate Fluxes During Brain Activation 149
- Parallel Glucose Utilization Assays Reveal Increased Glycolysis During Brain Activation 149
- Lactate is the Predominant Labeled Metabolite of Glucose Released From Brain 151
- Impact of Lactate Spreading and Release on Functional Imaging of Brain Activation 151

Astrocytic Lactate Trafficking Via Gap Junctions 152
- Dye Coupling 152
- Selectivity of Gap Junctional Trafficking of Molecules Involved in Glycolysis 154
- Lactate Uptake and Shuttling 155
- Glucose Shuttling 157
- Summary 157

Perivascular Routes for Metabolite Discharge From Activated Brain Structures 158

Influence of Noradrenaline on Astrocytic Lactate Fluxes 158
- Adrenergic Signaling and Aerobic Glycolysis 158
- β_2-Adrenergic Vagus Nerve Signaling by Adrenaline and Noradrenaline in Blood 159

✉Correspondence address
E-mail: gadienel@uams.edu

Noradrenergic Signaling and Astroglia. DOI: http://dx.doi.org/10.1016/B978-0-12-805088-0.00007-4
© 2017 Elsevier Inc. All rights reserved.

Excitatory and Inhibitory Effects of Lactate and Influence on Brain Noradrenaline Release 159
Influence of Noradrenaline on Astrocytic Metabolism 161
Conclusions 162
Abbreviations 163
References 163

ABSTRACT
Aerobic glycolysis is preferential upregulation of nonoxidative metabolism of glucose in the presence of adequate levels of oxygen. Aerobic glycolysis occurs during alerting, sensory stimulation, exercise, and pathophysiological conditions and involves increased lactate production and release. Lactate trafficking is mediated mainly by astrocytes; they take up lactate from extracellular fluid at higher rates and greater quantities than neurons. Astrocytes also disperse the lactate to other gap junction-coupled astrocytes to a much greater extent than shuttling lactate to nearby neurons. Astrocytic endfeet surround the vasculature and provide a route for discharge of lactate to the perivascular-lymphatic drainage system and to cerebral venous blood. Adrenergic regulation of aerobic glycolysis involves vagus nerve signaling to stimulate noradrenaline release from the locus coeruleus to brain. Noradrenaline influences many metabolic activities of astrocytes and gap junctional communication, and has a key role in governing the stoichiometry of glucose and oxygen consumption in activated brain.

Keywords: Adrenaline; aerobic glycolysis; astrocyte; gap junctional communication; glucose; glycogen; glycolysis; glycogenolysis; lactate; noradrenaline

INTRODUCTION

Aerobic Glycolysis

Aerobic glycolysis is a prevalent condition that becomes manifest during brain activation evoked by many types of stimuli, ranging from alerting, sensory stimulation, mental work, vigorous exercise, and abnormal conditions. The hallmark of aerobic glycolysis is preferential upregulation of nonoxidative metabolism of glucose in the presence of adequate levels and delivery of oxygen to brain, and it is characterized by enhanced glycolytic and pentose-phosphate shunt pathway (PPP) fluxes in unidentified cell types, as well as increased glycogen utilization and turnover in astrocytes.[1] These processes inherently involve lactate production and release from the activated brain cells because the rate of oxygen consumption (CMR_{O2}) would otherwise rise in proportion to glucose utilization (CMR_{glc}), and it does not (Fig. 7.1). In many studies, CMR_{O2} rises an average of $\sim$20% during activation, whereas CMR_{glc} rises by a much greater magnitude. Detailed studies of brain activation in awake rats have ruled out large contributions of biosynthetic activity or pool filling to the glucose consumed in excess of oxygen, leading to the conclusion that lactate release is the key factor in carrying glucose-derived carbon

Brain activation preferentially increases glycolysis and mobilizes glycogen

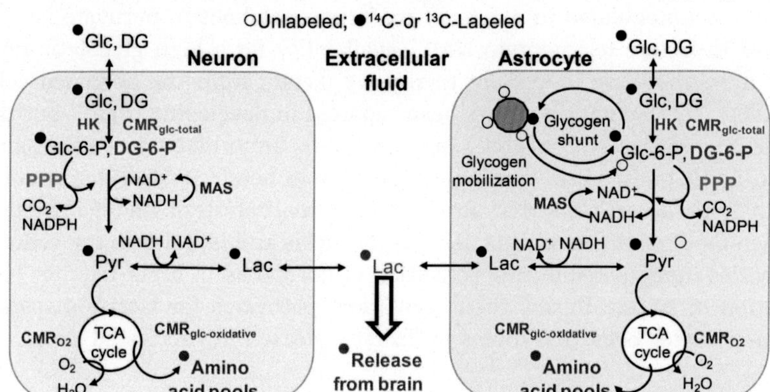

Activation: ↑CMR_{glc} > ↑CMR_{O2} ; $CMR_{glc\text{-}total}$ (hexokinase flux) > $CMR_{glc\text{-}oxidative}$ (TCA cycle flux)

- ➢CMR_{O2}/CMR_{glc} *falls* from resting value
- ➢*Small* rise in CMR_{O2}; *small* increase in TCA cycle-derived amino acid pool labeling
- ➢Lactate is derived mainly from blood glucose entering brain and metabolized
- ➢Rapid increase in lactate dispersal from activated cells; large lactate release to blood
- ➢Pentose phosphate pathway flux increases during activation

If release lactate: ↓ CMR_{O2}/CMR_{glc}, ↓ label trapping in amino acid pools
If stoichiometric lactate shuttling + oxidation: CMR_{O2} must match CMR_{glc} – ***it does not***

FIGURE 7.1 **Metabolic pathways involved in aerobic glycolysis.** Labeled glucose (Glc) and deoxyglucose (DG) are used to measure local rates of glucose utilization in brain. Because DG-6-phosphate (DG-6-P) is not metabolized further by the glycolytic pathway, this tracer measures the hexokinase (HK) reaction and the rate of total glucose utilization ($CMR_{glc\text{-}total}$). Glc-6-P is further metabolized and its rate of incorporation into TCA cycle-derived amino acids is used to calculate the rate of oxidative metabolism of glucose ($CMR_{glc\text{-}oxidative}$). The difference between rates calculated for DG and Glc give an estimate of loss of labeled products, mainly lactate, and of glycolytic upregulation. Assays of glucose (CMR_{glc}) and oxygen (CMR_{O2}) utilization can also be determined by arteriovenous differences, and represent global values. Astrocytes contain glycogen, and glycogen mobilization increases during brain activation. The PPP flux also increases during brain activation, releasing carbon one from glucose as CO_2 and generating NADPH that is mainly used for management of oxidative stress in normal adult brain. The stoichiometry of glucose and oxygen utilization is represented by the oxygen-glucose index, OGI = CMR_{O2}/CMR_{glc}, and under resting conditions OGI is close to 6.0, i.e., the theoretical maximum due to consumption of $6O_2$ per molecule of glucose oxidized. Under resting conditions, most of the cytoplasmic NADH produced by glycolysis is transferred to mitochondria by the malate-aspartate shuttle (MAS), thereby regenerating NAD^+ and producing pyruvate as oxidative substrate. Preferential upregulation of glycolysis during brain activation compared with oxidative metabolism leads to lactate production to regenerate NAD^+. The major contributor to aerobic glycolysis is rapid lactate dispersal and release from activated cells. The cellular origin of lactate production in brain in vivo is unknown, and translational conclusions drawn from lactate production by cultured neurons and astrocytes are speculative and unlikely to be correct due to the large extracellular volume of cultured cells and developmental differences compared with mature cells in adult brain. *Modified from Figure 2 of Dienel GA, Cruz NF. Contributions of glycogen to astrocytic energetics during brain activation. Metab Brain Dis. 2015;30(1):281–298 (Ref. 2), © 2014, Springer Science + Business Media New York, with permission of Springer.*

and glucose-derived label from activated tissue. When glycolysis is markedly upregulated in excess of oxidative metabolism, pyruvate is converted to lactate to maintain NAD^+ availability for a high glycolytic rate, and it is, therefore, necessary to remove lactate from the activated cells (Fig. 7.1). Aerobic glycolysis has been reported in developing brain,[3] but the functions of glucose consumed in excess of O_2 are unlikely to be the same as in adult brain. Also, regional differences in aerobic glycolysis in adult brain[4] may be artifacts that arose from normalization of metabolic data.[5] Understanding of the basis of aerobic glycolysis and its roles in the cellular activities upregulated during activation of normal adult brain requires elucidation of lactate fluxes, their regulation, pathways for lactate dispersal from activated cells, and routes for lactate release from activated tissue.

Lactate Release vs Lactate Shuttling-Oxidation

Two conflicting concepts are relevant to the predominant fate of lactate during brain activation, lactate release, and lactate oxidation, i.e., if lactate is released, it is not oxidized as supplemental fuel and vice versa. Lactate oxidation requires oxygen, and CMR_{O2} almost never rises to the same extent as CMR_{glc} during activation. The notion that lactate, generated by cultured astrocytes and released to the medium in response to exposure to extracellular glutamate, serves as an important neuronal oxidative fuel during excitatory neurotransmission was put forth by Pellerin and Magistretti[6] and is referred to as astrocyte–neuron lactate (ANL) shuttle. However, ANL shuttling coupled with local neuronal oxidation has never been demonstrated and quantified in brain in vivo, and the concept remains unproven after $>$20 years. Most claims of ANL shuttling are over-interpreted, and essential aspects of the tissue culture-based model remain unsubstantiated: (1) many laboratories have not replicated glutamate-evoked stimulation of CMR_{glc} and lactate release in cultured astrocytes—this behavior is not a robust phenotype of cultured astrocytes; (2) glutamate oxidation after its uptake can help support energetics because the glutamate-induced rise in astrocytic CMR_{O2} generates much more ATP than glycolysis; (3) cultured neurons and synaptosomes isolated from adult rodents can upregulate glycolysis, $CMR_{glc,}$ and CMR_{O2} by large amounts, indicating that neurons are capable of increasing glycolysis and do not need lactate as supplemental fuel due to putative deficits in glycolytic capacity. Furthermore, independent lines of evidence obtained in brain in vivo indicate that little, if any, of the lactate produced in brain during activation is used as a supplemental fuel for neurons, including the following: (1) upregulation of CMR_{glc} in excess of CMR_{O2} involves rapid dispersal and discharge of lactate from activated tissue; (2) most locally produced lactate is not oxidized; (3) blockade of glutamate transporters does not alter lactate release to extracellular space indicating that lactate production is independent of glutamate uptake by astrocytes; (4) phosphorylation of glucose by synaptic endings in vivo during activation increases in proportion to that of whole brain, demonstrating high capacity of neuronal glycolysis in vivo; and (5) the small rise in CMR_{O2} during activation places an upper limit on lactate oxidation, and CMR_{O2} includes oxidation of pyruvate, lactate, and glutamate. Suzuki et al.[7] concluded that ANL transport is required for memory formation, but they did not measure transport, unambiguously identify the endogenous source(s) of lactate, and take into account the complex compensatory metabolic responses of astrocytes to inhibition of glycogenolysis, their knockdown of lactate transporters was unlikely to have much effect on lactate transport, and their lactate dose used to rescue memory was in the pathological range that would impair neuronal firing by at least 50%; the roles of glycogen and lactate in memory are much more complex than portrayed. Magistretti, Oddo, and colleagues have also claimed that high-dose lactate supplementation is beneficial for patients with traumatic brain injury, but their reports are mis- and over-interpreted.[8] Evidence documenting the above points with more detailed discussion is available in recent reviews[1,9–12,59] and references cited therein.

Thematic Sequence

The following discussion first presents details of lactate generation and movements in brain in vivo during aerobic glycolysis, followed by fluxes of lactate into and among astrocytes, and adrenergic regulation of aerobic glycolysis and astrocytic metabolism.

LACTATE FLUXES DURING BRAIN ACTIVATION

Parallel Glucose Utilization Assays Reveal Increased Glycolysis During Brain Activation

Predominant upregulation of glycolysis during brain activation was first recognized and quantified in vivo in awake rats in the late 1980s when Collins et al.[13] and Lear and Ackermann[14,15] assayed local rates of glucose utilization (CMR_{glc}) in parallel assays with [^{14}C]deoxyglucose (DG) or [^{18}F]fluorodeoxyglucose (FDG) and [6-^{14}C]glucose. They found that registration of increases in CMR_{glc} with labeled glucose was much less than that with [^{14}C]DG during visual stimulation and [^{18}F]FDG in hippocampal seizures and concluded that large underestimates of calculated CMR_{glc} were due to rapid efflux of labeled lactate. The basis for this conclusion is illustrated in Fig. 7.1 in which [^{14}C]DG is shown to measure total CMR_{glc} at the hexokinase step, whereas labeling with [6-^{14}C]glucose registers all downstream metabolites of glucose-6-phosphate (Glc-6-P), mainly oxidative metabolism due to high labeling of tricarboxylic acid (TCA) cycle-derived amino acids. Note that if all labeled metabolites of glucose were quantitatively retained in tissue, values obtained with DG or FDG and glucose would be similar.

The [^{14}C]DG method[16] takes advantage of the limited metabolism of [^{14}C]DG after its phosphorylation by hexokinase to generate [^{14}C]DG-6-P, which is trapped in the cell in which it is formed. [^{18}F]FDG has similar properties, with intracellular trapping of [^{18}F]FDG-6-P. The rate of phosphorylation of [^{14}C]DG is converted to rate of glucose utilization by dividing by the lumped constant, the factor that accounts for kinetic differences in transport and phosphorylation between [^{14}C]DG and glucose. The value of the lumped constant in normal rats is ~0.48, i.e., two glucose molecules are phosphorylated per [^{14}C]DG molecule. Because DG and glucose compete for transport and phosphorylation, changes in glucose concentration need to be taken into account by using the appropriate value for the lumped constant, which is relatively stable within the normal range of blood and brain glucose levels.[17] Collins et al.[13] measured glucose levels and showed that they were stable during visual stimulation, and Ackermann and Lear[15] increased the value of the lumped constant by 20% to account for potential decreases in brain glucose level during seizures.

In contrast to [^{14}C]DG, [6-^{14}C]glucose is extensively metabolized beyond the Glc-6-P step, and label is incorporated into many intermediates in the PPP, glycolytic pathway, TCA cycle, and amino acid pools derived from various pathways in all brain cell types and into glycogen in astrocytes (Fig. 7.1). Labeling of TCA cycle-derived amino acids accounts for about half of the label in metabolites at 5 min after pulse labeling with [6-^{14}C]glucose during rest and sensory stimulation of awake rats, and lactate accounts for variable amounts of the label, more during stimulation or abnormal conditions.[18,19] Within 5 min lactate reaches its maximal specific activity (i.e., the ratio of labeled lactate to molar amount of unlabeled lactate), which is half that of [6-^{14}C]glucose because one pyruvate/lactate is labeled and one is not labeled.[20] Because lactate has the highest specific activity, it is the most diffusible of the glycolytic intermediates, and it carries half of the glucose molecule, it has a high impact on calculated CMR_{glc} and O_2-glucose stoichiometry, i.e., the oxygen-glucose index ($OGI = CMR_{O2}/CMR_{glc}$) and the oxygen-carbohydrate index ($OCI = CMR_{O2}/[CMR_{glc} + 0.5\ CMR_{lac}]$) that accounts for lactate metabolism. OGI has a theoretical maximal value of 6.0 because $6O_2$ are consumed per glucose oxidized. Diffusion of lactate down its concentration gradient from cells to extracellular fluid where it can be dispersed by bulk flow of interstitial and perivascular fluid[21] and released to blood and the lymphatic drainage system are processes that contribute to reduced registration of focal activation of metabolism when labeled glucose is the tracer. Also, release of unlabeled lactate contributes to underestimation of calculated CMR_{glc} and to reducing OGI.

Lactate is the Predominant Labeled Metabolite of Glucose Released From Brain

The above studies provided strong evidence for upregulation of glycolysis, incomplete trapping of labeled metabolites of glucose, and large underestimation of calculated CMR_{glc} using labeled glucose under activating conditions but did not prove that lactate release was the primary cause of low CMR_{glc}. We, therefore, used spreading cortical depression as an experimental model for brain activation because left–right differences could be quantitatively evaluated and arteriovenous differences could be used to identify and quantify metabolites released. We found that CMR_{glc} increased by 50% in the activated cortex when assayed with [^{14}C]DG and using appropriate values for the lumped constant, whereas it increased only 16% with [6-^{14}C]glucose.[19] Lactate accounted for a threefold greater percentage of the label in the activated cortex, and labeled lactate increased in direct proportion to unlabeled lactate, with a specific activity half that of brain glucose, i.e., the lactate was derived from blood--borne glucose. Next, efflux of labeled compounds was evaluated by analysis of paired samples of arterial and cerebral venous blood. Release of lactate was detectable within 2 min after pulse labeling with [6-^{14}C] glucose, it was continuous, and it accounted for 96% of the label released.[22] Release of both labeled and unlabeled lactate accounted for 22% of the labeled and unlabeled glucose entering the brain. Autoradiographic analysis of [^{14}C]DG labeling patterns in cerebral cortex showed that ^{14}C levels were highest in the most dorsal and ventral cortical laminae, whereas the laminar labeling differences were not detected with [6-^{14}C]glucose, suggesting rapid distribution of labeled metabolites within brain in addition to lactate release to blood. After microinjection of [U-^{14}C]lactate into brain, label diffusion was sufficient to cause loss of cortical laminar labeling patterns, i.e., it labeled a volume 17-fold greater than the injected volume and diffused up to 1.5 mm within 10 min.[22] Together, these studies established that lactate is the major labeled metabolite released from brain, and lactate trafficking within brain and release from brain reduces registration of focal metabolic activation. However, the quantity of lactate released to blood only accounted for ~50% of the underestimation of CMR_{glc} by [6-^{14}C]glucose, suggesting another major efflux route or diffusion within brain.

Impact of Lactate Spreading and Release on Functional Imaging of Brain Activation

Acoustic stimulation of the inferior colliculus, a midbrain auditory-processing structure, was used to evaluate physiological metabolic activation with labeled glucose and DG. The inferior colliculus was chosen because it has the highest metabolic rate in brain, highest capillary density, and highest blood flow rate,[23] so metabolic trafficking and efflux should be correspondingly high. Also, the inferior colliculus exhibits tonotopic organization in which selective groups of cells preferentially respond to tones of specific frequencies, and acoustic stimulation is readily achieved in awake rats. Tonotopic bands of metabolic activation were readily detected with [^{14}C]DG but not with [1-^{14}C]glucose.[24] Notably, band detection was enhanced by changing the position of the label to [6-^{14}C]glucose, by halothane anesthesia, and by inhibition of lactate transporters or astrocytic gap junctions. Microdialysis assays showed that extracellular lactate level doubled during acoustic stimulation, whereas extracellular glucose level was stable, indicating that the lumped constant did not change.

Diffusion of ^{14}C-labeled lactate from a microinfusion probe increased during acoustic stimulation, whereas that for labeled glucose fell and [^{14}C]DG was unchanged.[24] This means that [^{14}C]DG was taken up and phosphorylated in the same tissue volume during rest and activation, whereas spreading of lactate increased but the tissue volume labeled by glucose fell due to label loss. Inhibition of gap junctions by microinfusion of either α-glycyrrhetinic acid or oleamide prior to microinfusion of [1-^{14}C]glucose reduced the volume of labeled tissue by 35%–70%, indicating that metabolite diffusion within astrocytic gap junctional communication contributes to rapid dispersal of glucose metabolites. The calculated rate of the PPP increased 3.5-fold during acoustic stimulation,[10] indicating that $^{14}CO_2$ release also contributes to reduced registration of metabolic activation when [1-^{14}C]glucose is the tracer. Together, these experiments identified lactate transporters, lactate diffusion, and gap junctional communication as important contributing factors to underestimation of focal CMR_{glc} using labeled glucose.

ASTROCYTIC LACTATE TRAFFICKING VIA GAP JUNCTIONS

Dye Coupling

Connectivity of astrocytes by gap junctional channels was evaluated in slices of inferior colliculus from adult rats using Lucifer yellow to visualize coupled cells. Using a 5 min dye diffusion interval equal to the [^{14}C] glucose CMR_{glc} assay duration, up to ~12,000 cells were labeled after microinjection of Lucifer yellow into a single astrocyte (Fig. 7.2Aa).[25] Two striking characteristics of the pattern of Lucifer yellow labeling were dispersal from the center of the slice toward the meningial border and extensive perivascular labeling that was 1.7-fold higher than in the adjacent neuropil (Fig. 7.2Ab,c). Because the dye is trapped inside the cells, it reveals the size of the astrocytic syncytium but does not represent the actual extent of metabolite diffusion that will be influenced by local gradients, metabolism, and influx and efflux from cells. Also, capacity for

(A) Lucifer yellow spread from 1 astroycte in inferior colliculus slice

a. Heterogeneous, widespread labeling in 5 min

b. Dye spread to meningeal border

c. Perivascular labeling

Injection site

0.5 mm

(B) Selective transfer of metabolites through astrocytic gap junctions

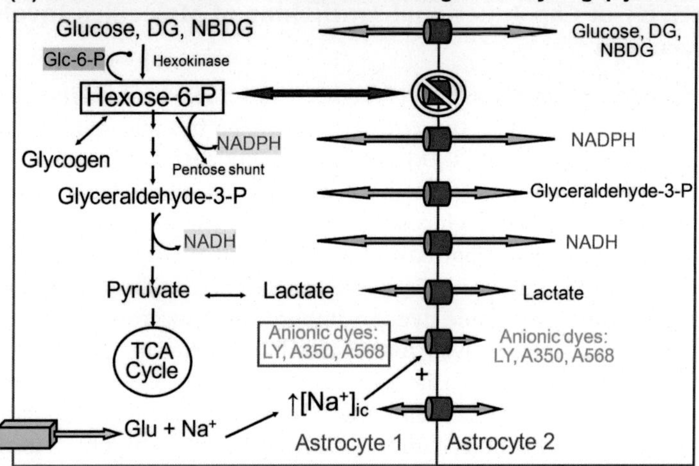

FIGURE 7.2 **Dye spread among gap junction-coupled astrocytes and selective trafficking of glycolytic metabolites and fluorescent dyes.** (A) Diffusion of Lucifer yellow from a single astrocyte in a slice of inferior colliculus from adult rat brain. (a) Widespread but heterogeneous diffusion in a radial direction from the center of the slice toward the meningial border to label up to 12,000 astrocytes within 5 min. *From Figure 5B of Ball KK, Gandhi GK, Thrash J, Cruz NF, Dienel GA.* Astrocytic connexin distributions and rapid, extensive dye transfer via gap junctions in the inferior colliculus: implications for [(14)C]glucose metabolite trafficking. *J Neurosci Res. 2007;85(15):3267–3283, © 2007 Wiley-Liss, Inc., with permission of John Wiley and Sons.* (b) Labeling of cell bodies (bright punctate spots) and cytoplasm (haze) near the meningial border (scale bar = 100 μm), with (c) strong perivascular labeling (scale bar = 25 μm). *From Figure 7a,b of Gandhi GK, Cruz NF, Ball KK, Dienel GA. Astrocytes are poised for lactate trafficking and release from activated brain and for supply of glucose to neurons. J Neurochem. 2009;111(2):522–536, © 2009 The Authors. Journal Compilation © 2009 International Society for Neurochemistry, with permission of John Wiley and Sons.* (B) Passage of three hexose-6-phophates (P), glucose-6-P, deoxyglucose-6-P (DG-6-P) and 2-(*N*-(7-nitrobenz-2-oxa-1,3-diazol-4-yl)amino)-2-deoxyglucose (2-NBDG-6-P), through astrocytic gap junctional channels is highly restricted, whereas the parent sugars, other phosphorylated compounds, anionic dyes, and lactate readily diffuse among coupled cells. *Data are from Gandhi GK, Cruz NF, Ball KK, Theus SA, Dienel GA. Selective astrocytic gap junctional trafficking of molecules involved in the glycolytic pathway: impact on cellular brain imaging. J Neurochem. 2009;110 (3):857–869, and the figure is modified from Figure 7 of Dienel GA. Fueling and imaging brain activation. ASN Neuro. 2012;4(5):267–321. art:e00093.doi:00010.01042/AN20120021, © 2012 The Author(s), with permission.*

lactate dispersal is anticipated to vary with physiological state because exposure of cultured astrocytes to extracellular K^+ or glutamate increases the extent of dye coupling, and activation of protein kinase C reduces coupling.[26,27] Elevated intracellular $[Na^+]$ also stimulates dye transfer among astrocytes,[28] and dye coupling is regulated by noradrenaline (NA).[29] Extensive labeling of astrocytes and their endfeet that surround the vasculature indicates that compounds derived from a single cell can be rapidly dispersed among many astrocytes and to their endfeet surrounding blood vessels.

Selectivity of Gap Junctional Trafficking of Molecules Involved in Glycolysis

Well-defined tonotopic bands of acoustic stimulus-evoked increase in CMR_{glc} were readily detected with $[^{14}C]$DG in both 5 and 45 min experiments, suggesting that $[^{14}C]$DG-6-P was highly restricted from passage through gap junctions because the bands did not dissipate over time as occurred with ^{14}C-labeled glucose. This finding was unexpected because hexose-6-Ps have been reported to be transferred through gap junctional channels.[30,31] Cultured astrocytes were microinjected with a combination of a gap junction-permeable dye, an impermeable macromolecule, and a test compound being assayed for transcellular diffusion. These types of experiments demonstrated that glucose, DG, and 2- or 6-NBDG (fluorescent glucose analogs) readily diffused among coupled cells, whereas passage of Glc-6-P, DG-6-P, and 2-NBDG-6-P was highly restricted (Fig. 7.2B).[28] Curiously, glyceraldehyde-3-P and larger phosphorylated compounds, NADPH and NADH, as well as other larger anionic dyes readily traversed through the channels, and intracellular sodium increased spreading diffusion of the reference permeant fluorescent dye. These findings raise very interesting questions as to how Glc-6-P, a feedback inhibitor of hexokinase, is restricted to the cell where it is formed; regulation is not simply due to phosphorylation, negative charge, or size. This is very important because the rate of first irreversible step of glucose utilization is governed by its product, indicating that metabolic regulation of astrocytic glucose utilization is achieved at the single-cell level. In contrast, upstream (glucose) and downstream (glyceraldehyde-3-P, lactate) metabolites and cofactors involved in oxidation–reduction reactions of the PPP and oxidative stress management (NADPH), and of glycolysis (NADH) are readily distributed within the syncytium (Fig. 7.2B). These results demonstrate that fluorescent dyes commonly used to assess gap junctional communication do not represent or predict the transfer of biological molecules. Furthermore, assays with $[^{14}C]$glucose-6-P[29] only measured label transfer, and ^{14}C-tracer diffusion was probably an artifact due to its metabolism to permeant compounds.

Lactate Uptake and Shuttling

To compare cellular rates of lactate uptake and dispersal, a sensitive, specific, real-time assay was devised using lactate oxidase and horseradish peroxidase to convert Amplex red to the fluorescent resorufin,[32] with a 1:1 stoichiometry for lactate and resorufin. The standard curves for fluorescence vs concentration were linear and identical in astrocytes and neurons, and resorufin was not gap junction permeable so it could not diffuse among coupled cells and interfere with assays. We first measured lactate uptake into astrocytes and neurons from an extracellular point source, i.e., a micropipette containing 0, 2, 10, 20, or 40 mmol/L L-lactate, a range chosen to span the Km values for monocarboxylic acid transporters (MCTs) in neurons (MCT2) and astrocytes (MCT1 and MCT4) (i.e., ~0.7, 4, and 28 mmol/L, respectively[33]). Initial rates and net uptake of lactate into astrocytes were 4.3-fold faster and 2.3-fold higher compared with neuronal uptake. Even at 2 mmol/L lactate, the level generally observed in activated brain, lactate uptake into astrocytes predominates, and as extracellular lactate level rises more enters astrocytes (Fig. 7.3A–D). Next, shuttling of lactate among coupled astrocytes was compared with lactate shuttling from an astrocyte to neuron. First, a reporter astrocyte or a neuron was impaled with a micropipette containing the reaction mixture and a dye to identify coupled astrocytes. Then a micropipette containing 0, 2, 5, or 10 mmol/L L-lactate was introduced into a coupled lactate-donor astrocyte located ~50 μm from the reporter astrocyte or neuron. The initial rate of lactate diffusion among astrocytes increased with concentration from 0 to 5 mmol/L, and was about twice that into neurons, whereas lactate transfer from an astrocyte to neuron did not vary over the concentration range of 2–10 mmol/L (Fig. 7.3E–H). The net transfer of lactate from one astrocyte to one coupled astrocyte was almost fivefold greater than to a neuron, and extensive gap junctional coupling among astrocytes indicates that dispersal of lactate within the syncytium vastly exceeds transfer to a nearby neuron. Thus, when an ANL gradient is imposed by insertion of lactate into an astrocyte, most of the lactate is dispersed to other astrocytes (Figs. 7.2 and 7.3).

In sharp contrast to these results, Mächler et al.[34] claimed that there is substantial lactate shuttling from astrocytes to neurons. However, their data interpretation is weak because, (1) they used an unusual anesthetic cocktail that probably preferentially altered astrocytic metabolism and created an artifactual ANL gradient and (2) ANL flux was emphasized while ignoring lactate efflux to blood and lactate diffusion among coupled astrocytes. The blood lactate level illustrated in their Fig. 6 in Ref. 34 is much lower than that in the astrocyte and similar to that in the neuron, yet Mächler et al. omitted discussion of lactate outflow to blood and gap junctional lactate trafficking.[1]

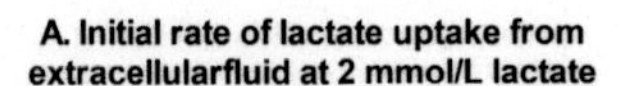

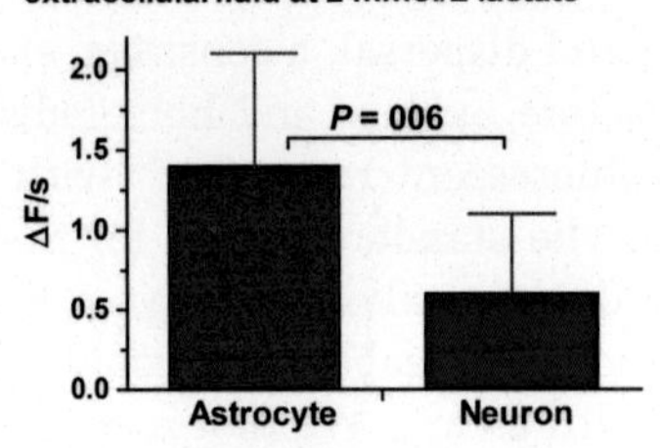

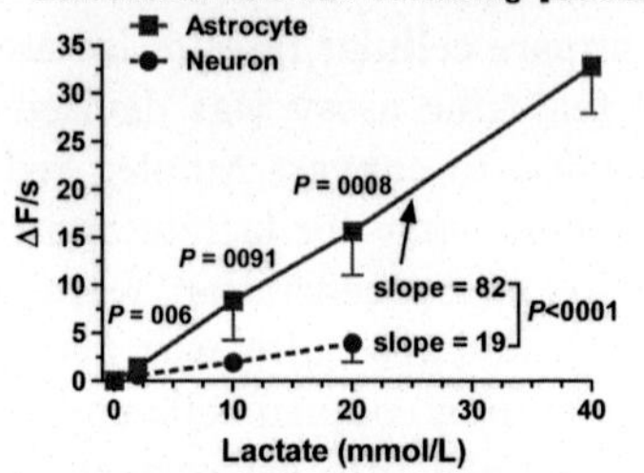

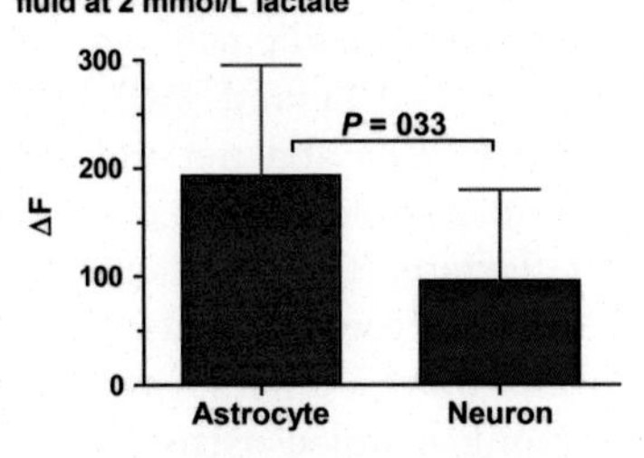

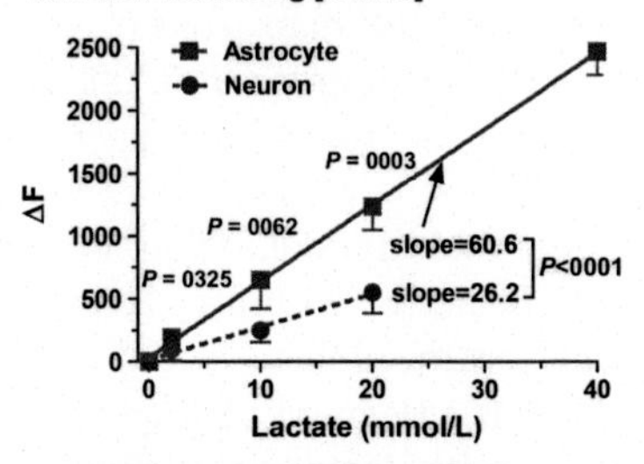

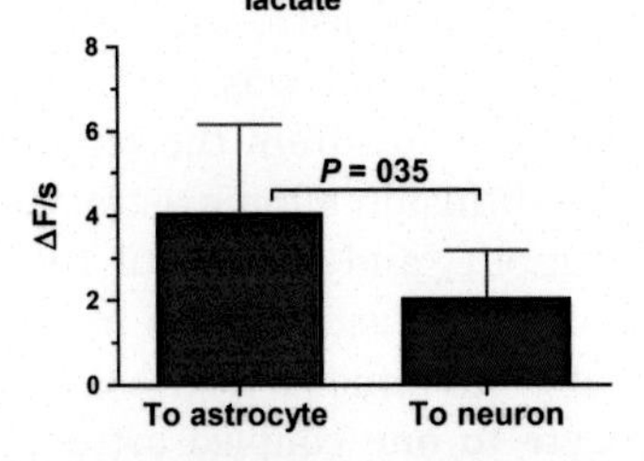

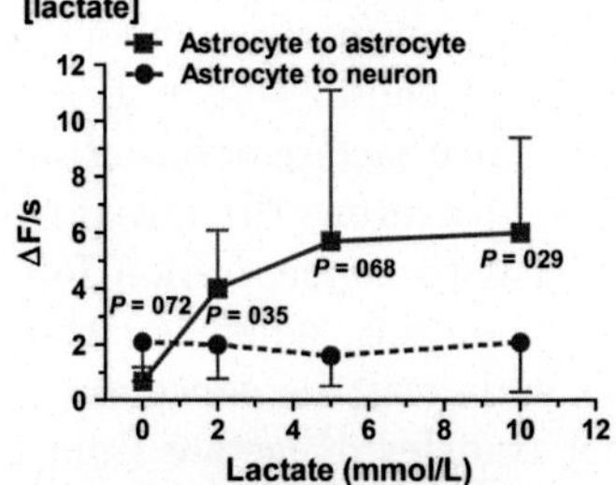

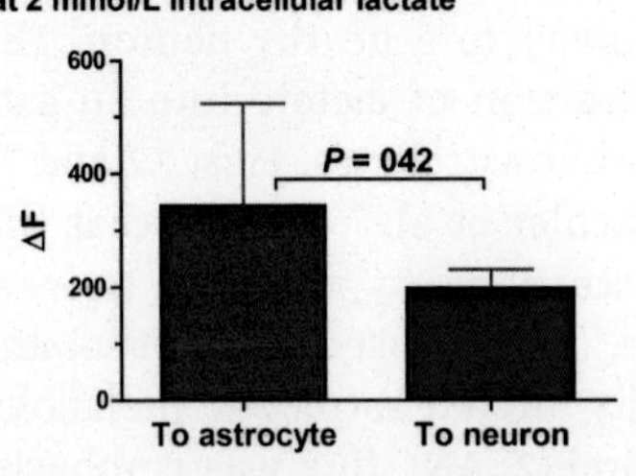

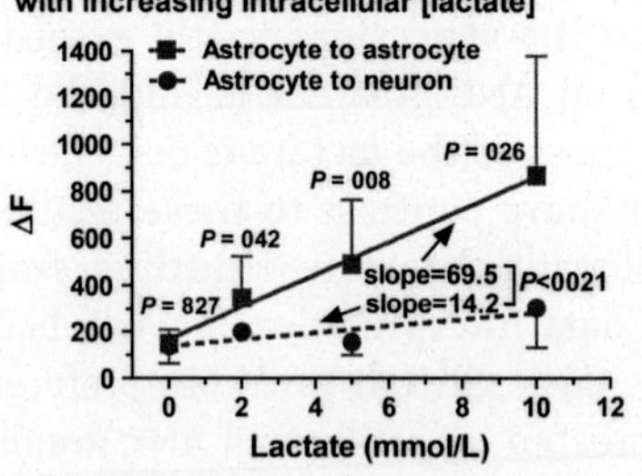

FIGURE 7.3 Lactate uptake and shuttling among astrocytes and neurons. Initial (A, B) and net (C, D) rates of lactate uptake from an extracellular point source into astrocytes and neurons at different lactate concentrations. Initial (E, F) and net (G, H) rates of lactate shuttling from an astrocyte to another astrocyte or an equidistant neuron at different lactate levels inserted into the donor astrocyte. *From Figures 4 and 5 of Gandhi GK, Cruz NF, Ball KK, Dienel GA. Astrocytes are poised for lactate trafficking and release from activated brain and for supply of glucose to neurons. J Neurochem. 2009;111(2):522–536, © 2009 The Authors. Journal Compilation © 2009 International Society for Neurochemistry, with permission of John Wiley and Sons.*

Glucose Shuttling

A similar assay using glucose oxidase was devised to evaluate glucose transfer from astrocytes to neurons because Rouach et al.,[35] reported that they could not detect transfer of a fluorescent glucose analog from astrocytes to neurons, implying that lactate is needed as a neuronal fuel. In fact, concentration-driven astrocyte-to-neuron glucose shuttling was similar in magnitude to direct neuronal glucose uptake.[32]

Summary

The cellular source(s) of lactate produced in activated brain in vivo remain to be identified. However, astrocytes are poised to rapidly take up and disperse lactate and other compounds from extracellular fluid surrounding activated neurons, and they can quickly discharge lactate from brain via their endfeet. Blood is an infinite sink for lactate due to its large volume and fast flow, and when blood lactate concentration is lower than that of astrocytes, lactate will be released down its concentration gradient to blood (Fig. 7.4).

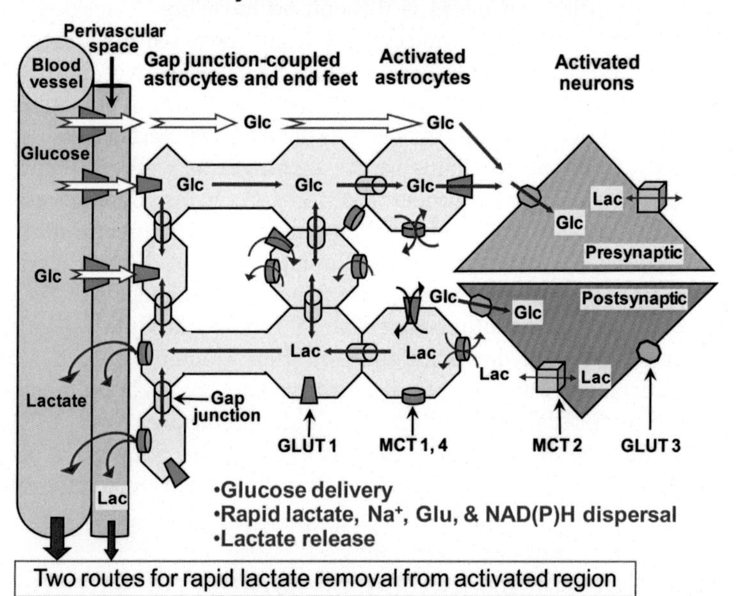

FIGURE 7.4 **Roles for gap junctional communication during brain activation.** Rapid uptake of metabolites and electrolytes by astrocytes adjacent to activated neurons and their dispersal into the astrocytic syncytium can reduce gradients of metabolites in the most activated region, reduce the energetic requirements placed on these cells, and facilitate release of lactate from endfeet to perivascular fluid and blood. *Modified from Figure 7c of Mächler P, Wyss MT, Elsayed M, et al. In vivo evidence for a lactate gradient from astrocytes to neurons. Cell Metab. 2016;23(1):94–102, © 2009 The Authors. Journal Compilation © 2009 International Society for Neurochemistry, with permission of John Wiley and Sons.*

PERIVASCULAR ROUTES FOR METABOLITE DISCHARGE FROM ACTIVATED BRAIN STRUCTURES

Release of lactate to blood sampled at the sagittal sinus during spreading depression accounted for ~50% of the underestimate of calculated CMR_{glc}, suggesting that there are other routes for lactate discharge from brain. Perivascular fluid flow delivers material contained in the interstitial fluid to lymph nodes by bulk flow and quickly distributes proteins and small molecules throughout the brain and to lymph nodes.[21] The perivascular efflux route was examined by microinfusion of either [1-^{14}C]glucose or the nonmetabolizable D-[^{14}C]lactate into extracellular fluid of the inferior colliculus of awake rats for 5 min, the duration of the typical [^{14}C]glucose metabolic assay.[36] The infused inferior colliculus contained ~60% of the recovered label in the [1-^{14}C]glucose infusions and meninges had 34%. On a weight basis, the meninges contained 2.3 times that of the inferior colliculus, indicating high labeling and rapid release of glucose, lactate and other metabolites to the meningial membranes. Seventeen percent of the D-[^{14}C] lactate was recovered in meninges vs 70% in the inferior colliculus. Together these results demonstrate rapid, sizeable removal of labeled compounds including lactate from extracellular fluid to the meninges.

Analysis of efflux pathways is difficult when using ^{14}C-labeled compounds, so fluorescent macromolecules that are restricted to extracellular fluid and perivascular spaces were used to visualize perivascular efflux routes from the inferior colliculus. Microinfusion of Evans blue-albumin or fluorescent-amyloid-β into awake rats labeled the perivascular space along vessels from the colliculus to the meninges as well as labeling of the circle of Willis, vessels under the olfactory bulbs, middle cerebral artery, and cervical lymph nodes.[36] The perivascular-lymphatic drainage system bypasses the cerebral venous drainage and can disperse and discharge metabolites released from astrocytic endfeet metabolites, as well as macromolecules and other material in interstitial fluid[21] (Fig. 7.4). Thus, lactate diffusion down its concentration gradients to other cells and to extracellular fluid, lactate dispersal within brain via bulk convective flow of interstitial and perivascular fluid, and lactate release to blood and the lymphatic drainage system contribute to the complexity of lactate "trafficking" within and from brain that underlies aerobic glycolysis.

INFLUENCE OF NORADRENALINE ON ASTROCYTIC LACTATE FLUXES

Adrenergic Signaling and Aerobic Glycolysis

Brain activation is frequently associated with adrenergic signaling from blood to brain. For example, catecholamine release to blood from the adrenal glands and sympathetic nerves increases during alerting, handling, and immobilization of experimental animals,[37] and during mental work[38] and vigorous exercise[39] in humans, and NA levels rise in human cerebrospinal fluid during exercise.[39] The strong link between β_2-adrenergic signaling and aerobic glycolysis was revealed by three lines of evidence, (1) prevention of the fall in OGI or OCI when alerted rats[40] or exercising humans[41,42] were pretreated with the nonspecific β-blocker, propranolol, (2) no effect on OCI of the specific β_1-blocker, metoprolol, in exercising humans,[43] and (3) reduced OCI in resting humans given an intravenous infusion of adrenaline.[44] Thus, increased plasma adrenaline levels are sufficient, but probably not the only factor, to alter the stoichiometry of brain carbohydrate and oxygen utilization, a conclusion consistent with findings from many older studies (reviewed by Ref. 1).

β_2-Adrenergic Vagus Nerve Signaling by Adrenaline and Noradrenaline in Blood

A wide range of stressors, including those commonly used in memory-evoking tasks (e.g., footshock, bitter taste, maze testing), involve catecholamine release to blood and signaling by adrenaline at β_2-adrenergic receptors (β_2-ARs) in the vagus nerve[45] to increase NA release throughout the brain from the locus coeruleus to elicit a broad range of effects on brain function. Thus, even though adrenaline and NA cannot cross the intact blood–brain barrier, they influence brain metabolism by stimulating NA release in brain (Fig. 7.5).

Excitatory and Inhibitory Effects of Lactate and Influence on Brain Noradrenaline Release

NA release is stimulated not only by vagus nerve signaling, but also by local release of lactate in the locus coeruleus.[46] Lactate release from astrocytes stimulates NA release from locus coeruleus neurons at normal lactate levels (EC_{50} $\sim$ 0.5 mmol/L) in a stereoselective manner without neuronal uptake of lactate. The receptor-mediated excitatory effects of lactate on neurons involve cyclic adenosine monophosphate (cAMP) and protein kinase A signaling. Exogenous L-lactate application to organotypic cultures or brain slices from 1- or 2-month-old rats also triggered NA release, and microinjection of L-lactate into the locus coeruleus of the intact rat caused cardiovascular and electroencephalogram changes consistent with excitatory actions of lactate on locus coeruleus neurons. Thus, activation of astrocytes in the locus coeruleus to cause their release of lactate has stimulatory effects on nearby neurons leading to increased NA release.

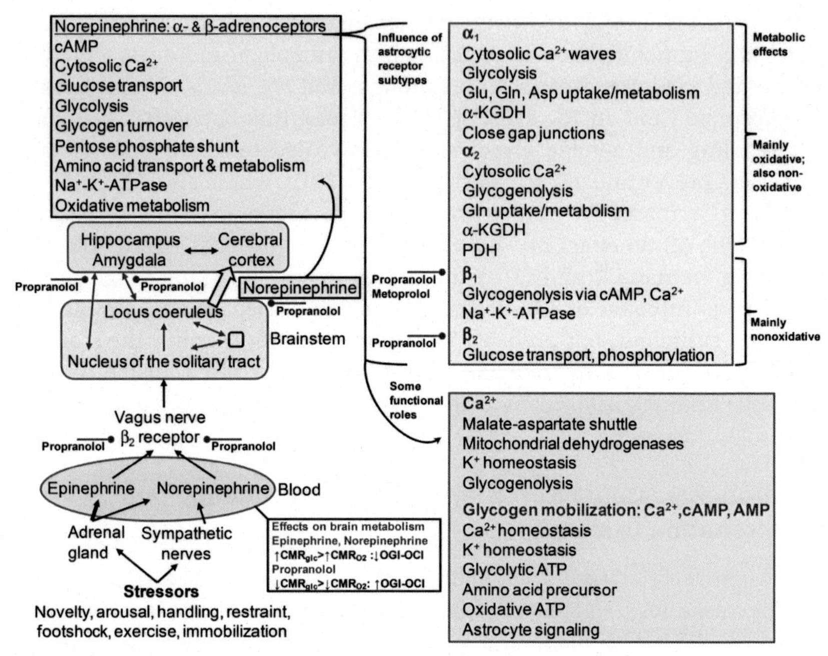

FIGURE 7.5 **Influence of adrenaline and NA on astrocytic metabolism.** Major pathways for adrenergic influence on brain metabolism and aerobic glycolysis. *From Dienel GA, Cruz NF. Aerobic glycolysis during brain activation: adrenergic regulation and influence of norepinephrine on astrocytic metabolism. J Neurochem. 2016;138(1):14–52, © 2016 The Authors. Journal Compilation © 2016 International Society for Neurochemistry, with permission of John Wiley and Sons.*

Lactate has also suppressive effects on neuronal signaling. Exogenous lactate acts via the hydroxycarboxylic acid-1 receptor that has an EC_{50} ~4.4 mmol/L for inhibition of spiking of both excitatory and inhibitory neurons in culture.[47] In brain slices, higher levels of lactate inhibit cAMP production via this receptor.[48] When lactate is infused into the hippocampus of intact rats via microdialysis probes, it inhibits neuronal firing by >85% at levels of ~12 mmol/L.[49] The above effects were observed in the presence of glucose in the culture or slice medium or in vivo, but brain slice experiments showed that replacement of glucose with lactate or pyruvate attenuated population field potentials and excitatory postsynaptic potentials even though ATP and phosphocreatine levels were normal.[50] Similarly, Okada and Lipton found that lactate did not support evoked population spikes when brain slices were rapidly prepared, and early long-term potentiation of population spikes was less robust with lactate in place of glucose; lactate did, however, maintain normal energy status.[51] On the other hand, slower slice preparation involved an unidentified calcium-uptake event that removed the dependence of neurotransmission on glucose and glycolysis, and lactate could now support evoked potentials. Studies with cultured neurons demonstrated the glucose dependence of *N*-methyl-D-aspartate-induced synaptic activity that involves Ca^{2+} regulation of the malate-aspartate shuttle and oxidative metabolism.[52] Clearly, lactate has strong effects on neuronal activity independent of its metabolism as fuel, and more work is required to elucidate the requirement for glycolysis in neuronal function. Thus, lactate dispersal among astrocytes and lactate release to interstitial fluid at some distance from the source of lactate can influence neuronal activity by interacting with receptors.

The use of high lactate levels for "therapeutic" purposes can also have other unanticipated effects on astrocytic and neuronal functions and metabolism. For example, lactate can disrupt metabolite trafficking within the astrocytic syncytium and influence astrocyte–neuron interactions. Lactate uptake into cultured neurons and astrocytes reduces intracellular pH due to cotransport of lactate with H^+.[53] Intracellular acidification reduces gap junctional communication, and a pH decrease from 7.4 to 6.5 in cultured astrocytes caused a large reduction in dye-coupled cells.[54] Intracellular acidification can also inhibit phosphofructokinase activity, and flooding tissue with large amounts of lactate will drive the lactate dehydrogenase reaction toward pyruvate plus NADH, thereby reducing availability of NAD^+ for glycolysis. These effects can reduce glycolytic flux and dispersal of lactate, electrolytes, and other compounds within the astrocytic syncytium.

Influence of Noradrenaline on Astrocytic Metabolism

Astrocytes respond to NA via its interactions with subtypes of α- and β-ARs to stimulate second messenger systems that have a strong influence on astrocytic functions. Noradrenergic stimulation of astrocytes evokes oscillatory changes in intracellular Ca^{2+} concentrations and slower, persistent cAMP levels, both of which can propagate to neighboring astrocytes by various mechanisms including gap junctional communication and extracellular pathways to induce many types of downstream effects, including morphological changes, gliotransmitter release, and changes in intracellular glucose level related to glycogenolysis.[55] In addition, NA has a wide spectrum of metabolic effects on astrocytes that can secondarily influence lactate fluxes (reviewed in Ref. 1, see also Chapter 2: Astroglial Adrenergic Receptor Signaling in Brain Cortex). The actions of NA on AR-subtypes can vary with species, brain region, developmental stage, and culture conditions. For example, in chick astrocytes, glycogenolysis is stimulated by NA via β_2-ARs, whereas in rodent astrocytes NA stimulates glycogen mobilization via β_1-ARs. In *mammalian* astrocytes, NA action on α-ARs stimulates cytosolic Ca^{2+} waves, glycolysis, uptake and metabolism of glutamate, aspartate, and glutamine, α-ketoglutarate dehydrogenase activity, and closure of gap junctions. Its actions on α-ARs stimulates cytosolic Ca^{2+} waves, glycogenolysis, glutamine uptake and metabolism, and α-ketoglutarate and pyruvate dehydrogenase activities (Fig. 7.5). In addition, incubation of cultured mouse astrocytes with NA plus yohimbine (α-AR antagonist) impaired glycogen synthesis, whereas clonidine (α-AR agonist) stimulated glycogenesis (see Figure 12B in Ref. 56). Thus, α-ARs stimulate both glycogen mobilization and re-synthesis (not illustrated in Fig. 7.5), i.e., the glycogen shunt pathway in which some Glc-6-P traverses through glycogen prior to reentering the glycolytic pathway. Interactions with β-ARs have strong effects on glycogenolysis, Ca^{2+} levels and Na^+-K^+-ATPase, whereas β-ARs influence glucose transport and phosphorylation. Taken together, the effects of α-ARs are mainly but not exclusively on oxidative metabolism, whereas those of β-ARs are predominantly on nonoxidative metabolism (Fig. 7.5). Furthermore, NA is taken up into and metabolized by astrocytes by monoamine oxidase to generate hydrogen peroxide,[57] which stimulates the PPP flux to generate NADPH and glycogenolysis to provide Glc-6-P as substrate for the PPP.[58] Together, these varied actions of NA on astrocytic receptors and its uptake and metabolism have a strong influence on nearly all aspects of astrocytic metabolism (Fig. 7.5). Blockade of β-AR-mediated effects but not α-AR-mediated effects of NA by propranolol has the overall effect of modulating nonoxidative metabolism of glucose more an oxidative metabolism, thereby increasing OGI or OCI or blocking its fall during brain activation.

CONCLUSIONS

Astrocytes are poised to generate, release, take up, disperse, discharge from brain, and release lactate to act on nearby noradrenergic neurons to stimulate release of NA that has a high impact on astrocytic metabolism, lactate fluxes, and lactate signaling. A large range of behavioral conditions, from very mild to noxious, cause adrenaline and NA levels to rise in blood, increase vagus signaling to the locus coeruleus, and enhance NA release throughout the brain. Astrocytes are major targets of NA, and this neurotransmitter is a key factor in manifestation of aerobic glycolysis. Increased nonoxidative metabolism associated with lactate dispersal through gap junctions and lactate release from brain are primary consequences of noradrenergic signaling. The cellular functions served by glycolytic metabolism remain to be established, but they are essential components of brain activation and have important roles in brain disorders.

ABBREVIATIONS

ANL	astrocyte–neuron lactate
AR	adrenergic receptor
CMR	cerebral metabolic rate for oxygen (CMR_{O2}), glucose (CMR_{glc}), or lactate (CMR_{lac})
DG	2-deoxy-D-glucose
DG-6-P	deoxyglucose-6-phosphate
EC_{50}	concentration that gives half-maximal excitatory response
FDG	2-fluoro-2-deoxy-D-glucose
FDG-6-P	fluorodeoxyglucose-6-phosphate
Glc	glucose
Glc-6-P	glucose-6-phosphate
Lac	lactate
MCT	monocarboxylic acid transporter
NA	noradrenaline
NAD^+ and NADH	oxidized and reduced forms, respectively, of nicotinamide adenine dinucleotide
$NADP^+$ and NADPH	oxidized and reduced forms, respectively, of nicotinamide adenine dinucleotide phosphate
2- or 6-NBDG	2- or 6-(*N*-(7-nitrobenz-2-oxa-1,3-diazol-4-yl)amino)-2-deoxyglucose
OCI	oxygen-carbohydrate index, $OCI = CMR_{O2}/(CMR_{glc} + 0.5\ CMR_{lac})$
OGI	oxygen-glucose index, $OGI = CMR_{O2}/CMR_{glc}$
PPP	pentose-phosphate shunt pathway
TCA	tricarboxylic acid

REFERENCES

1. Dienel GA, Cruz NF. Aerobic glycolysis during brain activation: adrenergic regulation and influence of norepinephrine on astrocytic metabolism. *J Neurochem.* 2016;138(1):14–52.
2. Dienel GA, Cruz NF. Contributions of glycogen to astrocytic energetics during brain activation. *Metab Brain Dis.* 2015;30(1):281–298.
3. Goyal Manu S., Hawrylycz M., Miller Jeremy A., Snyder Abraham Z., Raichle Marcus E. Aerobic glycolysis in the human brain is associated with development and neotenous gene expression. *Cell Metabolism.* **19**(1), 2014,49–57.
4. Vaishnavi SN, Vlassenko AG, Rundle MM, Snyder AZ, Mintun MA, Raichle ME. Regional aerobic glycolysis in the human brain. *Proc Natl Acad Sci USA.* 2010;107(41):17757–17762.
5. Hyder F, Herman P, Bailey CJ, et al. Uniform distributions of glucose oxidation and oxygen extraction in gray matter of normal human brain: no evidence of regional differences of aerobic glycolysis. *J Cereb Blood Flow Metab.* 2016;36(5):903–916.
6. Pellerin L, Magistretti PJ. Glutamate uptake into astrocytes stimulates aerobic glycolysis: a mechanism coupling neuronal activity to glucose utilization. *Proc Natl Acad Sci USA.* 1994;91(22):10625–10629.
7. Suzuki A, Stern SA, Bozdagi O, et al. Astrocyte-neuron lactate transport is required for long-term memory formation. *Cell.* 2011;144(5):810–823.
8. Dienel GA, Rothman DL, Nordström CH. Microdialysate concentration changes do not provide sufficient information to evaluate metabolic effects of lactate supplementation in brain-injured patients. *J Cereb Blood Flow Metab.* 2016;36(11):1844–1864.

9. Dienel GA. Brain lactate metabolism: the discoveries and the controversies. *J Cereb Blood Flow Metab.* 2012;32(7):1107–1138.
10. Dienel GA. Fueling and imaging brain activation. *ASN Neuro.* 2012;4(5):267–321:art: e00093.doi:00010.01042/AN20120021
11. Dienel GA. Astrocytic energetics during excitatory neurotransmission: What are contributions of glutamate oxidation and glycolysis? *Neurochem Int.* 2013;63(4):244–258.
12. Dienel GA. The metabolic trinity, glucose-glycogen-lactate, links astrocytes and neurons in brain energetics, signaling, memory, and gene expression. *Neurosci Lett.* 2017;10 (637):18–25. Available from: http://dx.doi.org/10.1016/j.neulet.2015.1002.1052.
13. Collins RC, McCandless DW, Wagman IL. Cerebral glucose utilization: comparison of [14C]deoxyglucose and [6-14C]glucose quantitative autoradiography. *J Neurochem.* 1987;49 (5):1564–1570.
14. Lear JL, Ackermann RF. Why the deoxyglucose method has proven so useful in cerebral activation studies: the unappreciated prevalence of stimulation-induced glycolysis. *J Cereb Blood Flow Metab.* 1989;9(6):911–913.
15. Ackermann RF, Lear JL. Glycolysis-induced discordance between glucose metabolic rates measured with radiolabeled fluorodeoxyglucose and glucose. *J Cereb Blood Flow Metab.* 1989;9(6):774–785.
16. Sokoloff L, Reivich M, Kennedy C, et al. The [^{14}C]deoxyglucose method for the measurement of local cerebral glucose utilization: theory, procedure, and normal values in the conscious and anesthetized albino rat. *J Neurochem.* 1977;28(5):897–916.
17. Dienel GA, Cruz NF, Mori K, Holden JE, Sokoloff L. Direct measurement of the lambda of the lumped constant of the deoxyglucose method in rat brain: determination of lambda and lumped constant from tissue glucose concentration or equilibrium brain/plasma distribution ratio for methylglucose. *J Cereb Blood Flow Metab.* 1991;11(1):25–34.
18. Dienel GA, Wang RY, Cruz NF. Generalized sensory stimulation of conscious rats increases labeling of oxidative pathways of glucose metabolism when the brain glucose-oxygen uptake ratio rises. *J Cereb Blood Flow Metab.* 2002;22(12):1490–1502.
19. Adachi K, Cruz NF, Sokoloff L, Dienel GA. Labeling of metabolic pools by [6-^{14}C]glucose during K(+)-induced stimulation of glucose utilization in rat brain. *J Cereb Blood Flow Metab.* 1995;15(1):97–110.
20. Dienel GA, Cruz NF. Exchange-mediated dilution of brain lactate specific activity: implications for the origin of glutamate dilution and the contributions of glutamine dilution and other pathways. *J Neurochem.* 2009;109(suppl 1):30–37.
21. Bradbury MWB, Cserr HF. Drainage of cerebral interstitial fluid and of cerebrospinal fluid into lymphatics. In: Johnston MG, ed. *Experimental biology of the lymphatic circulation.* New York: Elsevier; 1985:355–394.
22. Cruz NF, Adachi K, Dienel GA. Rapid efflux of lactate from cerebral cortex during K+-induced spreading cortical depression. *J Cereb Blood Flow Metab.* 1999;19(4):380–392.
23. Gross PM, Sposito NM, Pettersen SE, Panton DG, Fenstermacher JD. Topography of capillary density, glucose metabolism, and microvascular function within the rat inferior colliculus. *J Cereb Blood Flow Metab.* 1987;7(2):154–160.
24. Cruz NF, Ball KK, Dienel GA. Functional imaging of focal brain activation in conscious rats: impact of [(14)C]glucose metabolite spreading and release. *J Neurosci Res.* 2007;85 (15):3254–3266.
25. Ball KK, Gandhi GK, Thrash J, Cruz NF, Dienel GA. Astrocytic connexin distributions and rapid, extensive dye transfer via gap junctions in the inferior colliculus: implications for [(14)C]glucose metabolite trafficking. *J Neurosci Res.* 2007;85(15):3267–3283.

26. Enkvist MO, McCarthy KD. Astroglial gap junction communication is increased by treatment with either glutamate or high K+ concentration. *J Neurochem.* 1994;62(2):489–495.
27. Enkvist MO, McCarthy KD. Activation of protein kinase C blocks astroglial gap junction communication and inhibits the spread of calcium waves. *J Neurochem.* 1992;59 (2):519–526.
28. Gandhi GK, Cruz NF, Ball KK, Theus SA, Dienel GA. Selective astrocytic gap junctional trafficking of molecules involved in the glycolytic pathway: impact on cellular brain imaging. *J Neurochem.* 2009;110(3):857–869.
29. Giaume C, Marin P, Cordier J, Glowinski J, Premont J. Adrenergic regulation of intercellular communications between cultured striatal astrocytes from the mouse. *Proc Natl Acad Sci USA.* 1991;88(13):5577–5581.
30. Tabernero A, Giaume C, Medina JM. Endothelin-1 regulates glucose utilization in cultured astrocytes by controlling intercellular communication through gap junctions. *Glia.* 1996;16(3):187–195.
31. Finbow ME, Pitts JD. Permeability of junctions between animal cells. Intercellular exchange of various metabolites and a vitamin-derived cofactor. *Exp Cell Res.* 1981;131 (1):1–13.
32. Gandhi GK, Cruz NF, Ball KK, Dienel GA. Astrocytes are poised for lactate trafficking and release from activated brain and for supply of glucose to neurons. *J Neurochem.* 2009;111(2):522–536.
33. Manning Fox JE, Meredith D, Halestrap AP. Characterisation of human monocarboxylate transporter 4 substantiates its role in lactic acid efflux from skeletal muscle. *J Physiol.* 2000;529(Pt 2):285–293.
34. Mächler P, Wyss MT, Elsayed M, et al. In vivo evidence for a lactate gradient from astrocytes to neurons. *Cell Metab.* 2016;23(1):94–102.
35. Rouach N, Koulakoff A, Abudara V, Willecke K, Giaume C. Astroglial metabolic networks sustain hippocampal synaptic transmission. *Science.* 2008;322(5907):1551–1555.
36. Ball KK, Cruz NF, Mrak RE, Dienel GA. Trafficking of glucose, lactate, and amyloid-beta from the inferior colliculus through perivascular routes. *J Cereb Blood Flow Metab.* 2010;30(1):162–176.
37. Kvetnansky R, Goldstein DS, Weise VK, et al. Effects of handling or immobilization on plasma levels of 3,4-dihydroxyphenylalanine, catecholamines, and metabolites in rats. *J Neurochem.* 1992;58(6):2296–2302.
38. Madsen PL, Schmidt JF, Holm S, et al. Mental stress and cognitive performance do not increase overall level of cerebral O2 uptake in humans. *J Appl Physiol.* 1992;73 (2):420–426.
39. Dalsgaard MK, Ott P, Dela F, et al. The CSF and arterial to internal jugular venous hormonal differences during exercise in humans. *Exp Physiol.* 2004;89(3):271–277.
40. Schmalbruch IK, Linde R, Paulson OB, Madsen PL. Activation-induced resetting of cerebral metabolism and flow is abolished by beta-adrenergic blockade with propranolol. *Stroke.* 2002;33(1):251–255.
41. Gam CMB, Rasmussen P, Secher NH, Seifert T, Larsen FS, Nielsen HB. Maintained cerebral metabolic ratio during exercise in patients with β-adrenergic blockade. *Clin Physiol Funct Imaging.* 2009;29(6):420–426.
42. Larsen TS, Rasmussen P, Overgaard M, Secher NH, Nielsen HB. Non-selective beta-adrenergic blockade prevents reduction of the cerebral metabolic ratio during exhaustive exercise in humans. *J Physiol.* 2008;586(Pt 11):2807–2815.

43. Dalsgaard MK, Ogoh S, Dawson EA, Yoshiga CC, Quistorff B, Secher NH. Cerebral carbohydrate cost of physical exertion in humans. *Am J Physiol Regul Integr Comp Physiol.* 2004;287(3):R534–R540.
44. Seifert TS, Brassard P, Jorgensen TB, et al. Cerebral non-oxidative carbohydrate consumption in humans driven by adrenaline. *J Physiol.* 2009;587(Pt 1):285–293.
45. Schreurs J, Seelig T, Schulman H. β2-Adrenergic receptors on peripheral nerves. *J Neurochem.* 1986;46(1):294–296.
46. Tang F, Lane S, Korsak A, et al. Lactate-mediated glia-neuronal signalling in the mammalian brain. *Nat Commun.* 2014;5:3284.
47. Bozzo L, Puyal J, Chatton JY. Lactate modulates the activity of primary cortical neurons through a receptor-mediated pathway. *PLoS One.* 2013;8(8):e71721.
48. Lauritzen KH, Morland C, Puchades M, et al. Lactate receptor sites link neurotransmission, neurovascular coupling, and brain energy metabolism. *Cereb Cortex.* 2014;24(10):2784–2795.
49. Gilbert E, Tang JM, Ludvig N, Bergold PJ. Elevated lactate suppresses neuronal firing in vivo and inhibits glucose metabolism in hippocampal slice cultures. *Brain Res.* 2006;1117(1):213–223.
50. Cox DW, Bachelard HS. Partial attenuation of dentate granule cell evoked activity by the alternative substrates, lactate and pyruvate: evidence for a postsynaptic action. *Exp Brain Res.* 1988;69(2):368–372.
51. Okada Y, Lipton P. Glucose, oxidative energy metabolism, and neural function in brain slices—glycolysis plays a key role in neural activity. In: Gibson GE, Dienel GA, eds. *Brain Energetics. Integration of Molecular and Cellular Processes.* 3rd ed. Berlin: Springer-Verlag; 2007:17–39.
52. Bak LK, Obel LF, Walls AB, et al. Novel model of neuronal bioenergetics: post-synaptic utilization of glucose but not lactate correlates positively with Ca2+ signaling in cultured mouse glutamatergic neurons. *ASN Neuro.* 2012;4(3):151–160:art:e00083.doi:00010.01042/AN20120004
53. Nedergaard M, Goldman SA. Carrier-mediated transport of lactic acid in cultured neurons and astrocytes. *Am J Physiol.* 1993;265(2 Pt 2):R282-289.
54. Spray DC, Harris AL, Bennett MV. Gap junctional conductance is a simple and sensitive function of intracellular pH. *Science.* 1981;211(4483):712–715.
55. Vardjan N, Zorec R. Excitable astrocytes: Ca(2+)- and cAMP-regulated exocytosis. *Neurochem Res.* 2015;40(12):2414–2424.
56. Hertz L, Peng L, Dienel GA. Energy metabolism in astrocytes: high rate of oxidative metabolism and spatiotemporal dependence on glycolysis/glycogenolysis. *J Cereb Blood Flow Metab.* 2007;27(2):219–249.
57. Pelton 2nd EW, Kimelberg HK, Shipherd SV, Bourke RS. Dopamine and norepinephrine uptake and metabolism by astroglial cells in culture. *Life Sci.* 1981;28(14):1655–1663.
58. Rahman B, Kussmaul L, Hamprecht B, Dringen R. Glycogen is mobilized during the disposal of peroxides by cultured astroglial cells from rat brain. *Neurosci Lett.* 2000;290(3):169–172.
59. Dienel GA. Lack of appropriate stoichiometry: Strong evidence against an energetically important astrocyte-neuron lactate shuttle in brain. *J Neurosci Res.* 2017 Feb 2;. Available from: http://dx.doi.org/10.1002/jnr.24015.

Chapter 8

Dialogue Between Astrocytes and Noradrenergic Neurons Via L-Lactate

Anja G. Teschemacher[✉] **and Sergey Kasparov**[✉]
University of Bristol, Bristol, United Kingdom

Chapter Outline

The Noradrenaline-to-Astrocyte Signaling Axis 168
L-Lactate Release by Astrocytes 169
L-Lactate as a Gliotransmitter Feed Forward Signal to Noradrenergic Neurons? 172
Further Potential Signaling Roles of L-Lactate in the Brain 174
Conclusions 178
Abbreviations 178
Acknowledgments 178
References 178

ABSTRACT

Noradrenaline (NA) is one of the most powerful central neuromodulators and its effects are extremely diverse, ranging from control of appetite to the regulation of pain transmission and clearance of macromolecules from the brain. It is becoming increasingly clear that the many effects of NA cannot be explained by the simplistic idea of synaptic transmission and direct modulation of neurones in the target areas. Instead, the main target of NA released from the wide-spread noradrenergic varicosities are astrocytes, which via a variety of mechanisms then affect the surrounding neuronal networks. Here we focus on one specific pathway where the inter-cellular signaling may be carried out by lactate, which is otherwise known for its role in energy production by all cells of the body. Lactate can be almost instantly released by astrocytes in response to activation of the adjacent neurons and, in addition to being a potential substrate for ATP production, can via different mechanisms affect neurones. NA is a powerful stimulant of lactate production by astrocytes and, interestingly, lactate can act on noradrenergic neurones themselves and increase release of NA. This forms a positive feedback loop, which

[✉]Correspondence address
E-mail: anja.teschemacher@bristol.ac.uk; sergey.kasparov@bristol.ac.uk

Noradrenergic Signaling and Astroglia. DOI: http://dx.doi.org/10.1016/B978-0-12-805088-0.00008-6
© 2017 Elsevier Inc. All rights reserved.

could explain how activation of astrocytes can be coupled to the activity of neuronal networks, both, globally and locally. The search for the molecular mechanisms of lactate-mediated signaling is ongoing.

Keywords: Noradrenaline; astrocyte; glia; lactate; receptor

Noradrenaline (NA) is one of the most powerful neuromodulators in the mammalian brain. All NA released in the upper part of the front brain originates from a rather small and compact cluster of neurons located within the lower brainstem, the locus coeruleus (LC), alternatively known as the A6 noradrenergic cell group. These neurons are very unusual in almost every aspect of their anatomy and physiology. For example, many are connected by gap junctions, and their axons are thin and unmyelinated, with frequent large varicosities, and cover long distances to their targets in the cortex, hippocampus, or spinal cord. Intracellular recordings from these cells reveal that they typically have relatively depolarized membrane potentials at rest (~ -50 mV) and slow firing rates (1–2 Hz) which, upon stimulation, rise for a fraction of a second to 15–20 Hz, often followed by a period of lower activity. Further, they are subject to robust negative feedback via locally released NA acting via inhibitory α_2-adrenergic receptors (α_2-ARs).

THE NORADRENALINE-TO-ASTROCYTE SIGNALING AXIS

We were the first to characterize vesicular release of NA from central neurons using micro-amperometry and found that these neurons release, in addition to a population of small quanta, a substantial fraction of their NA by surprisingly large events which we could not explain at that time.[1] Indeed, the large NA release events suggested NA quanta, which could not be packaged into traditional small vesicles as they are usually associated with neurotransmission in central synapses releasing glutamate or γ-amino butyric acid (GABA). Recently, using advanced cryo-preservation techniques and electron microscopy we found that vesicular organelles containing NA in central noradrenergic neurons are indeed larger than was assumed, with sizes comparable to the granules in adrenal chromaffin cells.[2] As others before us, we have noticed that such NA-containing vesicles are hardly ever positioned next to structures, which could be identified as presynapse, implying that the majority of NA is not directed toward a specified cellular target, as per traditional synaptic transmission, but is instead released into extracellular space, i.e., in volume transmission mode. Moreover, release of such large quantities of NA implies a long signaling range for these events—multiple mechanisms for uptake and inactivation of catecholamines exist in the brain, but it is obvious that large local gradients of NA will facilitate its diffusion toward further targets. So if NA is not released in a targeted manner to signal from the axon of an LC neuron to a dedicated neuron in the frontal brain, what is then the target for NA released from these varicosities? Ample evidence indicates that the key recipients of NA signaling in the brain are astrocytes and this idea dates back more than 20 years.[3–5] Indeed, according to the mouse brain transcriptome, astrocytes express comparable or higher levels of α_{2A}- and α_{1B}-ARs, and much greater levels of β_1-ARs than other brain cells.[6] The latter is consistent with our unpublished transcriptomic data in the LC. All ARs are of the 7-transmembrane metabotropic type, which couple to intracellular signaling events through activation of different heterotrimeric G-proteins. Accordingly, NA, via α_2-, α_1-, and β_1-ARs, can recruit Gi-, Gq-, and Gs-protein signaling, respectively, and activate the downstream cascades in astrocytes. Several studies have demonstrated that Ca^{2+} signaling in astrocytes in vivo is under critical influence from NA released by LC fibers.[7,8] An interesting recent study suggested that in the mouse cortex, astrocytes are much more sensitive to NA than the local neurons.[9] It was further demonstrated that NA-mediated recruitment of astrocytes in cortical brain slices enabled synaptic plasticity and that the addition of a low concentration of NA was sufficient to induce long-term potentiation using a sub-threshold theta-burst stimulation. This effect was abolished when vesicular release in astrocytes was interrupted. Given the fundamental role of NA in control of neuronal network excitability, as exemplified for example by the role of NA in control of sleep-wake cycles, it appears that the NA-astrocyte signaling pathway is of paramount importance.

While the effects of NA on astrocytes have been studied extensively, it needs to be remembered that the absolute majority of these studies have been performed using cultured astrocytes, which are known to transform and change their physiology and even morphology in vitro. Nevertheless, it can be accepted as firmly established that NA recruits both, Ca^{2+} and cyclic adenosine monophosphate (cAMP)-mediated signaling mechanisms.[3,10,11] Because of the availability of high-quality tools for $[Ca^{2+}]_i$ imaging in live cells and, until relatively recently, the lack of equally good tools for monitoring intracellular cAMP dynamics, we know much more about NA-related Ca^{2+} signaling and comparatively little about the effects of NA on cAMP activity in integrated preparations. Importantly, both, the Gq-protein/phospholipase C/inositol trisphosphate/Ca^{2+} cascade and the Gs-protein/adenylate cyclase/cAMP pathway can lead to activation of glycogen breakdown in astrocytes. Therefore, since astrocytes are the main cell type in the brain where glycogen is found in measurable amounts, glycogenolysis will be one predictable outcome of astrocytic activation by NA.

L-LACTATE RELEASE BY ASTROCYTES

Glycolytic processing of glycogen and of newly imported glucose molecules inevitably leads to rapid buildup of pyruvate and, consequently, of

L-lactate in astrocytes. Neurons obviously also produce L-lactate but whether they produce as much as astrocytes and whether they release a significant amount of it into the extracellular space is still not completely clear. Recent in vivo imaging experiments with a genetically encoded lactate sensor, Laconic[12,13], have revealed that in spite of multiple transporters expressed by both neurons and astrocytes, L-lactate does not equilibrate across intra-/extracellular brain compartments and astrocytes evidently have a higher intracellular content of L-lactate than neurons.[13,14] At the same time, it is becoming clear that astrocytes can release L-lactate not only via monocarboxylate transporters (MCT) where the only factor affecting the speed of equilibration is the concentration gradient, but also via hemichannels[15] or other ion channels.[16] NA, via a variety of mechanisms, can trigger a rapid increase in both, synthesis and release of L-lactate from astrocytes.[17,18]

Extracellular concentration of L-lactate in resting and physiologically oxygenated brain has been estimated to vary between a few hundreds of micromolar and 1–2 mM.[19] L-Lactate levels in vivo are highly dynamic and faithfully follow sleep-wake cycles,[20–22] which are also well known to be paralleled by the activity of noradrenergic neurons and their projections. In spite of many years of research we still do not fully understand the physiological significance of L-lactate release from astrocytes.

One of the theories, widely known as "lactate shuttle hypothesis" proposes that L-lactate, following its release from astrocytes, is taken up by neurons and constitutes an essential component of neuronal energy supply, particularly at times when metabolic demand is high.[23–25] Two features put astrocytes into a unique position to fulfill this role—their ability to handle glucose flux across the blood–brain barrier, and the fact that they are the main cell type in the brain to store glycogen which constitutes energy reserve of the brain. The lactate shuttle hypothesis was initially based on a number of theoretical arguments, including differential expression of lactate dehydrogenase (LDH) and MCTs between neurons and astrocytes. There is evidence to suggest that the LDH isoenzyme 5 that favors L-lactate synthesis from pyruvate is enriched in astrocytes. Neurons, on the other hand, express mainly LDH isoenzyme 1 that is faster in oxidizing L-lactate. In theory, this distribution should support utilization of astrocyte-derived L-lactate as an energy substrate in neurons.[26] Interestingly, L-lactate production within astrocytes might even be compartmentalized—at least there is plenty of glycogen phosphorylase in perivascular end feet of astrocytes[27] and it is hard to imagine that L-lactate produced there would be traveling back to the somata of the astrocytes via the miniscule processes of these cells. Consequently L-lactate production and release in these areas could be rather directed toward the extracellular space around blood vessels and endothelial cells.

Transport of L-lactate across neuronal and glial cell membranes is supported mainly by MCT that symport monocarboxylates (such as L-lactate and pyruvate) and protons in 1:1 stoichiometry. Of the multiple types of MCTs that have been identified, only MCT1–MCT4 transport L-lactate.[28] MCT4, the major MCT type in astrocytes, has a relatively low affinity but high transport rate.[29,30] In contrast, the high affinity transporter MCT2 is thought to be more neuron-specific.[31–33] The intermediate affinity MCT1 was also found in cultured astrocytes but this is currently controversial as some studies suggest that it localizes specifically to blood vessels.[32]

Of course, since the activity and reaction direction of LDH enzymes and MCTs are gradient-dependent, L-lactate is produced, and may be released, not only by astrocytes but also by neurons. However, with astrocytes being the primary glucose source, and neuronal transporters saturating at relatively low L-lactate levels, neurons are likely to be poor exporters of L-lactate.[34] The differential distribution of (predominantly astrocytic) MCT4 and (predominantly neuronal) MCT2 further predicts that they facilitate the flux of L-lactate from astrocytes toward neurons. This situation would be somewhat comparable to skeletal muscle, where MCT4 is abundantly found on the glycolytic muscle fibers which export L-lactate while MCT1 is localized in oxidative muscle cells which import L-lactate.[35]

Several studies noted a decrease in intracellular pH after increasing extracellular L-lactate concentration in astrocyte cultures[36] or oocytes expressing MCT1.[37] This is probably a result of cotransport of protons along with L-lactate molecules by MCT.

During periods of increased neuronal activity in particular, glutamate spill-over from synapses can be taken up by astrocytes in a Na^+-dependent process and lead to glycolysis and production of L-lactate which can then be provided to the neurons.[38] Indeed, L-lactate of astrocytic origin does rapidly appear in the brain during activated states.[30,36]

Considering that astrocytes also mediate the supply of glucose, which is abundant in cerebrospinal fluid, why should neurons prefer L-lactate to glucose? In fact, a number of publications reason that, in the presence of physiological concentrations of glucose, neurons should not need to rely on alternative fuels such as L-lactate.[39] To complement this argument, it has been suggested that increased glycogenolysis in astrocytes during periods of elevated local brain activity fulfills rather the function of sustaining the astrocytes themselves and, in preventing their own rising demand for glucose, leaving more glucose for the metabolically challenged neurons.[30]

On the other hand, there is some data to suggest that increased rates of glycolysis such as would be required to support sustained levels of high activity might be toxic to neurons.[40] This work also reported that neurons limit glycolysis by actively degrading 6-phosphofructo-2-kinase/fructose-2,6-biphosphatase 3, a critical enzyme in the glucose consumption pathway, and instead actively use the pentose phosphate pathway. Therefore the ability of neurons to produce substantial levels of L-lactate would be limited. However, this warrants further investigation as, to our knowledge, no later study confirmed the existence of neuronal limitation of the speed of glycolysis.

In terms of physiological relevance, a number of studies have demonstrated that interfering with L-lactate synthesis or transport at critical time points during learning paradigms has a negative impact on memory formation, thus clearly an increase in L-lactate supply on demand is important.[41–44] Interestingly, a rise of levels of the D-isomer of L-lactate in plasma as a result of intestinal pathology has a negative effect on memory.[45] However, many of these effects of L-lactate cannot be fully and easily ascribed to its caloric value for neurons following its import from the extracellular space.

In summary, while the presence of glycogen stores mainly in astrocytes, the differential distribution of transporters, a putative limit of glycolysis in neurons, and other findings are generally consistent with the idea of a "lactate shuttle," i.e., that astrocytes make L-lactate available to neurons as a source of energy, there is currently no consensus on how biologically important this additional resource really is. Therefore, can L-lactate production, release and increase in the extracellular space in the brain have other than simply metabolic functions?

L-LACTATE AS A GLIOTRANSMITTER FEED FORWARD SIGNAL TO NORADRENERGIC NEURONS?

As discussed above, there are a range of mechanisms by which noradrenergic neurons can activate astrocytes, including L-lactate production and release. But what is the significance of astrocytic activation for noradrenergic neurons—can the astrocytes signal back to these neurons and modulate release of NA? We tested this hypothesis using organotypic cultured slices from the LC area which were co-transfected with two adenoviral vectors, one targeting astrocytes and transducing them with a Channelrhodopsin-2 mutant, the other targeting LC noradrenergic neurons for expression of a red fluorescent protein to aid visualization.[46] This approach enabled us to selectively activate astrocytes by optogenetics, and carry out whole cell recordings from LC neurons. Following about 60 s of intermittent optogenetic stimulation of astrocytes with blue light we recorded depolarizations and increased action potential activity in the neurons. These effects were blocked by treatments that prevented formation of L-lactate, for example by the LDH inhibitor oxamate, or by inhibiting glycogen metabolism by 1,4-dideoxy-1,4-imino-D-arabinitol.

Interestingly, optogenetic stimulation caused acidification in cultured astrocytes, which also could be prevented by blocking glycogen metabolism, indicating a buildup of acidic products of glycolysis. Incidentally, this process probably explains acidification of astrocytes in the experiments recently published by Beppu et al.[47] Under physiological conditions, the proton gradient is outward and opening of a H^+ conductance as such, using Channelrhodopsin-2, cannot lead to any significant intracellular acidification. It is also worth remembering that transport of L-lactate following its gradient inevitably leads to acidification even when pH-neutral L-lactate solutions are used because of the cotransport of H^+ by MCT. This needs to be taken into account when interpreting experiments using extracellular L-lactate application. We do not yet know to what extent changes in pH may have contributed to L-lactate effects in many published studies.

We reasoned that the delayed depolarization of LC neurons mentioned above may be resulting from L-lactate release by stimulated astrocytes.[46] Indeed, we observed a concentration-dependent excitatory effect using exogenously applied L-lactate (0.2–6 mM). The apparent EC_{50} for the depolarization was calculated as 680 μM. Co-application of D-lactate abolished depolarization of noradrenergic neurons evoked by either optogenetic activation of astrocytes or by application of L-lactate. Interestingly, blockers of AMPA and NMDA glutamate receptors and of purinergic P2Y1 receptors did not prevent the stimulatory effect, arguing in favor of a direct effect of L-lactate on these neurons. Several lines of evidence argued against the notion that the functional role of L-lactate for LC neurons was as an energy source. First, all experiments were performed in the presence of an excess of glucose (extracellular glucose 5.5 mM; internal pipette solution 5 mM glucose plus 2 mM ATP). Second, pyruvate did not mimic the L-lactate action—no depolarizations were observed in LC neurons. Third, when added to the internal pipette solution, L-lactate did not evoke membrane depolarizations but when added into the bath solution it still excited LC neurons. Finally, an MCT blocker 4-CIN was ineffective, arguing against the requirement of L-lactate uptake for depolarizations in LC neurons. Interestingly, inhibition of adenylate cyclase and protein kinase A, but not of protein kinase C, abolished the effect of L-lactate, consistent with an L-lactate-evoked cAMP-mediated signaling pathway in LC neurons (Fig. 8.1). We then employed fast scan voltammetry in organotypic brain slices, and found that LC neurons release NA in response to optogenetic activation of astrocytes. Exogenous L-lactate application also caused NA release in organotypic and acute slices in a concentration-dependent manner and these responses were mediated by the same adenylate cyclase-dependent cascade.

These observations implied that L-lactate should be able to activate the LC in vivo. We tested this in anaesthetized rats. Microinjection of L-lactate into the LC induced an increase in the power of the high frequency bands of the electroencephalogram, consistent with increased noradrenergic activity in the cortex. A transient increase in arterial blood pressure was also observed, possibly reflecting an episode of arousal.

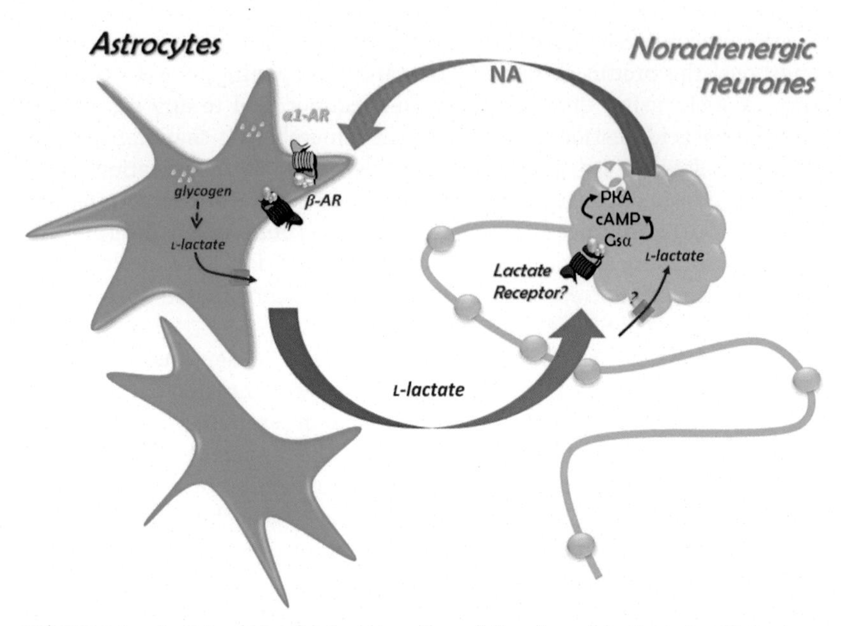

FIGURE 8.1 L-Lactate production by astrocytes and its release into the extracellular space is stimulated by noradrenaline (NA) via activation of astrocytic adrenergic receptors (ARs). Extracellular L-lactate is excitatory to noradrenergic neurons through a mechanism likely to involve a metabotropic receptor. This scenario creates a specific positive feedback loop between astrocytes and noradrenergic neurons and may therefore engage astrocytes as amplifiers of localized and global noradrenergic modulation of neural network activity.

At present the identity of the molecular mechanisms behind these effects of L-lactate on noradrenergic neurons is unknown. As discussed below, two G-protein-coupled receptors (GPCRs), which are sensitive to L-lactate, have been previously described but our data are more consistent with the existence of a yet unidentified excitatory metabotropic receptor for L-lactate (Fig. 8.1). In any case, whatever the molecular nature of this L-lactate signaling cascade, it is likely to support a positive feedback loop between noradrenergic axons and astrocytes and may, in theory, couple local noradrenergic input to the activity and metabolic status of neurons and astrocytes in brain areas targeted by LC projections.

FURTHER POTENTIAL SIGNALING ROLES OF L-LACTATE IN THE BRAIN

Scanning of the human and other genomes has made it possible to identify open reading frames for putative GPCRs. Guided by their highly conserved structural determinants including their seven transmembrane domain structure with extracellular amino- and intracellular carboxy-terminals, it was found that they occupy a large proportion of mammalian genomes. They represent extremely valuable pharmacological targets since numerous currently used clinical drugs block, activate or otherwise modulate these types of receptors. Up until the identification of their physiological ligands, putative GPCRs are classified as orphan receptors. GPR81 (or hydroxycarboxylic acid receptor 1; HCA1) was an orphan receptor until 2008–09, when two groups reported that L-lactate acts as a natural ligand and agonist of HCA1.[48,49] The monocarboxylates α-hydroxybutyrate, glycolate, α-hydroxyisobutyrate, and γ-hydroxybutyrate were also recognized as HCA1 agonists. In contrast, L-lactate was unable to activate GPR109a (HCA2) and GPR109b (HCA3), which are highly homologous to HCA1.

In a Chinese hamster ovary cell line CHO-K1-based HCA1 expression system, L-lactate concentration-dependently increased binding of ^{35}S-GTPγS ($EC_{50} \approx 1.3$ mM) and inhibited forskolin-stimulated cAMP production.[48] L-Lactate also activated Ca^{2+} mobilization in CHO cells stably co-expressing HCA1 with the G_{qi5} protein, which promiscuously couples various GPCRs to the phospholipase C/inositol trisphosphate pathway. Pretreatment with pertussis toxin in order to inactivate G_i-protein signaling completely abolished the effect of L-lactate-mediated activation of HCA1 in a ^{35}S-GTPγS binding assay. To note, in the same assay the stereoisomer D-lactate also activated HCA1 but with a higher EC_{50} (~3 mM). These experiments clearly demonstrated that L-lactate stimulates the G_i-coupled HCA1 pathway and therefore causes inhibition of cAMP-mediated intracellular signaling events. HCA1 is highly expressed in adipose tissue and was originally proposed as a putative target for treating hypercholesterolemia.[48,49] Initial studies suggested that HCA1 is expressed at negligible levels in the brain when compared to adipocytes, and a later study estimated about 100-fold lower levels of HCA1 expression in mouse brain than in adipose tissue.[50] The most prominent expression, at mRNA and protein levels, was reported for Purkinje cells in the cerebellum, pyramidal cells of the hippocampus, and neurons within the dentate hilus and at a lower level in the neocortex. Immunoreactivity for HCA1 was localized mostly in neurons and to a lesser extent in vascular endothelial cells and astrocytes.[50] Primary cultures of mouse neocortical neurons also showed functional HCA1 expression.[51] In human tissue, HCA1 mRNA expression was detected in pituitary gland but not in frontal, temporal, and occipital lobes of the cortex, nor in forebrain, caudate nucleus, nucleus accumbens, and hippocampus.[52] However, the published transcriptomes of various cell types in the mouse and human brain suggest miniscule levels of HCA1 expression in the brain http://bioinf.nl:8080/GOAD2/databaseSelectServlet, http://web.stanford.edu/group/barres_lab/cgi-bin/geneSearch.py?geneNameIn=HCAR1.[6,53] Our own (as yet unpublished) transcriptomic data from the LC area also show no measurable expression of HCA1. Therefore, expression of HCA1 in the brain remains a controversial issue and it appears unlikely that the effects of L-lactate on NA release in our experiments could in any way be mediated by this receptor. However, in spite of its very low level of expression it might be functional in some frontal brain structures. For example, Ca^{2+} imaging experiments in primary cortical cultures showed that L-lactate inhibited the frequency of intracellular Ca^{2+} transients by about 50% in principal as well as GABAergic interneurons in a concentration-dependent manner ($IC_{50} \sim 5$ mM).[51] This effect of L-lactate was evoked in the presence of 5 mM glucose, and could not be mimicked by the application of pyruvate or by high (10 mM) or low (0.5 mM) concentrations of glucose. Furthermore, inactivation of G_i-protein signaling abolished the inhibitory effect of 10 mM L-lactate on intracellular Ca^{2+} transient frequency.

To summarize, HCA1 may be present in some parts of the brain at low levels. Given that all studies published to date consistently required applications of very high L-lactate concentrations for activation of HCA1 (5 mM or more), it may play a role under conditions of ischemia or neuronal hyperactivity, as for example during seizures. There is currently no evidence to suggest that HCA1 plays a role in the communication between astrocytes and the noradrenergic system under physiological conditions.

While the search for the explanations of central effects associated with L-lactate continues, recent publications implicated yet another L-lactate-sensitive receptor that may be involved in local signal transduction in the carotid body.[54] Chang et al. were able to demonstrate that the orphan receptor OR51E2 (Olfr78 in mouse, Olr59 in rat) is highly expressed on glomus cells and is activated by L-lactate, causing increased carotid sinus nerve activity.[54] OR51E2 belongs to the family of olfactory GPCRs but is expressed not only in the olfactory bulb, but also in the prostate, specific autonomic ganglia, the juxtaglomerular apparatus, and the brainstem.[55–57] Importantly, knock-out of OR51E2 eliminated activation of glomus cells in response to hypoxia, as well as L-lactate-induced respiratory responses.[54] Given the indications for localized expression of OR51E2 orthologs in parts of the brain, a possible contribution of these proteins to L-lactate signaling in the brain should be considered. In our own transcriptome from the LC area, OR51E2 is present at low levels but appears neither in the astrocytic nor in the noradrenergic neuron fraction, so could potentially be involved in local signaling in an indirect fashion.

In addition to its metabolic and signaling roles, a recent study proposed that L-lactate may modulate NMDA receptor function via its redox potential.[58] The study found that L-lactate drives expression of the plasticity-related immediate early genes Arc, c-Fos, and Zif268 in cortical neurons in a time- and concentration-dependent manner. L-Lactate (20 mM), in the presence of high glucose concentrations (25 mM), increased immediate early gene expression four to eightfold with a maximum increase detected after 60 min. The effect of L-lactate was neither mimicked by application of glucose (20 mM, an equicaloric level to 10 mM L-lactate), nor by pyruvate, nor by D-lactate (10 mM). However, it could be blocked by the MCT inhibitor UK5099 and by the NMDA receptor antagonist MK801. Further, L-lactate (10 and 20 mM) enhanced inward currents evoked by co-application of glutamate and glycine, an effect, which was also abolished by MK801. This data suggest that the effect of L-lactate may be mediated by a change in the redox state within the neurons because metabolism of the imported L-lactate leads to production of pyruvate and an increase in the ratio of reduced/oxidized nicotinamide adenine dinucleotide ($NADH/NAD^+$ ratio). However, the effects were only significant from a minimal L-lactate concentration of 2.5 mM and most of the experiments were performed using 20 mM L-lactate, which is outside of a working range that could explain the effect of L-lactate on noradrenergic transmission.[46] These concentrations are also an order of magnitude higher than the physiological range of physiological brain L-lactate concentrations.[19] Therefore, a modulation of gene expression in neurons via L-lactate-evoked NMDA current potentiation may be relevant following central hyperexcitation such as with epileptiform activity. Of note, one earlier study reported for hippocampal neurons that very high concentrations of L-lactate (250—500 mM) when applied via microdialysis were inhibitory rather than excitatory and so may counteract neuronal overactivation.[59] Further work is needed to establish the physiological relevance of the modulation of NMDA receptors via L-lactate's effect on the neuronal redox state.

As discussed above the transfer of L-lactate from astrocytes to neurons is clearly possible. However, not all evidence supports the idea that, while glucose is available, such a transfer would actually confer a benefit to neurons. Nevertheless, if neurons, in fact, did use astrocyte-derived L-lactate to support their adenosine triphosphate (ATP) production, this could resemble the situation in the pancreas where the ATP/adenosine 5′-diphosphate ratio in β-cells is monitored by K_{ATP} channels. These channels close in response to a rise in cytoplasmic ATP levels when the supply of energy substrates is high, and this leads to membrane depolarization.[60,61] K_{ATP} channels are present in the brain[62] and may be involved in the effects of L-lactate on certain populations of neurons (for review see Ref. 19). Our transcriptome data suggest the presence of low levels of the relevant K_{ATP} channel subunits, $K_{ir}6.1$ and $K_{ir}6.2$, in noradrenergic neurons in the LC. However, in our recordings, LC neurons depolarized when exposed to extracellular L-lactate, without the prerequisite of importing it.[46] A recent publication suggested that, in addition to the modulation of K_{ATP} channels, uptake of large concentrations of L-lactate in neurons could facilitate release of ATP into the extracellular space, which then may cause neuronal excitation by activating purinergic receptors.[63]

CONCLUSIONS

Some of the evidence discussed above supports the idea of a signaling role of L-lactate in the brain and bidirectional L-lactate-mediated communication between astrocytes and noradrenergic neurons and their varicosities. Such a mechanism could potentially explain how thin, unmyelinated and highly varicose axons projecting from the LC over very long distances to the frontal brain structures might operate in the "local response mode" in all areas where NA release occurs. L-Lactate levels in vivo are highly dynamic and faithfully follow sleep-wake cycles that are also well known to be paralleled by the activity of noradrenergic neurons and their projections.[20–22] By recruiting additional signaling mechanisms mediated by NA locally, activity triggered in the neuronal networks by synaptic inputs may be modulated. In this way, by tuning the activity of the LC, such as during cycles of sleep and arousal, the brain may exert differential control over the computations performed by individual neuronal networks: when local neurons and astrocytes release L-lactate which facilitates release of LC-derived NA, which then evokes further release of L-lactate, this could create a very efficient mechanism for coordination of the energy homeostasis and excitability of various cellular compartments of the brain.

ABBREVIATIONS

AR	adrenergic receptor
ATP	adenosine triphosphate
cAMP	cyclic adenosine monophosphate
CHO	Chinese hamster ovary cells
GABA	γ-amino butyric acid
GPCR or GPR	G-protein-coupled receptors
HCA	hydroxycarboxylic acid receptor
LC	locus coeruleus
LDH	lactate dehydrogenase
MCT	monocarboxylate transporters
NA	noradrenaline (norepinephrine)

ACKNOWLEDGMENTS

A.G.T. and S.K. are supported by BBSRC grants BB/L019396/1, BB/K009192/1, and MRC grant MR/L020661/1.

REFERENCES

1. Chiti Z, Teschemacher AG. Exocytosis of norepinephrine at axon varicosities and neuronal cell bodies in the rat brain. *FASEB J.* 2007;21:2540–2550.

2. Kourtesis I, Kasparov S, Verkade P, Teschemacher AG. Ultrastructural correlates of enhanced norepinephrine and neuropeptide Y cotransmission in the spontaneously hypertensive rat brain. *ASN Neuro.* 2015;7. Available from: http://dx.doi.org/10.1177/1759091415610115.
3. Stone EA, Ariano MA. Are glial cells targets of the central noradrenergic system? A review of the evidence. *Brain Res - Brain Res Rev.* 1989;14:297–309.
4. Hertz L. Autonomic control of neuronal-astrocytic interactions, regulating metabolic activities, and ion fluxes in the CNS. *Brain Res Bull.* 1992;29:303–313.
5. Hertz L, Lovatt D, Goldman SA, Nedergaard M. Adrenoceptors in brain: cellular gene expression and effects on astrocytic metabolism and [Ca(2+)]i. *Neurochem Int.* 2010;57:411–420.
6. Zhang Y, Chen K, Sloan SA, et al. An RNA-sequencing transcriptome and splicing database of glia, neurons, and vascular cells of the cerebral cortex. *J Neurosci.* 2014;34:11929–11947.
7. Ding F, O'Donnell J, Thrane AS, et al. α_1-Adrenergic receptors mediate coordinated Ca^{2+} signaling of cortical astrocytes in awake, behaving mice. *Cell Calcium.* 2013;54:387–394.
8. Bekar LK, He W, Nedergaard M. Locus coeruleus alpha-adrenergic-mediated activation of cortical astrocytes in vivo. *Cereb Cortex.* 2008;18:2789–2795.
9. Pankratov Y, Lalo U. Role for astroglial alpha1-adrenoreceptors in gliotransmission and control of synaptic plasticity in the neocortex. *Front Cell Neurosci.* 2015;9:230. Available from: http://dx.doi.org/10.3389/fncel.2015.00230.
10. Schimmer BP, Schimmer BP. Effects of catecholamines and monovalent cations on adenylate cyclase activity in cultured glial tumor cells. *Biochim Biophys Acta.* 1971;252:567–573.
11. Kon C, Breckenridge BM. Potentiation of guanosine 3':5'-monophosphate accumulation in C-6 glial tumor cells. *J Cyclic Nucleotide Res.* 1979;5:31–41.
12. San Martin A, Ceballo S, Ruminot I, Lerchundi R, Frommer WB, Barros LF. A genetically encoded FRET lactate sensor and its use to detect the warburg effect in single cancer cells. *PLoS ONE.* 2013;8:e57712.
13. Mächler P, Wyss MT, Elsayed M, et al. In vivo evidence for a lactate gradient from astrocytes to neurons. *Cell Metab.* 2016;23:94–102.
14. Kasparov S. Are astrocytes the pressure-reservoirs of lactate in the brain?. *Cell Metab.* 2016;23:1–2.
15. Karagiannis A, Sylantyev S, Hadjihambi A, Hosford PS, Kasparov S, Gourine AV. Hemichannel-mediated release of lactate. *J Cereb Blood Flow Metab.* 2015.
16. Sotelo-Hitschfeld T, Niemeyer MI, Machler P, et al. Channel-mediated lactate release by K^+-stimulated astrocytes. *J Neurosci.* 2015;35:4168–4178.
17. Magistretti PJ, Sorg O, Yu N, Martin JL, Pellerin L. Neurotransmitters regulate energy metabolism in astrocytes: implications for the metabolic trafficking between neural cells. *Dev Neurosci.* 1993;15:306–312.
18. Subbarao KV, Hertz L. Stimulation of energy metabolism by alpha-adrenergic agonists in primary cultures of astrocytes. *J Neurosci Res.* 1991;28:399–405.
19. Mosienko V, Teschemacher AG, Kasparov S. Is L-lactate a novel signaling molecule in the brain?. *J Cereb Blood Flow Metab.* 2015;35:1069–1075.
20. Cocks JA. Change in the concentration of lactic acid in the rat and hamster brain during natural sleep. *Nature.* 1967;215:1399–1400.

21. Shram N, Netchiporouk L, Cespuglio R. Lactate in the brain of the freely moving rat: voltammetric monitoring of the changes related to the sleep-wake states. *Eur J Neurosci*. 2002;16:461–466.
22. Li B, Freeman RD. Neurometabolic coupling between neural activity, glucose, and lactate in activated visual cortex. *J Neurochem*. 2015;135:742–754.
23. Gladden LB. Lactate metabolism: a new paradigm for the third millennium. *J Physiol*. 2004;558:5–30.
24. Magistretti PJ, Sorg O, Yu N, Martin JL, Pellerin L. Neurotransmitters regulate energy metabolism in astrocytes: implications for the metabolic trafficking between neural cells. *Dev Neurosci*. 1993;15:306–312.
25. Pellerin L, Pellegri G, Bittar PG, et al. Evidence supporting the existence of an activity-dependent astrocyte-neuron lactate shuttle. *Dev Neurosci*. 1998;20:291–299.
26. Bittar PG, Charnay Y, Pellerin L, Bouras C, Magistretti PJ. Selective distribution of lactate dehydrogenase isoenzymes in neurons and astrocytes of human brain. *J Cereb Blood Flow Metab*. 1996;16:1079–1089.
27. Pfeiffer B, Elmer K, Roggendorf W, Reinhart PH, Hamprecht B. Immunohistochemical demonstration of glycogen phosphorylase in rat brain slices. *Histochemistry*. 1990;94:73–80.
28. Halestrap AP. The monocarboxylate transporter family—structure and functional characterization. *IUBMB Life*. 2012;64:1–9.
29. Dimmer KS, Friedrich B, Lang F, Deitmer JW, Broer S. The low-affinity monocarboxylate transporter MCT4 is adapted to the export of lactate in highly glycolytic cells. *Biochem J*. 2000;350(Pt 1):219–227.
30. Gandhi GK, Cruz NF, Ball KK, Dienel GA. Astrocytes are poised for lactate trafficking and release from activated brain and for supply of glucose to neurons. *J Neurochem*. 2009;111:522–536.
31. Rafiki A, Boulland JL, Halestrap AP, Ottersen OP, Bergersen L. Highly differential expression of the monocarboxylate transporters MCT2 and MCT4 in the developing rat brain. *Neuroscience*. 2003;122:677–688.
32. Bergersen L, Rafiki A, Ottersen OP. Immunogold cytochemistry identifies specialized membrane domains for monocarboxylate transport in the central nervous system. *Neurochem Res*. 2002;27:89–96.
33. Debernardi R, Pierre K, Lengacher S, Magistretti PJ, Pellerin L. Cell-specific expression pattern of monocarboxylate transporters in astrocytes and neurons observed in different mouse brain cortical cell cultures. *J Neurosci Res*. 2003;73:141–155.
34. Dienel GA. Brain lactate metabolism: the discoveries and the controversies. *J Cereb Blood Flow Metab*. 2012;32:1107–1138.
35. Bergersen LH. Is lactate food for neurons? Comparison of monocarboxylate transporter subtypes in brain and muscle. *Neuroscience*. 2007;145:11–19.
36. Nedergaard M, Goldman SA. Carrier-mediated transport of lactic acid in cultured neurons and astrocytes. *Am J Physiol*. 1993;265:R282–R289.
37. Broer S, Rahman B, Pellegri G, et al. Comparison of lactate transport in astroglial cells and monocarboxylate transporter 1 (MCT 1) expressing Xenopus laevis oocytes. Expression of two different monocarboxylate transporters in astroglial cells and neurons. *J Biol Chem*. 1997;272:30096–30102.
38. Pellerin L, Magistretti PJ. Glutamate uptake into astrocytes stimulates aerobic glycolysis: a mechanism coupling neuronal activity to glucose utilization. *Proc Natl Acad Sci USA*. 1994;91:10625–10629.

39. Hall CN, Klein-Flugge MC, Howarth C, Attwell D. Oxidative phosphorylation, not glycolysis, powers presynaptic and postsynaptic mechanisms underlying brain information processing. *J Neurosci.* 2012;32:8940–8951.
40. Herrero-Mendez A, Almeida A, Fernandez E, Maestre C, Moncada S, Bolanos JP. The bioenergetic and antioxidant status of neurons is controlled by continuous degradation of a key glycolytic enzyme by APC/C-Cdh1. *Nat Cell Biol.* 2009;11:747–752.
41. Gibbs ME, Hertz L. Inhibition of astrocytic energy metabolism by D-lactate exposure impairs memory. *Neurochem Int.* 2008;52:1012–1018.
42. Gibbs ME, Lloyd HG, Santa T, Hertz L. Glycogen is a preferred glutamate precursor during learning in 1-day-old chick: biochemical and behavioral evidence. *J Neurosci Res.* 2007;85:3326–3333.
43. Newman LA, Korol DL, Gold PE. Lactate produced by glycogenolysis in astrocytes regulates memory processing. *PLoS ONE.* 2011;6:e28427.
44. Suzuki A, Stern SA, Bozdagi O, et al. Astrocyte-neuron lactate transport is required for long-term memory formation. *Cell.* 2011;144:810–823.
45. Hanstock TL, Mallet PE, Clayton EH. Increased plasma D-lactic acid associated with impaired memory in rats. *Physiol Behav.* 2010;101:653–659.
46. Tang F, Lane S, Korsak A, et al. Lactate-mediated glia-neuronal signalling in the mammalian brain. *Nat Commun.* 2014;5:3284.
47. Beppu K, Sasaki T, Tanaka KF, et al. Optogenetic countering of glial acidosis suppresses glial glutamate release and ischemic brain damage. *Neuron.* 2014;81:314–320.
48. Cai TQ, Ren N, Jin L, et al. Role of GPR81 in lactate-mediated reduction of adipose lipolysis. *Biochem Biophys Res Commun.* 2008;377:987–991.
49. Liu C, Wu J, Zhu J, et al. Lactate inhibits lipolysis in fat cells through activation of an orphan G-protein-coupled receptor, GPR81. *J Biol Chem.* 2009;284:2811–2822.
50. Lauritzen KH, Morland C, Puchades M, et al. Lactate receptor sites link neurotransmission, neurovascular coupling, and brain energy metabolism. *Cereb Cortex.* 2014;24 (10):2784–2795.
51. Bozzo L, Puyal J, Chatton JY. Lactate modulates the activity of primary cortical neurons through a receptor-mediated pathway. *PLoS ONE.* 2013;8:e71721.
52. Lee DK, Nguyen T, Lynch KR, et al. Discovery and mapping of ten novel G protein-coupled receptor genes. *Gene.* 2001;275:83–91.
53. Holtman IR, Noback M, Bijlsma M, et al. Glia Open Access Database (GOAD): a comprehensive gene expression encyclopedia of glia cells in health and disease. *Glia.* 2015;63:1495–1506.
54. Chang AJ, Ortega FE, Riegler J, Madison DV, Krasnow MA. Oxygen regulation of breathing through an olfactory receptor activated by lactate. *Nature.* 2015;527:240–244.
55. Conzelmann S, Levai O, Bode B, et al. A novel brain receptor is expressed in a distinct population of olfactory sensory neurons. *Eur J Neurosci.* 2000;12:3926–3934.
56. Yuan TT, Toy P, McClary JA, LIN RJ, Miyamoto NG, Kretschmer PJ. Cloning and genetic characterization of an evolutionarily conserved human olfactory receptor that is differentially expressed across species. *Gene.* 2001;278:41–51.
57. Pluznick JL, Protzko RJ, Gevorgyan H, et al. Olfactory receptor responding to gut microbiota-derived signals plays a role in renin secretion and blood pressure regulation. *Proc Natl Acad Sci USA.* 2013;110:4410–4415.
58. Yang J, Ruchti E, Petit JM, et al. Lactate promotes plasticity gene expression by potentiating NMDA signaling in neurons. *Proc Natl Acad Sci USA.* 2014;111:12228–12233.

59. Gilbert E, Tang JM, Ludvig N, Bergold PJ. Elevated lactate suppresses neuronal firing in vivo and inhibits glucose metabolism in hippocampal slice cultures. *Brain Res.* 2006;1117:213–223.
60. Matsuo M, Kimura Y, Ueda K. KATP channel interaction with adenine nucleotides. *J Mol Cell Cardiol.* 2005;38:907–916.
61. Yokoshiki H, Sunagawa M, Seki T, Sperelakis N. ATP-sensitive K+ channels in pancreatic, cardiac, and vascular smooth muscle cells. *Am J Physiol.* 1998;274:C25–C37.
62. Karschin C, Ecke C, Ashcroft FM, Karschin A. Overlapping distribution of K(ATP) channel-forming Kir6.2 subunit and the sulfonylurea receptor SUR1 in rodent brain. *FEBS Lett.* 1997;401:59–64.
63. Jourdain P, Allaman I, Rothenfusser K, Fiumelli H, Marquet P, Magistretti PJ. L-Lactate protects neurons against excitotoxicity: implication of an ATP-mediated signaling cascade. *Sci Rep.* 2016;6:21250.

Chapter 9

Noradrenergic System and Memory: The Role of Astrocytes

Manuel Zenger[1#], Sophie Burlet-Godinot[1,2#], Jean-Marie Petit[1,2§,✉] and Pierre J. Magistretti[1,3§,✉]

[1]*École Polytechnique Fédérale de Lausanne, Lausanne, Switzerland,* [2]*Centre Hospitalier Universitaire Vaudois, Prilly, Switzerland,* [3]*King Abdullah University of Science and Technology, Thuwal, Kingdom of Saudi Arabia*

Chapter Outline

Introduction 184
Brain Noradrenergic System and Its Weight on Cerebral Energy Metabolism 184
- Noradrenergic Pathways and Receptors 184
- The Specific Action of Noradrenaline on Glycogen Metabolism 187

Noradrenaline and Memory 188
- Noradrenaline Action on Synaptic Plasticity: Neurons as Targets 188
- Noradrenaline Action on Synaptic Plasticity: Astrocytes as Targets 189
- Role of Noradrenaline in Memory Paradigms 189

Modulation of Astrocytic Energy Metabolism by Noradrenaline: Impact on Memory 190
- Brain Energy Metabolism and Memory 190
- The Central Role of Glycogen 191
- Influence of the Sleep–Wake Cycle 193

Conclusions 194
Abbreviations 194
References 195

ABSTRACT

There is ample evidence indicating that noradrenaline plays an important role in memory mechanisms. Noradrenaline is thought to modulate these processes through activation of adrenergic receptors in neurons. Astrocytes that form essential partners for synaptic function, also express alpha- and beta-adrenergic receptors. In astrocytes, noradrenaline triggers metabolic actions such as the glycogenolysis leading to an increase in L-lactate formation and release. L-Lactate can be used by neurons as a source of energy during memory tasks and can also

✉Correspondence address
E-mail: jean-marie.petit@epfl.ch; pierre.magistretti@kaust.edu.sa
#These authors contributed equally to this work.
§Co-last authors.

Noradrenergic Signaling and Astroglia. DOI: http://dx.doi.org/10.1016/B978-0-12-805088-0.00009-8
© 2017 Elsevier Inc. All rights reserved.

induce transcription of plasticity genes in neurons. Activation of α-adrenergic receptors can also trigger gliotransmitter release resulting of intracellular calcium waves. These gliotransmitters modulate the synaptic activity and thereby can modulate long-term potentiation mechanisms. In summary, recent evidences indicate that noradrenaline exerts its memory-promoting effects through different modes of action both on neurons and astrocytes.

Keywords: Hippocampus; learning; L-lactate; glycogen; sleep

INTRODUCTION

The brain noradrenergic system is mobilized in response to stress, novelty exposure, or sensory challenge. Each of these conditions could be present during a learning task and consequently explain why noradrenaline (NA) is involved in learning and memory mechanisms. At the cellular levels, NA is classically thought to act on memory formation and consolidation through its modulatory role on the neuronal long-term potentiation (LTP). However, based on anatomical and physiological data, synaptic mechanisms need to be thought in terms of the "tripartite synapse" in which astrocyte process that ensheath the synapse plays an active role in synapse functioning. Moreover, astrocytes express adrenergic receptors (ARs) through which NA can induce bursts of intracellular calcium (Ca^{2+}) and second messenger activation leading to metabolic changes. These data point to a potential role of NA on memory mechanisms through its action on astrocytes.

The aim of this review is to present data supporting the involvement of astrocytes in the memory-promoting action of NA and to put them into perspective of general principles of synaptic transmission.

BRAIN NORADRENERGIC SYSTEM AND ITS WEIGHT ON CEREBRAL ENERGY METABOLISM

Noradrenergic Pathways and Receptors

The mammalian central nervous system (CNS) contains several distinct groups of noradrenergic neurons that were first identified by Dahlström and Fuxe in 1964 and labeled groups A1–A7. These groups are distributed from the rostral pons to the caudal medulla.[1] More recently, based on developmental gene expression distribution, Robertson et al. defined four subpopulations of NA neurons that differ in their anatomical localization, their axon morphology, and their projections pattern.[2]

The most noticeable group of NA neurons lies in the locus coeruleus (LC). This compact cluster of NA neurons sends extensive and widespread projections via immensely ramified axons throughout the neuroaxis, from neocortex to spinal cord. More precisely, depending on their location in the LC, neurons provide noradrenergic innervation to the olfactory bulb, the cortex, in particular to the medial prefrontal region that exhibits the highest density of NA varicosities[3]; the cerebellum, the hippocampus, the amygdala, the thalamus, the hypothalamus as well as the solitary nucleus and both ventral and dorsal horns of the spinal cord (for review see Refs. 4,5).

LC neurons exhibit distinct phasic and tonic modes of firing, which differ in both the pattern of spike discharge and NA release properties.[6,7] Tonic firing of LC—NA neurons respond to diurnal variation, which is related to the state of cortical arousal. Phasic firing of LC—NA is observed during voluntary tasks that involve attention. Both modes of firing interact and their balance allows optimal performance during processes that require attention.[6,8] The noradrenergic system is also a key mediator in the stress response and has been implicated in relapse to drug addiction.[9]

With as few as 10% of release sites being associated with conventional synaptic contacts made by NA containing axons, NA release is mainly achieved via en passant synapses or nonjunctional varicosities into the surrounding neuropil in a nonsynaptic or paracrine mode. In addition to impact ensembles of neurons within noradrenergic terminal fields by nonspecific diffusion, NA released can also target multiple cell types in particular astrocytes and microglia, by volume transmission.[10,11]

After release, NA interacts with different types and subtypes of ARs that are widely distributed in many brain regions, expressed on distinct cell types and have distinct pharmacological profiles. To date, they are classified in three distinct subclasses, β-, α_1-, and α_2-ARs and each subclass comprises of several subtypes (see Table 9.1).

Thus far, the existence of 2 "typical" (β_1 and β_2) and one atypical β-AR (β_3, insensitive to the most commonly used antagonist) is generally accepted. Three distinct subtypes of α_1-AR (α_{1A}, α_{1B}, and α_{1D}) and three subtypes of α_2-AR (α_{2A}, α_{2B}, and α_{2C}) have also been localized in the brain. While a certain degree of regional overlap exists, each AR subtype has a distinct anatomical distribution in the brain as demonstrated by mRNA in situ hybridization. In rat CNS, there is a high to very high level of α_{1A}-AR mRNA in the olfactory system, some nuclei of the hypothalamus, some motor nuclei, and the spinal cord. α_{1B}-AR mRNA is highly expressed in the thalamus, the dorsal and median raphe nuclei as well as the pineal gland. α_{1D}-AR mRNA is predominantly expressed in the olfactory bulb, the reticular nucleus of the thalamus, the hippocampal formation, and the cerebral cortex.[12] Concerning α_2-AR subtypes, α_{2A}-AR mRNA is highly expressed in the LC, some hypothalamic nuclei, in the hippocampal region and cerebral cortex in rat but not in mouse; α_{2B}-AR mRNA is marginally expressed in the brain except in the thalamus and the olfactory system; α_{2C}-AR mRNA is intensely expressed in the olfactory system, the hippocampal region, and the basal ganglia.[13,14] The β_1- and β_3-AR mRNAs are both present in the olfactory system, the hippocampus, and the cerebral cortex (in particular in layers I and II), but

TABLE 9.1 ARs in Summary

Receptors	Subtype		Intracellular pathway	Subcellular Localization	Refs.
Alpha	α_1	α_{1A}[a]	Activation of phospholipase C	Postsynaptic and astrocytes	17,18
		α_{1B}		Postsynaptic and astrocytes	17,18
		α_{1D}		Postsynaptic and astrocytes	17,18
	α_2	α_{2A}	Inhibition of cAMP production from ATP	Presynaptic, postsynaptic and astrocytes	19,29,30
		α_{2B}		Presynaptic	29
		α_{2C}		Presynaptic	29,31
Beta	β_1		Activation of cAMP-dependent pathway by activating adenylyl cyclase	Postsynaptic and astrocytes	32,33
	β_2			Presynaptic, postsynaptic and astrocytes	32–35
	β_3			Presynaptic and postsynaptic	32,33

[a]*α_{1A} was previously also described as α_{1C}, but turned out to be the same as α_{1A}.*[24,36,37]

only β_1-AR mRNA is expressed in the striatum. Finally, β_2 mRNA expression is the highest in olfactory bulb, the piriform cortex, the hippocampus, the thalamic interlaminar nuclei, and the cerebellar cortex.[15,16]

These different types of receptor also differ in their cellular localization in the brain. Indeed, α_1-AR is predominantly situated postsynaptically on nonnoradrenergic nerve terminals, but also distributed on presynaptic membranes and in glial cells in some cerebral structures.[17,18] The three α_2-AR subtypes are found postsynaptically. However, in the hippocampus, α_{2A}-ARs were described as mostly presynaptic and also present on astrocytes.[19] In the cerebellar cortex, β-AR show a postsynaptic localization on neurons.[20] In the dentate gyrus, β-ARs were also localized on dendritic profiles, occasionally on presynaptic profiles and frequently detected on astrocytes.[21] Indeed, β_1-AR and predominantly β_2-AR have been described in astrocytes from the gray mater, whereas astrocytes from the white mater expressed exclusively β_2-AR.[22,23] In vitro, essentially β_1- and β_2-, as well as α_1-AR are expressed in astrocytes in culture, also when cocultured with neurons.[24–26]

All the ARs are members of the seven transmembrane domain, G protein-coupled receptor superfamily. In general, β-AR is coupled via G_s

to adenylyl cyclase and their activation results in an increase in cyclic adenosine monophosphate (cAMP) levels. In contrast, α_2-AR is generally coupled to G_i proteins, thus reducing intracellular cAMP production by inhibiting some adenylyl cyclase isoforms. Finally, α_1-AR is commonly coupled to the phosphatidyl inositol/protein kinase C (PKC) intracellular pathway via G_q proteins. Moreover, G proteins can also interact with voltage-sensitive Ca^{2+}channels, which can be, depending of the type of ARs activated, either stimulated (through α_1- and β-AR activation) or inhibited (through α_2-AR activation) (for review see Refs. 27,28).

In summary, there are different types of ARs for NA; ARs are G protein-coupled receptors, which are essentially activated by NA, but also by adrenaline (see Table 9.1).

The Specific Action of Noradrenaline on Glycogen Metabolism

NA promotes glycogenolysis (breakdown of glycogen to glucose and L-lactate) in the cerebral cortex[38,39] (Fig. 9.1). A similar metabolic regulation has been described for vasoactive intestinal peptide (VIP).[38] This action of NA and VIP occurs in astrocytes as glycogen is almost exclusively present in this cell type in the brain.[40,41] The glycogenolytic effect of NA is mediated by β-ARs (G protein-coupled receptors) that increase the intracellular levels of cAMP.[42,43] The effect of VIP is likewise mediated by activation of the cAMP signaling cascade leading to an activation of protein kinase A (PKA). PKA phosphorylates phosphorylase kinase, which in turn activates glycogen phosphorylase (GP).[44] In addition, glycogen synthase (GlyS) is deactivated by PKA-mediated phosphorylation.

Given the anatomical features of the neuronal systems containing VIP and NA in the cortex, we have proposed that VIP bipolar neurons can regulate energy availability in cortical columns while the noradrenergic system, which spans across cortically distinct areas can metabolically prime the whole cortex.[38]

In addition to the immediate effect of NA, a delayed effect has been described. In a time frame of up to 9 h after the application of NA to astrocyte cultures, glycogen resynthesis is observed leading to 6–10 times higher glycogen levels as compared to levels determined before the application of NA.[45] The delayed action is also cAMP-dependent, but requires transcription factors of the cytosine–cytosine–adenosine–adenosine–thymidine motif recognition/enhancer binding protein family (termed C/EBP). More precisely, the C/EBP-β and the C/EBP-δ are required; both are also induced by NA.[46]

NA also regulates a key gene encoding protein important for the glycogen synthesis, a protein targeting to glycogen that is one of the protein phosphatase-1 glycogen targeting subunits.[47,48]

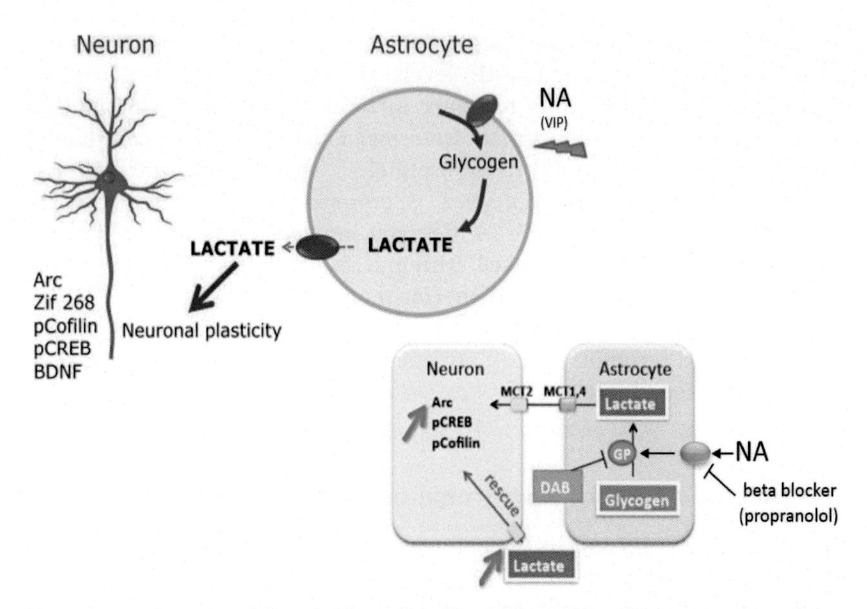

FIGURE 9.1 Summary of the role of noradrenaline in the context of astrocyte-neuron interactions. **Top panel**: The breakdown of glycogen in astrocytes promotes the release of lactate that can be taken up by neurons. Lactate can induce the expression of several plasticity genes in neurons. **Bottom panel**: Memory consolidation is impaired when the activity of glycogen phosphorylase (enzyme catalyzing the breakdown of glycogen) is inhibited (blocking the production of lactate in astrocytes). Exogenous lactate can rescue this inhibition. Similarly, monocarboxylate transporter (MCT)1 and MCT2 downregulations inhibit memory consolidation. Only the MCT1 and MCT4 downregulations are rescued by lactate, indicating the necessity of lactate transport into neurons for memory consolidation. If β-adrenergic receptor is blocked (by β-blockers such as propranonol), memory consolidation is also impaired, and is rescued by coadminitsration of lactate. *NA*, noradrenaline; *DAB*, 1,4-dideoxy-1,4-imino-D-arabinitol; *GP*, glycogen phosphorylase; *VIP*, vasointestinal peptide. *Modified from Boury-Jamot B, Carrard A, Martin JL, Halfon O, Magistretti PJ, Boutrel B. Disrupting astrocyte-neuron lactate transfer persistently reduces conditioned responses to cocaine.* Mol Psychiatry. *2016;21(8):1070–1076.*

Glucose homeostasis is also regulated by glucocorticoids, which include cortisol in human and corticosterone in rodents.[49] The glucocorticoids significantly reduce the glycogen synthesis provoked by NA.

NORADRENALINE AND MEMORY

Noradrenaline Action on Synaptic Plasticity: Neurons as Targets

As mentioned earlier, NA acts through β- and α-ARs. These receptors are expressed by hippocampal neurons, which is consistent with the dense innervation of these cells by noradrenergic fibers originating mainly from the LC (cf. Section "Noradrenergic Pathways and Receptors"). The hippocampus is a hub for episodic memory trace formation and recall.[50] At the cellular levels, memory formation is thought to involve LTP as well as long-term depression (LTD), two mechanisms of synaptic plasticity explaining the "tagging" of synapses related to a specific neuronal input. Since NA is released when emotional arousal and/or attention are mobilized,[51] it is not surprising that adrenergic agonists or antagonists interact with encoding and recall through their direct neuronal actions.[52] Indeed, a large body of evidences strongly suggest that NA, acting specifically through β-ARs, facilitates[53–57] and is even required for LTD mechanisms in hippocampus[58,59] and in amygdala.[60]

Noradrenaline Action on Synaptic Plasticity: Astrocytes as Targets

In addition to these neuronal effects, NA can act indirectly on synaptic plasticity via the astrocytes. Release of gliotransmitters, such as glutamate, ATP, or D-serine, can modulate synaptic strength.[61,62] Indeed, increases in intracellular astrocytic Ca^{2+} induce the vesicular release of these gliotransmitters and therefore, reflect astrocyte activation.[63–65] Recently, using transgenic mice expressing a Ca^{2+} indicator in astrocytes coupled with in vivo two-photon imagery, Paukert et al. showed that locomotion promotes a widespread activation of astrocyte networks over the cerebellar cortex.[66] Interestingly, this global stimulation of astrocytes was blocked by local application of an α_1-AR antagonist whereas β-AR antagonist had no effect.[66] In line with this, Pankratov and Lalo showed that activation of the astrocytic α_1-ARs triggers an intracellular increase of Ca^{2+} in hippocampal astrocytes promoting the release of ATP.[67] The release of ATP by astrocytes can induce a purinergic-receptor mediated current in pyramidal neurons and then facilitate LTP.[67] Taken together, these experiments indicate that adrenergic activation (likely through α_1-AR) of astrocytes, followed by intracellular Ca^{2+} increase and release of gliotransmitters, can promote LTP.

Role of Noradrenaline in Memory Paradigms

Among all different memory types, working memory reflects the ability to keep an information transiently to guide understanding, thinking, and planning. Working memory includes both storage and processing functions. Since the early seventies, evidences for a major role of the prefrontal cortex (PFC), in this memory processing have been accumulated.[68,69]

In this context, since PFC receives noradrenergic input from the LC, numerous studies have shown that NA has a critical beneficial influence on PFC functions, particularly through its postsynaptic action on α_2-AR.

Indeed, specific α_2-AR agonists, like guanfacine, improve the performance task that involves working memory. Studies emphasized the role of α_{2A}-AR activation and suggested that its beneficial action could be mediated by the inhibition of cAMP signaling in the PFC.[70,71] However, increased release of NA, as the one occurring in stress exposure conditions, would stimulate α_1-AR (which have a lower affinity to NA than α_2-AR) and impair working memory performance by suppressing delay-related firing in the PFC. Since this negative effect is reversed by pretreatment with lithium and PKC inhibitors, it has been suggested that it was mediated by the activation of the phosphatidyl inositol/PKC pathway. Finally, limited evidence suggests that β-AR activation could also influence PFC functions and working memory. Surprisingly, β_1- and β_2-AR that may be coupled to the same pathway, seem to have opposite effect on working memory. Indeed, endogenous activation of β_1-AR impairs PFC cognitive function whereas β_2-AR stimulation has a beneficial effect on working memory in aging animals (for review see Ref. 72). A direct participation of astrocytes in this specific type of memory has not been shown yet. However, a collaboration of astrocytes in this process cannot be excluded since it has been shown that a specific loss of astrocytes induced by local microinjection of the astrocyte toxin L-aminoadipate in the medial PFC would impair working memory[73] and induce depression-like behaviors in rats.[74]

MODULATION OF ASTROCYTIC ENERGY METABOLISM BY NORADRENALINE: IMPACT ON MEMORY

Brain Energy Metabolism and Memory

There are several lines of evidence indicating a positive effect of glucose ingestion on cognitive performance (for review see Ref. 75). As a matter of fact, hippocampal extracellular glucose, measured by microdialysis, exhibits a transient decrease of 30% at the beginning of cognitive tasks involving spatial working memory.[76] Such a decrease depends on the brain area since there is a simultaneous increase in striatum.[77] When the local extracellular glucose deficit is corrected by means of glucose infusion, the cognitive performances are improved.[78] Glucose may also enhance cognitive abilities through stimulation of glycolysis since pyruvate, the end-product of the glycolysis, exerts similar effects.[79] Levels of L-lactate, another end-product of glycolysis, were measured by Gold et al. in the extracellular space during a four-arm spontaneous alternation task. Interestingly, they observed an increase in L-lactate concentration that has been confirmed when extracellular levels of L-lactate and glucose were measured with a better time resolution (1 second) using biosensors. These opposite variations between glucose and L-lactate are in agreement with the massive release of L-lactate from astrocytes that is coupled with the uptake of glutamate during focal excitatory stimulation. This extracellular L-lactate is used, in addition to glucose, as energy substrate by the active synapses. These mechanisms constitute the so-called "astrocyte-neuron lactate shuttle" (ANLS).[40,80,81] The inhibition of long-term memory formation in a fear conditioning test by blocking the L-lactate transport from astrocytes to neurons using local antisense RNA against lactate transporter MCT1, MCT4, and MCT2 supports this view.[82]

The Central Role of Glycogen

Glycogenolysis Induced by β-AR Mediates Noradrenaline Effects on Memory

As described above (cf. Section "The Specific Action of Noradrenaline on Glycogen Metabolism"), glycogen, which constitutes the sole reserve of glucose in the brain,[83,84] is regulated by the balance between a synthesis and degradation that are respectively regulated by the activity of the GlyS and the GP. In brain slices of mice as well as in chicken, monoamines (NA, serotonin, dopamine, and histamine) induce a rapid glycogenolysis occurring within minutes following application.[38,41,85] For NA, this regulation is induced by β_2-ARs acting through G_s and cAMP mediated mechanisms.

In a series of experiments, the group of Marie Gibbs has shown the implication of NA-mediated glycogen regulation in memory process. In a discriminated bead task led in 1-day-old chicken, Gibbs et al. performed intracerebral injection of 1,4-dideoxy-1,4-imino-D-arabinitol (DAB) (100 pmol), an inhibitor of GP, which does not act on glycolysis, at different time points before and after training. When DAB was injected 5 min before training or 30 and 55 min posttraining, memory was impaired indicating that glycogenolysis was required for the acquisition and consolidation phases.[85] The memory consolidation observed following administration of β_2-AR agonist zinterol, which induces glycogenolysis, also supports the positive effects of glycogen mobilization on memory mechanisms.[86]

Using a model of fear conditioning in rats, Suzuki et al. also showed that blockade of glycogenolysis by intra-hippocampal injections of DAB or isofagomine (another inhibitor of the GP) injected 15 min before the test impaired the long-term memory.[82]

Astrocytic L-Lactate Derived From Glycogen is Crucial for Memory Mechanisms

Several lines of evidence indicate that astrocytic L-lactate is derived from glycogen[87–89] and the different experiments showing an active role for glycogen in memory mechanisms also support this view. In the paradigm used by Gibbs et al., where memory was blocked by DAB administration, intra-cerebral injections of L-lactate before or just after training were able to rescue memory.[90] Similarly, Suzuki et al. showed that intra-cerebral injection of L-lactate reverses the negative effects on fear conditioning induced by DAB or isofagomine administration in rats.[82] These results suggest that glycogen-derived L-lactate release from astrocytes is required for memory processing in neurons. Interestingly, glycogenolysis inhibition by DAB impaired in vivo LTP in hippocampus CA1 area of anesthetized rats. This points out the dependence of LTP mechanisms with respect to astrocyte L-lactate release.

Very recently, Gao et al. highlighted the role of astrocytic β_2-AR in the consolidation of a fear-based memory.[91] More precisely, injection into the dorsal hippocampus of propranolol, a nonspecific β_1- and β_2-AR antagonist, 15 min before fear conditioning had no effect on short-term memory tested 1 h after but impaired long-term memory evaluated 1 and 7 days after training. These results were reproduced following injection of a β_2-AR specific antagonist (ICI 118,551) but not after the injection of a selective β_1-AR antagonist (betaxolol).[91] The release of L-lactate measured in the same paradigm by microdialysis during 90 min (starting 10 min before the training) was also blocked by ICI 118,551 but not by betaxolol. These results indicate that β_2-AR subtypes are critical for the action of NA on fear-related memory, which involved a L-lactate release from astrocytes. Interestingly the impairment of memory consolidation caused by the β-AR antagonist propranolol in this task was fully reversed by co administration of L-lactate in the hippocampus, thus further pointing at a key role of NA-evoked glycogenolysis in memory consolidation.

In the same line, recent results by Boury-Jamot et al. indicate that glycogen-derived L-lactate is required for acquisition as well as for long-term conditioning in a cocaine-induced conditioned place preference (CPP) test.[92] As a matter of fact, long-term conditioning was impaired when glycogen mobilization was blocked by injection of DAB into the baso-lateral amygdala 15 min before acquisition sessions or 15 min before posttest sessions. When L-lactate was coadministered with DAB, drug memory was rescued. These results have been confirmed by a very similar study where micro-injections of DAB into the amygdala markedly impaired cocaine-induced CPP.[93] In this last study the drug-associated context memory was also impaired by the blockade of the expression of MCT1 or MCT2 L-lactate transporters by local injection of antisense oligonucleotides. This effect was reversed by the L-lactate injections but only when memory deficiencies were induced by MCT1 mRNA antisense but not those induced by MCT2 mRNA antisense, which impair the neuronal use of L-lactate by neurons.[93] These results are in agreement with the memory-promoting effects of L-lactate, which likely involve the stimulation of the ERK pathway and of the transcription.[92,94]

Influence of the Sleep–Wake Cycle

Sleep and Memory

The inhibition of sensory inputs to the cortex and the decrease in neuromodulatory activity during slow wave sleep (SWS) make this period ideal for off-line memory consolidation process. Indeed, learning and memory are markedly altered by sleep loss.[95-97] Different mechanisms for memory consolidation take place during SWS and rapid eye movement (REM) sleep. Spindles, slow waves and hippocampal ripples, three features of the SWS are specifically related to consolidation process of declarative and procedural memories involving perceptual skills. More precisely, hippocampal CA1 pyramidal neurons display bursts of high-frequency discharge (ripples, 100–200 Hz) that are temporally coupled with cortical spindles.[96] This phenomenon reflects the hippocampal-cortical dialog at the basis of memory consolidation, a view supported by the increase in ripple activity as well as in spindles density observed after learning.[98]

Noradrenaline and Sleep-Related Memory Mechanisms

As we mentioned above (see Section "Noradrenergic Pathways and Receptors") the noradrenergic neurons (particularly in the LC) participate in sleep–wake alternation with a tonic firing with bursts of activity during wakefulness and a global and marked decrease in discharge rate during sleep period. However, LC neurons exhibit discrete phases of activity throughout SWS.[99] In this context the participation of NA release on the sleep-related memory mechanisms has been investigated. Gais et al. tested this hypothesis using an odor discrimination task, which is known to be improved by posttraining sleep in humans. They observed a decrease in memory performance following blockade of NA activity by clonidine and an opposite effect following administration of riboxetine, an inhibitor of the NA reuptake.[100] However, memory consolidation was impaired, when LC was electrically stimulated at high frequency (50–100 Hz) during the ripple activity in rats that underwent a spatial discrimination task 1 h before.[101] These experiments suggest that the noradrenergic system plays a positive role in sleep-related memory consolidation mechanisms but that the timing of NA release is of crucial importance for this effect. It should be noted that REM sleep, which corresponds to periods of LC silencing, did not interfere with these effects.

The Role of Astrocytes in Sleep-Related Memory Mechanisms

To our knowledge the participation of astrocytes on the effects of NA on sleep-related memory consolidation mechanisms was not really investigated. The emerging role of astrocytes in sleep regulation[63,102] and in neuro-metabolic coupling associated to vigilance states should instigate new experiments to define their involvement in sleep-related memory consolidation mechanisms.

CONCLUSIONS

The positive impact of NA in cortical and hippocampal memory mechanisms is supported by a large amount of data in rodents as well as in human. This memory-promoting effect is the result of different mechanisms. In addition to its "traditional" action on ARs, which modulate synaptic plasticity through neurons, a growing number of studies indicates that NA also promotes memory through astrocytes. This can be achieved by different ways. In response to neuronal activity, astrocytes release gliotransmitters, which can ultimately promote LTP. By another way, NA also promotes astrocytic glycogenolysis, which in turn triggers the release of L-lactate that can be used as additional source of energy by neurons (ANLS) during memory tasks. In addition, the release of L-lactate induces the expression of several plasticity genes known to be involved in memory mechanisms.

In conclusion, NA is an important neurotransmitter that plays a central role in the regulation of the energy metabolism. In the brain, it is a key mediator implicated in the memory formation and retrieval, as well as in sleep-related memory mechanisms notably by its actions through astrocytes. The strive for new therapeutic treatments targeting astrocytes (or directly the ANLS) could also be of interest for diseases related to NA dysfunction.[103]

ABBREVIATIONS

ANLS astrocyte-neuron lactate shuttle
AR adrenergic receptor
Ca^{2+} calcium
cAMP cyclic adenosine monophosphate
CNS central nervous system
DAB 1,4-dideoxy-1,4-imino-D-arabinitol
GlyS glycogen synthase
GP glycogen phosphorylase
LC locus coeruleus
LTD long-term depression
LTP long-term potentiation
MCT monocarboxylate transporter
NA noradrenaline/norepinephrine
PFC prefrontal cortex
PKA protein kinase A
PKC protein kinase C

REM rapid eye movement
SWS slow wave sleep
VIP vasointestinal peptide

REFERENCES

1. Dahlstroem A, Fuxe K. Evidence for the existence of monoamine-containing neurons in the central nervous system. I. Demonstration of monoamines in the cell bodies of brain stem neurons. *Acta Physiol Scand Suppl.* 1964;232(suppl):231–255.
2. Robertson SD, Plummer NW, de Marchena J, Jensen P. Developmental origins of central norepinephrine neuron diversity. *Nature Neurosci.* 2013;16(8):1016–1023.
3. Agster KL, Mejias-Aponte CA, Clark BD, Waterhouse BD. Evidence for a regional specificity in the density and distribution of noradrenergic varicosities in rat cortex. *J Comp Neurol.* 2013;521(10):2195–2207.
4. Berridge CW, Waterhouse BD. The locus coeruleus–noradrenergic system: modulation of behavioral state and state-dependent cognitive processes. *Brain Res Rev.* 2003;42(1):33–84.
5. Foote SL, Bloom FE, Aston-Jones G. Nucleus locus ceruleus: new evidence of anatomical and physiological specificity. *Physiol Rev.* 1983;63(3):844–914.
6. Aston-Jones G, Cohen JD. Adaptive gain and the role of the locus coeruleus-norepinephrine system in optimal performance. *J Comp Neurol.* 2005;493(1):99–110.
7. Florin-Lechner SM, Druhan JP, Aston-Jones G, Valentino RJ. Enhanced norepinephrine release in prefrontal cortex with burst stimulation of the locus coeruleus. *Brain Res.* 1996;742(1–2):89–97.
8. Howells FM, Stein DJ, Russell VA. Synergistic tonic and phasic activity of the locus coeruleus norepinephrine (LC-NE) arousal system is required for optimal attentional performance. *Metab Brain Dis.* 2012;27(3):267–274.
9. Espana RA, Schmeichel BE, Berridge CW. Norepinephrine at the nexus of arousal, motivation and relapse. *Brain Res.* 2016;1641(Pt B):207–216.
10. Fuxe K, Agnati LF, Marcoli M, Borroto-Escuela DO. Volume transmission in central dopamine and noradrenaline neurons and its astroglial targets. *Neurochem Res.* 2015;40(12):2600–2614.
11. Beaudet A, Descarries L. The monoamine innervation of rat cerebral cortex: synaptic and nonsynaptic axon terminals. *Neuroscience.* 1978;3(10):851–860.
12. Day HEW, Campeau S, Watson Jr SJ, Akil H. Distribution of α1a-, α1b- and α1d-adrenergic receptor mRNA in the rat brain and spinal cord. *J Chem Neuroanat.* 1997;13(2):115–139.
13. Scheinin M, Lomasney JW, Hayden-Hixson DM, et al. Distribution of α2-adrenergic receptor subtype gene expression in rat brain. *Mol Brain Res.* 1994;21(1–2):133–149.
14. Wang R, Macmillan LB, Fremeau Jr RT, Magnuson MA, Lindner J, Limbird LE. Expression of α2-adrenergic receptor subtypes in the mouse brain: evaluation of spatial and temporal information impaires by 3 kb OF 5′ regulatory sequence for the α2a A-receptor gene in transgenic animals. *Neuroscience.* 1996;74(1):199–218.
15. Nicholas AP, Pieribone VA, Hökfelt T. Cellular localization of messenger RNA for beta-1 and beta-2 adrenergic receptors in rat brain: an in situ hybridization study. *Neuroscience.* 1993;56(4):1023–1039.
16. Summers RJ, Papaioannou M, Harris S, Evans BA. Expression of β3-adrenoceptor mRNA in rat brain. *Brit J Pharmacol.* 1995;116(6):2547–2548.

17. Nakadate K, Imamura K, Watanabe Y. Cellular and subcellular localization of alpha-1 adrenoceptors in the rat visual cortex. *Neuroscience*. 2006;141(4):1783–1792.
18. Mitrano DA, Schroeder JP, Smith Y, et al. Alpha-1 adrenergic receptors are localized on presynaptic elements in the nucleus accumbens and regulate mesolimbic dopamine transmission. *Neuropsychopharmacology*. 2012;37(9):2161–2172.
19. Milner TA, Lee A, Aicher SA, Rosin DL. Hippocampal α2A-adrenergic receptors are located predominantly presynaptically but are also found postsynaptically and in selective astrocytes. *J Comp Neurol*. 1998;395(3):310–327.
20. Strader CD, Pickel VM, Joh TH, et al. Antibodies to the beta-adrenergic receptor: attenuation of catecholamine-sensitive adenylate cyclase and demonstration of postsynaptic receptor localization in brain. *Proc Natl Acad Sci USA*. 1983;80(7):1840–1844.
21. Milner TA, Shah P, Pierce JP. β-Adrenergic receptors primarily are located on the dendrites of granule cells and interneurons but also are found on astrocytes and a few presynaptic profiles in the rat dentate gyrus. *Synapse*. 2000;36(3):178–193.
22. Cash R, Raisman R, Lanfumey L, Ploska A, Agid Y. Cellular localization of adrenergic receptors in rat and human brain. *Brain Res*. 1986;370(1):127–135.
23. Aoki C. β-Adrenergic receptors: astrocytic localization in the adult visual cortex and their relation to catecholamine axon terminals as revealed by electron microscopic immunocytochemistry. *J Neurosci*. 1992;12(3):781–792.
24. Lerea LS, McCarthy KD. Neuron-associated astroglial cells express beta- and alpha 1-adrenergic receptors in vitro. *Brain Res*. 1990;521(1–2):7–14.
25. Juric DM, Loncar D, Carman-Krzan M. Noradrenergic stimulation of BDNF synthesis in astrocytes: mediation via alpha1- and beta1/beta2-adrenergic receptors. *Neurochem Int*. 2008;52(1-2):297–306.
26. Kimelberg HK. Receptors on astrocytes—what possible functions? *Neurochem Int*. 1995;26(1):27–40.
27. Insel PA. Adrenergic receptors, G proteins, and cell regulation: implications for aging research. *Exp Gerontol*. 1993;28(4–5):341–348.
28. Reznikoff GA, Manaker S, Rhodes CH, Winokur A, Rainbow TC. Localization and quantification of beta-adrenergic receptors in human brain. *Neurology*. 1986;36(8):1067–1073.
29. Birnbaumer L. Expansion of signal transduction by G proteins. The second 15 years or so: from 3 to 16 alpha subunits plus betagamma dimers. *Biochim Biophys Acta*. 2007;1768 (4):772–793.
30. Lee A, Rosin DL, Van Bockstaele EJ. alpha2A-adrenergic receptors in the rat nucleus locus coeruleus: subcellular localization in catecholaminergic dendrites, astrocytes, and presynaptic axon terminals. *Brain Res*. 1998;795(1–2):157–169.
31. Lee A, Rosin DL, Van Bockstaele EJ. Ultrastructural evidence for prominent postsynaptic localization of alpha2C-adrenergic receptors in catecholaminergic dendrites in the rat nucleus locus coeruleus. *J Comp Neurol*. 1998;394(2):218–229.
32. Zaugg M, Schaub MC. Beta3-adrenergic receptor subtype signaling in senescent heart: nitric oxide intoxication or "endogenous" beta blockade for protection? *Anesthesiology*. 2008;109(6):956–959.
33. Nature Reviews Drug Discovery GQP. The state of GPCR research in 2004. *Nat Rev Drug Discov*. 2004;3(7):577–626.
34. Dong JH, Chen X, Cui M, Yu X, Pang Q, Sun JP. beta2-adrenergic receptor and astrocyte glucose metabolism. *J Mol Neurosci*. 2012;48(2):456–463.
35. Cohen Z, Molinatti G, Hamel E. Astroglial and vascular interactions of noradrenaline terminals in the rat cerebral cortex. *J Cereb Blood Flow Metab*. 1997;17(8):894–904.

36. Schwinn DA, Page SO, Middleton JP, et al. The alpha 1C-adrenergic receptor: characterization of signal transduction pathways and mammalian tissue heterogeneity. *Mol Pharmacol.* 1991;40(5):619–626.
37. Langer SZ. Nomenclature and state of the art on alpha1-adrenoceptors. *Eur Urol.* 1998;33 (suppl 2):2–6.
38. Magistretti PJ, Morrison JH, Shoemaker WJ, Sapin V, Bloom FE. Vasoactive intestinal polypeptide induces glycogenolysis in mouse cortical slices: a possible regulatory mechanism for the local control of energy metabolism. *Proc Natl Acad Sci USA.* 1981;78 (10):6535–6539.
39. Magistretti PJ, Morrison JH. Noradrenaline- and vasoactive intestinal peptide-containing neuronal systems in neocortex: functional convergence with contrasting morphology. *Neuroscience.* 1988;24(2):367–378.
40. Magistretti PJ, Allaman I. A cellular perspective on brain energy metabolism and functional imaging. *Neuron.* 2015;86(4):883–901.
41. Sorg O, Magistretti PJ. Characterization of the glycogenolysis elicited by vasoactive intestinal peptide, noradrenaline and adenosine in primary cultures of mouse cerebral cortical astrocytes. *Brain Res.* 1991;563(1–2):227–233.
42. Seeds NW, Gilman AG. Norepinephrine stimulated increase of cyclic AMP levels in developing mouse brain cell cultures. *Science.* 1971;174(4006):292.
43. Cummins CJ, Lust WD, Passonneau JV. Regulation of glycogenolysis in transformed astrocytes in vitro. *J Neurochem.* 1983;40(1):137–144.
44. Bollen M, Keppens S, Stalmans W. Specific features of glycogen metabolism in the liver. *Biochem J.* 1998;336(Pt 1):19–31.
45. Sorg O, Magistretti PJ. Vasoactive intestinal peptide and noradrenaline exert long-term control on glycogen levels in astrocytes: blockade by protein synthesis inhibition. *J Neurosci.* 1992;12(12):4923–4931.
46. Cardinaux JR, Magistretti PJ. Vasoactive intestinal peptide, pituitary adenylate cyclase-activating peptide, and noradrenaline induce the transcription factors CCAAT/enhancer binding protein (C/EBP)-beta and C/EBP delta in mouse cortical astrocytes: involvement in cAMP-regulated glycogen metabolism. *J Neurosci.* 1996;16(3):919–929.
47. Allaman I, Pellerin L, Magistretti PJ. Protein targeting to glycogen mRNA expression is stimulated by noradrenaline in mouse cortical astrocytes. *Glia.* 2000;30(4):382–391.
48. Allaman I, Pellerin L, Magistretti PJ. Glucocorticoids modulate neurotransmitter-induced glycogen metabolism in cultured cortical astrocytes. *J Neurochem.* 2004;88 (4):900–908.
49. McMahon M, Gerich J, Rizza R. Effects of glucocorticoids on carbohydrate metabolism. *Diab/Metab Res Rev.* 1988;4(1):17–30.
50. Battaglia FP, Benchenane K, Sirota A, Pennartz CM, Wiener SI. The hippocampus: hub of brain network communication for memory. *Trends Cogn Sci.* 2011;15(7):310–318.
51. Aston-Jones G, Rajkowski J, Cohen J. *Locus coeruleus and regulation of behavioral flexibility and attention. Prog Brain Res.* 126. 2000165–182.
52. O'Dell TJ, Connor SA, Gelinas JN, Nguyen PV. Viagra for your synapses: enhancement of hippocampal long-term potentiation by activation of beta-adrenergic receptors. *Cell Signal.* 2010;22(5):728–736.
53. Gelinas JN, Banko JL, Hou L, et al. ERK and mTOR signaling couple β-adrenergic receptors to translation initiation machinery to gate induction of protein synthesis-dependent long-term potentiation. *J Biol Chem.* 2007;282(37):27527–27535.

54. Gelinas JN, Nguyen PV. β-Adrenergic receptor activation facilitates induction of a protein synthesis-dependent late phase of long-term potentiation. *J Neurosci.* 2005;25 (13):3294–3303.
55. Nguyen PV, Duffy SN, Young JZ. Differential maintenance and frequency-dependent tuning of LTP at hippocampal synapses of specific strains of inbred mice. *J Neurophysiol.* 2000;84(5):2484–2493.
56. Qian H, Matt L, Zhang M, et al. β2-Adrenergic receptor supports prolonged theta tetanus-induced LTP. *J Neurophysiol.* 2012;107(10):2703.
57. Winder DG, Martin KC, Muzzio IA, et al. ERK plays a regulatory role in induction of LTP by theta frequency stimulation and its modulation by β-adrenergic receptors. *Neuron.* 1999;24(3):715–726.
58. Hansen N, Manahan-Vaughan D. Locus coeruleus stimulation facilitates long-term depression in the dentate gyrus that requires activation of β-adrenergic receptors. *Cereb Cortex.* 2015;25(7):1889–1896.
59. Lemon N, Aydin-Abidin S, Funke K, Manahan-Vaughan D. Locus coeruleus activation facilitates memory encoding and induces hippocampal LTD that depends on β-adrenergic receptor activation. *Cereb Cortex.* 2009;19(12):2827–2837.
60. Clem RL, Huganir RL. Norepinephrine enhances a discrete form of long-term depression during fear memory storage. *J Neurosci.* 2013;33(29):11825–11832.
61. Gordon GRJ, Iremonger KJ, Kantevari S, Ellis-Davies GCR, MacVicar BA, Bains JS. Astrocyte-mediated distributed plasticity at hypothalamic glutamate synapses. *Neuron.* 2009;64(3):391–403.
62. Pougnet J-T, Toulme E, Martinez A, Choquet D, Hosy E, Boué-Grabot E. ATP P2X receptors downregulate AMPA receptor trafficking and postsynaptic efficacy in hippocampal neurons. *Neuron.* 2014;83(2):417–430.
63. Halassa MM, Haydon PG. Integrated brain circuits: astrocytic networks modulate neuronal activity and behavior. *Annu Rev Physiol.* 2010;72(1):335–355.
64. Sahlender DA, Savtchouk I, Volterra A. What do we know about gliotransmitter release from astrocytes? *Philos Trans Royal Soc B: Biol Sci.* 2014;369:1654.
65. Wenker I. An active role for astrocytes in synaptic plasticity? *J Neurophysiol.* 2010;104 (3):1216–1218.
66. Paukert M, Agarwal A, Cha J, Doze Van A, Kang Jin U, Bergles Dwight E. Norepinephrine controls astroglial responsiveness to local circuit activity. *Neuron.* 2014;82(6):1263–1270.
67. Pankratov Y, Lalo U. Role for astroglial α1-adrenoreceptors in gliotransmission and control of synaptic plasticity in the neocortex. *Front Cell Neurosci.* 2015;9:230.
68. Fuster JM, Alexander GE. Neuron activity related to short-term memory. *Science.* 1971;173(3997):652–654.
69. Goldman-Rakic PS. Regional and cellular fractionation of working memory. *Proc Natl Acad Sci USA.* 1996;93(24):13473–13480.
70. Franowicz JS, Kessler LE, Borja CMD, Kobilka BK, Limbird LE, Arnsten AFT. Mutation of the α2A-adrenoceptor impairs working memory performance and annuls cognitive enhancement by guanfacine. *J Neurosci.* 2002;22(19):8771–8777.
71. Wang M, Ramos BP, Paspalas CD, et al. Alpha2A-adrenoceptors strengthen working memory networks by inhibiting cAMP-HCN channel signaling in prefrontal cortex. *Cell.* 2007;129(2):397–410.
72. Ramos BP, Arnsten AFT. Adrenergic pharmacology and cognition: focus on the prefrontal cortex. *Pharmacol Therap.* 2007;113(3):523–536.

73. Lima A, Sardinha VM, Oliveira AF, et al. Astrocyte pathology in the prefrontal cortex impairs the cognitive function of rats. *Mol Psychiatry.* 2014;19(7):834–841.
74. Banasr M, Duman RS. Glial loss in the prefrontal cortex is sufficient to induce depressive-like behaviors. *Biol Psychiatry.* 2008;64(10):863–870.
75. Messier C. Glucose improvement of memory: a review. *Eur J Pharmacol.* 2004;490(1–3):33.
76. McNay EC, Fries TM, Gold PE. Decreases in rat extracellular hippocampal glucose concentration associated with cognitive demand during a spatial task. *Proc Natl Acad Sci USA.* 2000;97(6):2881–2885.
77. McNay EC, McCarty RC, Gold PE. Fluctuations in brain glucose concentration during behavioral testing: dissociations between brain areas and between brain and blood. *Neurobiol Learn Mem.* 2001;75(3):325–337.
78. Stefani MR, Gold PE. Intra-septal injections of glucose and glibenclamide attenuate galanin-induced spontaneous alternation performance deficits in the rat. *Brain Res.* 1998;813(1):50–56.
79. Krebs DL, Parent MB. Hippocampal infusions of pyruvate reverse the memory-impairing effects of septal muscimol infusions. *Eur J Pharmacol.* 2005;520(1–3):91–99.
80. Pellerin L, Magistretti PJ. Glutamate uptake into astrocytes stimulates aerobic glycolysis: a mechanism coupling neuronal activity to glucose utilization. *Proc Natl Acad Sci USA.* 1994;91(22):10625–10629.
81. Magistretti PJ. Imaging brain aerobic glycolysis as a marker of synaptic plasticity. *Proc Natl Acad Sci USA.* 2016;113(26):7015–7016.
82. Suzuki A, Stern Sarah A, Bozdagi O, et al. Astrocyte-neuron lactate transport is required for long-term memory formation. *Cell.* 2011;144(5):810–823.
83. Brown AM. Brain glycogen re-awakened. *J Neurochem.* 2004;89(3):537–552.
84. Phelps CH. Barbiturate-induced glycogen accumulation in brain. An electron microscopic study. *Brain Res.* 1972;39(1):225–234.
85. Gibbs ME, O'Dowd BS, Hertz E, Hertz L. Astrocytic energy metabolism consolidates memory in young chicks. *Neuroscience.* 2006;141(1):9–13.
86. Gibbs ME. Role of glycogenolysis in memory and learning: regulation by noradrenaline, serotonin and ATP. *Front Integrat Neurosci.* 2016;9:70.
87. Evans RD, Brown AM, Ransom BR. Glycogen function in adult central and peripheral nerves. *J Neurosci Res.* 2013;91(8):1044–1049.
88. Sickmann H, Schousboe A, Fosgerau K, Waagepetersen H. Compartmentation of lactate originating from glycogen and glucose in cultured astrocytes. *Neurochem Res.* 2005;30(10):1295.
89. Tekkök SB, Brown AM, Westenbroek R, Pellerin L, Ransom BR. Transfer of glycogen-derived lactate from astrocytes to axons via specific monocarboxylate transporters supports mouse optic nerve activity. *J Neurosci Res.* 2005;81(5):644–652.
90. Gibbs ME, Lloyd HGE, Santa T, Hertz L. Glycogen is a preferred glutamate precursor during learning in 1-day-old chick: biochemical and behavioral evidence. *J Neurosci Res.* 2007;85(15):3326–3333.
91. Gao V, Suzuki A, Magistretti PJ, et al. Astrocytic beta2-adrenergic receptors mediate hippocampal long-term memory consolidation. *Proc Natl Acad Sci USA.* 2016;113(30):8526–8531.
92. Boury-Jamot B, Carrard A, Martin JL, Halfon O, Magistretti PJ, Boutrel B. Disrupting astrocyte-neuron lactate transfer persistently reduces conditioned responses to cocaine. *Mol Psychiatry.* 2016;21(8):1070–1076.

93. Zhang Y, Xue Y, Meng S, et al. Inhibition of lactate transport erases drug memory and prevents drug relapse. *Biol Psychiatry.* 2016;79(11):928–939.
94. Yang J, Ruchti E, Petit J-M, et al. Lactate promotes plasticity gene expression by potentiating NMDA signaling in neurons. *Proc Natl Acad Sci USA.* 2014;111 (33):12228–12233.
95. Killgore WDS. Effects of sleep deprivation on cognition. In: Kerkhof GA, Van Dongen HPA, eds. *Progress in Brain Research.* Vol 185. Amsterdam: Elsevier; 2010:105–129.
96. Rasch B, Born J. About sleep's role in memory. *Physiol Rev.* 2013;93:681–766.
97. Petit JM, Gyger J, Burlet-Godinot S, Fiumelli H, Martin JL, Magistretti PJ. Genes involved in the astrocyte-neuron lactate shuttle (ANLS) are specifically regulated in cortical astrocytes following sleep deprivation in mice. *Sleep.* 2013;36(10):1445–1458.
98. Eschenko O, Ramadan W, Mölle M, Born J, Sara SJ. Sustained increase in hippocampal sharp-wave ripple activity during slow-wave sleep after learning. *Learn Mem.* 2008;15(4):222–228.
99. Aston-Jones G, Bloom FE. Activity of norepinephrine-containing locus coeruleus neurons in behaving rats anticipates fluctuations in the sleep-waking cycle. *J Neurosci.* 1981;1(8):876–886.
100. Gais S, Rasch B, Dahmen JC, Sara S, Born J. The memory function of noradrenergic activity in non-REM sleep. *J Cogn Neurosci.* 2011;23(9):2582–2592.
101. Novitskaya Y, Sara SJ, Logothetis NK, Eschenko O. Ripple-triggered stimulation of the locus coeruleus during post-learning sleep disrupts ripple/spindle coupling and impairs memory consolidation. *Learn Mem.* 2016;23(5):238–248.
102. Petit JM, Magistretti PJ. Regulation of neuron-astrocyte metabolic coupling across the sleep-wake cycle. *Neuroscience.* 2016;323:135–156.
103. Elsayed M, Magistretti PJ. A new outlook on mental illnesses: glial involvement beyond the glue. *Front Cell Neurosci.* 2015;9:468.

Chapter 10

Hippocampal Noradrenaline Regulates Spatial Working Memory in the Rat

Rosario Gulino[1], Anna Kostenko[2], Gioacchino de Leo[2], Serena Alexa Emmi[2], Domenico Nunziata[2] and Giampiero Leanza[2,✉]
[1]University of Catania, Catania, Italy, [2]University of Trieste, Trieste, Italy

Chapter Outline

Introduction 202
Methods 203
Subjects and Experimental Design 203
Lesion and Transplantation Surgery 204
Behavioral Tests 204
Morris Water Maze 205
Postmortem Analyses 206
Microscopic Analysis and Quantitative Evaluation 207
Results 208
General Observations 208
Behavioral Analyses 209
Morphological Analyses 212
Effects of the Lesion and of Transplants 212
Discussion 214
Effects of the Anti-DBH-Saporin Lesion 214
Effects of Transplants 215
Conclusions 216
Abbreviations 216
Acknowledgments 217
References 217

ABSTRACT
Degeneration of noradrenergic neurons in the locus coeruleus and loss of fiber terminals in the neocortical and hippocampal target regions represent prominent and early features of Alzheimer's disease, however, whether these events are causally linked to cognitive decline in Alzheimer's disease is still unclear. In the present study, the noradrenergic contribution to the regulation of hippocampus-dependent spatial learning and memory was investigated. Postnatal day 4 rats underwent selective immunolesioning of hippocampal noradrenergic afferents and, 4 days later, the bilateral intrahippocampal implantation of locus coeruleus

✉Correspondence address
E-mail: gleanza@units.it

Noradrenergic Signaling and Astroglia. DOI: http://dx.doi.org/10.1016/B978-0-12-805088-0.00010-4
© 2017 Elsevier Inc. All rights reserved.

noradrenergic neuroblasts. Starting from 4 weeks and up to about 9 months post-surgery, sensory-motor and spatial navigation abilities were evaluated, followed by postmortem tissue analyses. All animals in the Control, Lesion, and Lesion + Transplant groups exhibited normal sensory-motor function and were equally efficient in the reference memory version of the water maze task, whereas working memory abilities were seen consistently impaired in the Lesion-only rats. Notably, the noradrenergic reinnervation promoted by the grafted progenitors reinstated a fairly normal working memory performance, suggesting a primary role for coeruleo-hippocampal noradrenergic inputs in the maintenance of specific aspects of cognition.

Keywords: Alzheimer's disease; noradrenaline; locus coeruleus; immunolesion; transplantation; water maze; working memory; rat

INTRODUCTION

The pontine locus coeruleus (LC) nucleus is the main source of noradrenergic inputs to the entire central nervous system, and has been implicated in a variety of physiological functions, including sleep/wakefulness, attention and alertness, anxiety and depression, and cognitive performance.[1–4] Accumulating evidence[5,6] has shown that the normal anatomy and functions of this nucleus undergo profound alterations in Alzheimer's disease (AD). In fact, extensive loss of noradrenergic LC neurons, and a marked decrease of cortical noradrenaline (NA) levels are common pathological findings in autoptic specimens from AD patients. They have been observed closely associated to the severity of the cognitive impairments, when present, and may occur at the very early stages of the disease.[7–12] Overall, these data would substantiate an important contribution of noradrenergic loss to the pathophysiology of AD. However, a general consensus on the functional role of the noradrenergic projection system in the regulation of specific cognitive domains has not yet been achieved, which is essential, if noradrenergic therapies are to be proposed in a clinical setting.[6,13] Likewise the development of preclinical models of cognitive impairments based on noradrenergic dysfunction have yielded conflicting results, partly due to the poor selectivity of the procedures used to target LC neurons.[14–16] A promising lesioning tool to address this issue is the antidopamine-beta-hydroxylase-saporin (anti-DBH-saporin).[17] This immunotoxin is based on the conjugation of saporin, a powerful ribosome-inactivating lectin from the plant *Saponaria officinalis*[18,19], to a monoclonal antibody raised against dopamine-beta-hydroxylase (DBH), a biosynthetic enzyme that, in addition to its main cytosolic localization, is also expressed onto the membrane surface of noradrenergic neurons.[20,21] Due to its structure, the saporin cannot enter the cell,[22] but when coupled to a carrier molecule (such as an antibody) it is able to specifically bind a surface protein (in this case DBH); the toxin can easily gain access to the cytosol and bind to the ribosome, interfering with protein synthesis, and inducing cell death.[23]

Using a series of hippocampus-dependent spatial navigation tasks in rats lesioned with anti-DBH-saporin, we have recently observed working memory deficits whose severity correlated with the extent of hippocampal noradrenergic depletion, whereas reference memory abilities were largely unaffected (Coradazzi et al., unpublished results). The hippocampus may therefore represent a candidate area where noradrenergic inputs converge to play a role in the regulation of working memory function. If so, restoration of hippocampal noradrenergic neurotransmission by, e.g., locally implanted noradrenergic-rich tissue should promote an amelioration of the cognitive performance disrupted by the lesion, but this issue has not been addressed yet.

Thus, in order to further substantiate the possible contribution of hippocampal noradrenergic innervation to aspects of spatial learning and memory, the present study examined the effects of selective hippocampal noradrenergic depletions upon performance in hippocampus-dependent spatial navigation tasks, and the possible restorative effects of intrahippocampal grafts of LC-derived neural progenitors.

METHODS

Subjects and Experimental Design

A total of 32 equally distributed male and female Sprague-Dawley rats from three different litters (provided by the animal facility at the University of Trieste) maintained in high efficiency, particulate air-filtered, double-decker cage units (Tecniplast, Italy), were used. The pups were randomly allocated into four groups: unoperated controls ($n = 5$), vehicle-injected ($n = 5$), lesioned controls ($n = 10$), and lesioned and grafted ($n = 12$). Litters (one per cage) were fostered by the mothers until weaning at 21 days of age and then housed in groups of four under standard conditions of light, temperature and humidity with ad libitum access to food and water. Evaluation of growth properties and motor skills for all animals has begun at about 4 weeks postsurgery and carried out monthly up to about 32 weeks of age, when the animals underwent the sequential administration of tests specifically designed to evaluate sensory-motor, as well as spatial navigation abilities. Upon conclusion of the behavioral testing, at about 36 weeks postlesioning and grafting, all the animals were perfused and the brains were processed for immunohistochemistry and quantitative microscopic analyses.

All efforts were made to minimize the number of rats used and their suffering. Experimental procedures were approved by the Ethical Committee at the University of Trieste and were performed following the Italian Guidelines for Animal Care (D.L. 116/92 and 26/2014) which are in compliance with the European Communities Council Directives (2010/63/EU), and the "Guide for the care and use of Laboratory animals" (8th ed. 2011) issued by the National Institutes of Health.

Lesion and Transplantation Surgery

Selective noradrenergic denervation of the developing hippocampus was performed on 4-day-old (postnatal day, PD, 4) pups under hypothermic anesthesia. The anti-DBH-saporin immunotoxin (purchased from Advanced Targeting Systems, CA, USA) was used at a dose of 0.08 μg dissolved in sterile phosphate-buffered saline (PBS). The immunotoxin or the vehicle solution alone was injected using a 1 μL microsyringe in a volume of 1 μL/side into the hippocampal formation at the following coordinates (in mm, relative to bregma and outer skull surface): $AP = -1.0$, $L = \pm 1.4$, $V = -2.2$, allowing 3 min for diffusion before the cannula was retracted.

Four days later (at PD8), a noradrenergic-rich cell suspension was prepared as previously described[24,25] following a modified protocol based on the cell suspension technique.[26] Briefly the developing LC was bilaterally dissected from the dorsolateral pons of 13- to 14-day-old donor rat fetuses (crown-to-rump length, 9–13 mm) of the same strain as the graft recipients, and collected in cold Dulbecco's modified Eagle medium (DMEM, Life Technologies, Italy). The tissue was mechanically dissociated with no prior enzymatic digestion, and then suspended in DMEM at a final volume of 4 μL/LC, to have about 30,000 cells/μL with an 85%–90% viability, as assessed by the trypan blue exclusion method. Transplantation surgery was performed on randomly selected lesioned pups, deeply reanesthetized by cooling them on ice and carefully fixed a hypothermic miniaturized device (Stoelting Co., Wood Dale, IL, USA), with atraumatic ear bars and tooth bar. The skull was exposed, and small holes were produced in the cranial bone with a dental drill. Using a 10 μL Hamilton microsyringe, fixed on the stereotaxic apparatus, 2.0 μL of the noradrenaline-rich cell suspension were bilaterally injected into the hippocampal formation, at the following coordinates (in mm): $AP = -3.5$; $L = \pm 2.2$; $V = 2.6$, using the same references as above. The suspension was injected as a single deposit, at a speed of 1 μL/2 min, with a further 3 min allowed for diffusion. After each surgery, the animals were allowed to fully recovery and reacquire normal body temperature, prior to being returned to the mothers.

Behavioral Tests

All testings were consistently carried out between 9:00 am and 3:00 pm. In order to assess unspecific motor disturbances possibly induced by the immunotoxin, simple motor tests of limb strength and coordination were administered to all animals every fourth week starting from about 4 weeks postlesion.[27] Briefly, locomotive form and support were assessed after placing the rat onto a wooden ramp 80 cm long, which was connected to the animal's home cage and was maintained either horizontal or inclined at a 45 degrees angle. An inclined (75 degrees) 80 cm × 30 cm framed grid made of coarse-mesh chicken wire was also used, where the rats were placed head-down, being requested to reverse the direction and climb onto it.

Morris Water Maze

Spatial navigation abilities were analyzed using the Morris water maze (MWM), a test originally developed by Morris[28,29] and later established and widely used as a popular tool to assess spatial learning and memory and the role of different neural systems in rodents. The test is known to be sensitive to the effects of hippocampal damage,[30] and relies on the water aversion exhibited by the rats, which strongly motivates them to swim and search for an escape route using the spatial cues outside the pool, with no need for reinforcers.

The test apparatus consisted of a circular pool, 140 cm in diameter and 50 cm deep filled to a depth of 35 cm with room temperature (20°C) water. Four equally spaced points (conventionally indicated as North, South, East, and West) served as start locations, also dividing the tank into four quadrants. The tank was located in a corner of a room containing many external cues that could be used by the animals for orientation. A circular platform (10 cm in diameter) was anchored to the bottom of the pool with its top 2 cm below (and thus invisible) the water surface, onto which the animal could climb to escape. Four annuli were defined as a circular area in the middle of each quadrant, corresponding to the site where the platform would have been, if placed in that quadrant.

The animals first received a free 60 s swim to become familiar with the pool environment, followed by a 3 days cued learning session, during which the platform was made visible by a striped flag and its position changed randomly on each of four daily trials. This test is performed in order to verify the specificity of the lesions and to exclude the possible occurrence of noncognitive (e.g., visual) impairments that may interfere with the correct execution of the test. Two different paradigms, designed to assess reference and working memory abilities, were adopted.

In the reference memory version of the water maze task, the animals were given four trials a day over five consecutive days, with a 30 s intertrial interval. On each trial the animal was released into the pool from one of the starting points, and then given 60 s to reach the platform, constantly kept in the southwest (SW) quadrant, and climb onto it. If the submerged platform was not found, the rat was gently guided by the experimenter and left on it for approximately 30 s, prior to starting with the next trial. The latency to find the hidden platform, the distance swum and the swim speed were recorded by a computer-based video tracking system. On the last day of testing, after the fourth trial, the platform was removed and a spatial probe trial commenced, during which the animal was allowed to swim freely for 60 s. In this test, normally used to evaluate the efficiency of the previous learning, the distance swum and the collisions with the annuli in each of the four quadrants were recorded.

Two days after completing the reference memory test, the animals underwent the working memory version of the task, consisting in a radial arm water maze (RAWM), with six plexiglass swim alleys (50 cm length × 20 cm wide) radiating out of an open central area. Approximately at the end of one of the arms (referred to as the goal arm) was a circular platform (10 cm in diameter) located 2.0 cm below the surface onto which the animals could escape. The platform remained in the same arm for the five trials given within a day, and moved pseudo randomly to a new arm on each of five consecutive testing days. For any given trial, the animal was placed into the designated start arm, and given up to 60 s to locate the hidden platform with a 30 s intertrial time. If the platform was not located within the given time, the animal was guided to the goal arm by the experimenter and allowed to remain on the platform for 30 s prior to being placed at the next start position. Entering an incorrect arm (i.e., an arm that did not contain the platform) was counted as an entry error. For each trial, the latency to reach the platform and the number of arm selection errors prior to locating the goal arm were recorded. Moreover, the difference between latency or error scores on trials 1 and 2, expressed as a percentage of the respective scores recorded on trial 1, provided an additional measure of working memory performance (savings).

Postmortem Analyses

Upon conclusion of the behavioral testing, at about 36 weeks postlesioning and grafting, the animals were deeply anesthetized (sodium pentobarbital, 6 mg/100 g i.p.) and perfused via the ascending aorta with 50–100 mL of room temperature saline followed by 250–300 mL ice-cold 4% paraformaldehyde in PBS (pH 7.4). The brains were removed, immersed in the same fixative for 2 h and then kept at 4°C in a phosphate-buffered 20% sucrose solution until they had sunk. Using a freezing microtome (Leitz Welzlar), 40 μm sections were cut from the level of the forebrain, through the frontoparietal cortex and the hippocampus, to the brainstem regions containing the LC/subcoeruleus (SubC) complex and collected into six series. One series was processed free-floating for DBH immunohistochemistry, following a standard avidin-biotin ABC procedure.[25] Briefly, after quenching endogenous peroxidase activity by treatment with 3% hydrogen peroxide and 10% methanol in 0.02 M potassium phosphate-buffered saline (KPBS, pH 7.4), the sections were preincubated for 1 h in 5% normal horse serum (Immunological Sciences, Rome, Italy) and 0.3% Triton (Merck, Darmstadt, Germany) in KPBS, and then exposed overnight to a mouse monoclonal anti-DBH antibody (1:2,000; Merck Millipore, Darmstadt, Germany), diluted in 2% normal horse serum and 0.3% Triton. The sections were rinsed with KPBS and incubated for 2 h in a 1:200 dilution of biotinylated horse anti-mouse IgG (Vector Laboratories, Burlingame, USA) in 2% normal serum and 0.3% Triton in KPBS, followed by a 1-h incubation in avidin-biotin-peroxidase solution (ABC kit, Vector Laboratories, Burlingame, USA). The staining was visualized using 0.05% 3,3′-diaminobenzidine (Sigma) as a chromogen and 0.01% hydrogen peroxide in KPBS for 2–5 min. The sections were mounted on chromalum-gelatin-coated slides, dehydrated through steps in ascending alcohol concentrations, clarified in xylene and coverslipped using DPX (Sigma).

To ensure consistency during morphometric analyses (see below), tissue processing and stainings were carried out under identical conditions, using all relevant sections at a time. Control tissue specimens for non-specific labeling were also used where the primary antibody had been omitted. All steps were performed at room temperature.

Microscopic Analysis and Quantitative Evaluation

All analyses were carried out on taped slides by investigators blinded to the groups identity. The extent of the noradrenergic neuronal depletion induced by lesion was estimated using an unbiased stereological quantification method based on the optical fractionator principle.[31] DBH-immunoreactive neurons in the LC were counted bilaterally from the level of the ventral portion of the parabrachial nucleus, rostrally, to the genu of the facial nerve, caudally. The counting included also the more scattered noradrenergic neurons in the SubC region, just ventral to the LC proper.[1,32] The sampling system consisted of an Olympus BH2 microscope (fitted with an $X-Y$ motorized stage and a microcator to measure distances in the Z-axis) interfaced with a color video camera (Sony) and a personal computer. The CAST GRID® software (Olympus Denmark A/S, Albertslund, Denmark) was used to delineate the selected regions at 4 × magnification, as well as to generate unbiased counting frames which were moved randomly and systematically until the entire delineated area was sampled. Using a 100 × oil objective, unambiguously positive cells were identified and counted after excluding guard volumes from both section surfaces, in order to prevent problems of lost caps.

Changes in the density of hippocampal noradrenergic innervation, possibly induced by either the immunotoxin treatment or grafted cells were assessed by densitometry, using the Image 1.61 image analysis software,[33] as previously described.[25] Briefly, measuring fields of consistent size (0.5 mm in diameter) were selected bilaterally along similar rostrocaudal levels from three different sections in the CA1 and CA3 subfields of the dorsal hippocampus and the dentate gyrus (DG). Background density, as determined in a structure normally lacking the DBH-positive reaction product (i.e., the corpus callosum), was subtracted from each measurement. Relative levels of DBH-positive optical density in the analyzed regions were separately measured in the left and right hemisphere, and then compared with each other to check for possible side-differences. Data were expressed as arbitrary units, averaging the values measured from the three sections.

RESULTS

General Observations

All animals, irrespective of their treatment, increased in body weight, and exhibited fairly normal sensory-motor functioning when evaluated in both bridge and grid tests at about 32 weeks of age (Table 10.1). These observations, together with the finding of no differences in cue learning or swim speed among the groups (see below) indicated that no changes in sensory-motor activity were produced by the lesion that would account for water maze performance. Rats with vehicle injections did not differ from the intact animals on any of the behavioral or morphological parameters analyzed. These animals were therefore combined into a single control group ($n = 10$) for all analyses and illustrations.

TABLE 10.1 Motor Performance

Group	Equilibrium Time on Ramp (%)	Latency to Cross Ramp (s)	Latency to Reverse on Grids (s)	Number of Falls in Grids
Control ($n = 10$)	98.0 ± 1.2	7.4 ± 0.7	6.8 ± 1.3	2.2 ± 0.9
Lesioned ($n = 10$)	97.2 ± 0.7	6.7 ± 0.4	6.4 ± 0.8	2.6 ± 0.6
LES + TRPL ($n = 12$)	97.3 ± 0.8	6.7 ± 0.5	6.6 ± 0.8	2.9 ± 0.7

Motor tests consisted in the evaluation of postural and locomotive form onto an 80 cm-long wooden ramp (maintained either horizontal or with a 45 degrees inclination), or an inclined (75 degrees) grid, both connected to the animals' home cage. Parameters to be analyzed were the % balance time onto the ramp and the time required to cross it, as well as the latency to reverse direction and the number of falls when placed onto a grid. Numbers represent the mean of four determinations ± SEM.

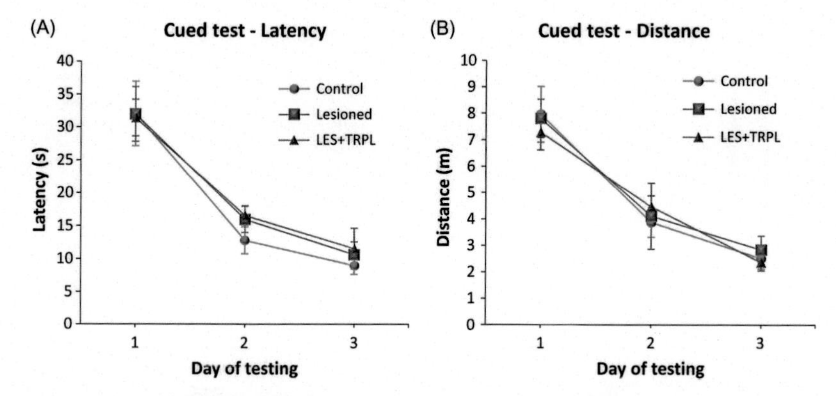

FIGURE 10.1 Performance in the cued version of the MWM test. All animals rapidly learned to locate the visually cued platform (which was moved to a new quadrant on each of the four daily trials) and exhibited similar escape latencies (A) and swim distances (B). Each point represents the mean value ± SEM for the block of four trials administered each day.

Behavioral Analyses

Morris Water Maze

Fig. 10.1 illustrates the performance of the groups in a cued paradigm, in which the platform was signaled and its position varied on each of the four daily trials. This test was adopted to possibly rule out lesion-induced sensory impairments that would affect navigation search in the pool. In general, animals in all groups improved their performance over time (repeated measure ANOVA, effect of day on latency, $F_{2,58} = 83.55$; on distance, $F_{2,58} = 54.48$; both $P < 0.001$), and did not differ from each other (main group effect on latency $F_{2,29} = 0.73$; on distance, $F_{2,29} = 0.67$; group $\times$ day on latency, $F_{4,58} = 0.56$; on distance, $F_{4,58} = 0.88$; all n.s.).

Mean latencies and swim distances required to find the platform in the place test, are shown in Fig. 10.2A,B. All animals initially required about 20–25 s and 5–6 m to locate the hidden platform, but improved significantly thereafter (repeated measure ANOVA, effect of day on latency, $F_{4,116} = 37.19$; on distance, $F_{4,116} = 33.38$; both $P < 0.001$). The learning rates were virtually identical throughout the 5 days of testing and the groups did not differ from each other (main group effect on latency $F_{2,29} = 1.63$; on distance, $F_{2,29} = 1.45$; group $\times$ day on latency, $F_{8,116} = 0.83$; on distance, $F_{8,116} = 0.90$; all n.s.). Swim speed, monitored as a general measure of motor ability during the execution of the navigation tasks, did not differ between groups and averaged 0.2–0.3 m/s across the days of testing.

During the spatial probe trial on day 5, when the platform was removed and a free swim was allowed (Fig. 10.2C–E), all animals swam primarily in the training (SW), quadrant (effect of quadrant on swim distance, $F_{3,87} = 36.86$; on annulus crossings, $F_{3,87} = 47.72$; both $P < .001$),

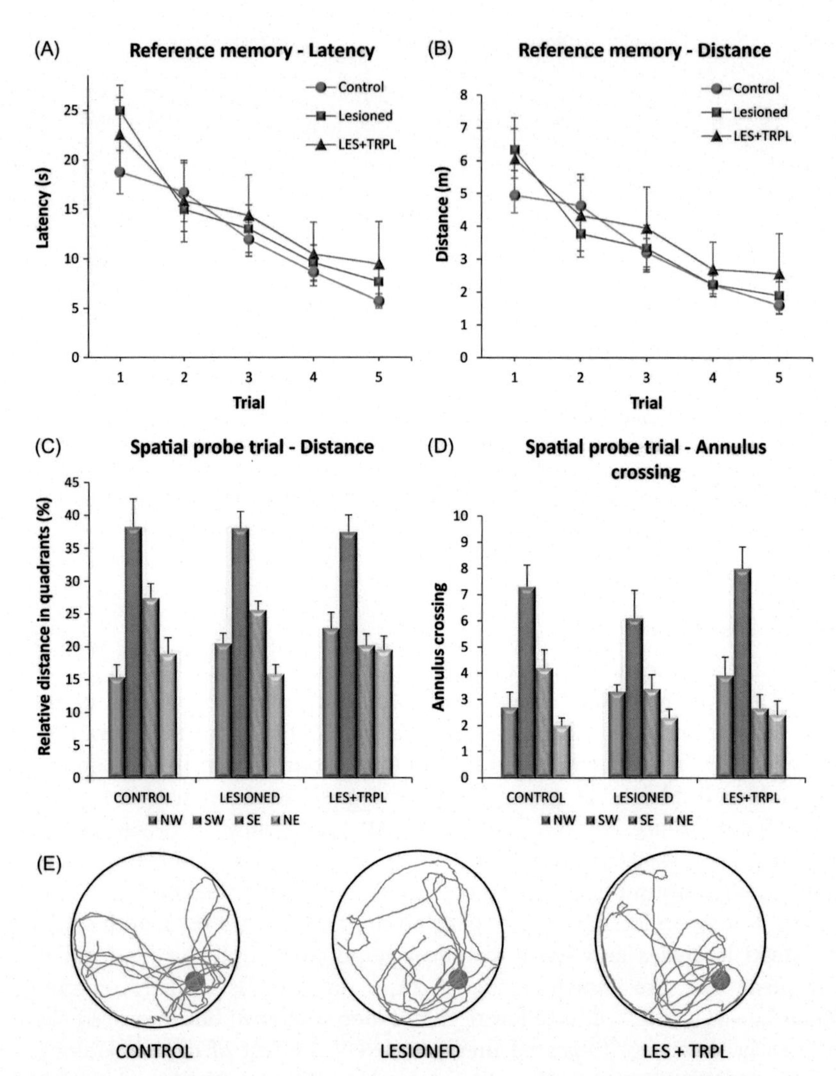

FIGURE 10.2 **Performance in the reference memory version of the MWM test.** All animals exhibited equally efficient performance and took similar escape latencies (A) and swim distances (B) to locate the submerged platform during the acquisition of the navigation task. Each point represents the mean value for the block of four trials on each of five consecutive days. Lower diagrams illustrate the mean relative distance swum (C) and the average number of annulus crossings (D) in each quadrant during the spatial probe trial, upon removal of the escape platform, as tested at about 32 weeks postgrafting. In (E) the actual swim paths taken by representative rats from the different groups are illustrated. All animals exhibited a pronounced bias for the original platform site and swam primarily in the training (SW) quadrant, indicating that the selective noradrenergic lesion has no effects upon reference memory abilities.

and exhibited an equally pronounced bias for the original platform site (main group effect on distance, $F_{2,29} = 1.42$; on annulus crossings, $F_{2,29} = 1.25$; group × quadrant interaction on distance, $F_{6,87} = 1.74$: on annulus crossings, $F_{6,87} = 2.06$: all n.s.). No group difference was observed in the total number of annulus crossings or swim speed (one-way ANOVA, main group effect $F_{2,29} = 4.10$ and $F_{2,29} = 2.82$, respectively, both n.s.) suggesting an equally active search behavior in all animals.

Radial Arm Water Maze

The performances of the groups in the RAWM testing are shown in Fig. 10.3A–C. In this task the platform was moved to a new quadrant

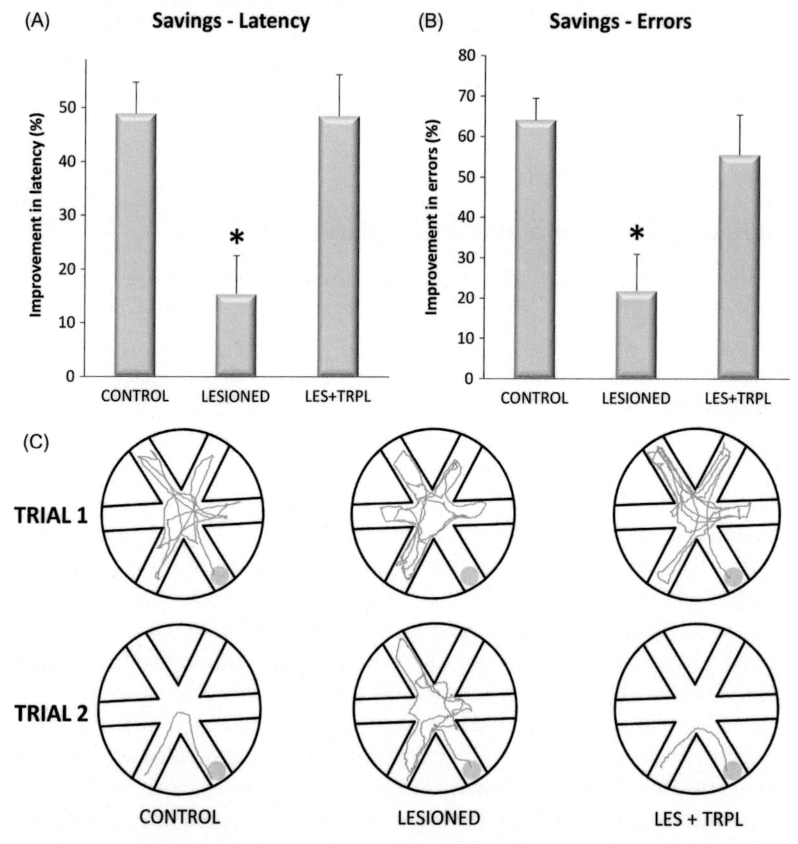

FIGURE 10.3 Working memory performance in the Radial Arm Water Maze task. Data are illustrated as percentage improvement (savings) between trials 1 and 2 for latency (A) and errors (B). In (C) the actual swim paths taken by representative rats from the different groups are illustrated. Note the dramatic impairments exhibited by the lesioned animals, compared to control animals, and the near complete functional recovery promoted by the intrahippocampal LC tissue grafts. The asterisk indicates significant difference from the control and the transplanted groups at $P < .01$.

every day, thus the animals had to relearn its position within the five trials of each testing day by developing a new search strategy. In general, all animals exhibited a relatively longer latency and a higher number of arm selection errors (i.e., entering an arm not containing the platform) in the first trial of each day, but both these measures were seen markedly reduced already in the second trial. Time and error savings, calculated as a percentage of improvement between trials 1 and 2 provided an estimation of learning efficiency in the RAWM task. Under these conditions, rats in the control and the transplanted groups were seen to reduce by about 50%–60% the latency and entry errors required to reach the platform. Conversely the lesioned animals did not learn to relocate the platform across trials as efficiently, their percent improvement being significantly lower than that exhibited by the control or the transplanted rats ($P < .01$ for both measures) and never exceeding 20% (Fig. 10.3A–B). The detailed swim paths obtained on the fifth day of training from representative animals in the three experimental groups are shown in Fig. 10.3C.

MORPHOLOGICAL ANALYSES

Effects of the Lesion and of Transplants

The bilateral infusion of the anti-DBH-saporin immunotoxin into the hippocampal formation of developing (PD4) rats produced a marked depletion of DBH-immunoreactive neurons in the LC (Fig. 10.4A–C). The neuronal loss was particularly prominent in the dorsal-central portion of the nucleus, where cells project mainly to rostral targets such as neocortex and hippocampus, whereas spared neurons were mainly scattered more caudally and ventrally, in the SubC nucleus, whose cells project to cerebellum and spinal cord.[34] As estimated by stereology (Table 10.2), the noradrenergic neuronal depletion averaged about 75%–77%, and was consistently observed in all lesioned animals, irrespective of the presence of transplanted progenitors in the hippocampus (one-way ANOVA followed by Fisher's PLSD post hoc test; $P < .01$ vs control). Likewise, in the lesioned animals, an almost complete loss of noradrenergic innervation was evident in all hippocampal subfields, where only sparse immunoreactive fibers could be detected, as opposed to control (compare, e.g., D with E in Fig. 10.4). A modest fiber loss could be detected also in the overlying medial cortex (not shown). Notably, all animals receiving embryonic noradrenergic-rich tissue bilaterally in the hippocampus exhibited surviving grafts that occurred as either small masses of heavily DBH-stained tissue within the hippocampal fissure (Fig. 10.4F) or small clusters of DBH-immunoreactive cells with clear neuronal morphology and rich arborization (not shown). In general the grafted cells appeared healthy, correctly positioned and well-integrated within the

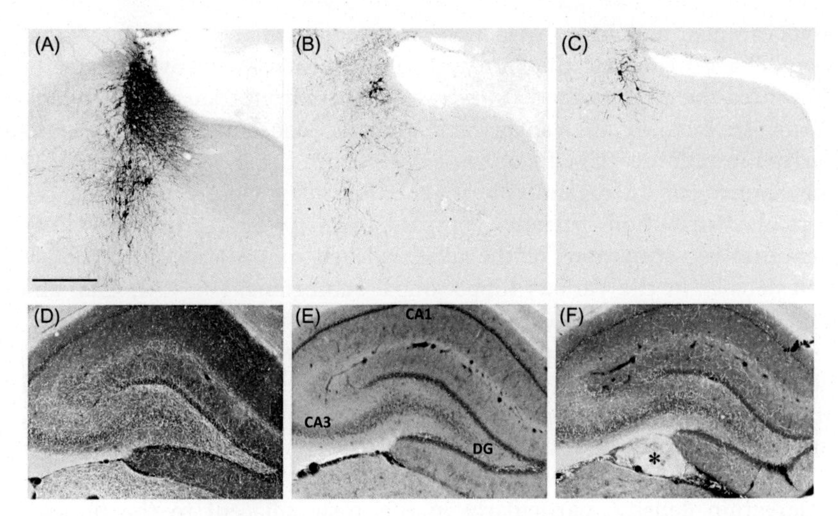

FIGURE 10.4 Representative examples of dopamine-beta-hydroxylase (DBH) immunostaining illustrating, on the coronal plane, the distribution of noradrenergic neurons in the LC/SubC (A–C), as well as the pattern, in dark field, of DBH-immunoreactive terminal innervation in the hippocampus (D–F) of control (A,D), lesioned (B,E), and lesioned and transplanted animals (C,F), at about 36 weeks after the intrahippocampal injection of the anti-DBH-saporin immunotoxin and implantation of fetal LC tissue, performed at PD4 and PD8, respectively. Note the marked loss of DBH-positive neurons in the LC/SubC similarly exhibited by the lesion-only (B) and the transplanted animals (C), compared to the control animals (A). Note also (in dark field) the near total loss of DBH-positive fibers in the hippocampus of lesioned animals (E) compared to control (D) and the near normal pattern of noradrenergic innervation reinstated by the transplanted LC tissue (F). *CA*, cornu ammonis of the hippocampus; *DG*, dentate gyrus. The asterisk in (F) indicates the location of the graft deposit. Scale bar in (A): 500 μm.

TABLE 10.2 Stereological and Densitometric Estimates of Dopamine-ß-Hydroxylase-Immunoreactive Neurons and Fiber Density in the LC and Hippocampal Terminal Regions, Respectively

Group	DBH-ir Neurons in LC/SubC	DBH-ir Fibers in CA1	DBH-ir Fibers in CA3	DBH-ir Fibers in DG
Control ($n = 10$)	1711.9 ± 41.0	66.3 ± 5.7	63.8 ± 1.2	62.5 ± 1.9
Lesioned ($n = 10$)	443.0 ± 30.5*	15.7 ± 1.7*	17.9 ± 1.2*	17.8 ± 1.3*
LES + TRPL ($n = 12$)	459.1 ± 28.8*	50.8 ± 1.8	54.4 ± 2.7	59.3 ± 2.4

Numbers represent the estimated total number of DBH-immunoreactive neurons in the LC, and the standardized relative density scores (± SEM) of DBH-positive innervation in each of the target areas in the various subdivisions of the hippocampal formation. Asterisks indicate significant difference from control group ($P < .001$).

hippocampus, and in no case was any obvious glial scar detected that would have disrupted the integrity of the host tissue environment or prevented the outgrowth of DBH-positive fibers from transplanted neuroblasts. In fact, in all transplanted animals, dense networks of graft-derived immunoreactive fibers were detected in the hippocampus, where they reinstated an organotypic innervation often closely matching the normal distribution pattern (compare, e.g., D and F in Fig. 10.4). Quantitative estimations of the relative levels of DBH-positive innervation density in the various hippocampal subregions, i.e., CA1, CA3, and DG are shown in Table 10.2. Statistical comparisons (two-way ANOVA followed by Fisher's PLSD post hoc test) confirmed a dramatic 75%–80% fiber loss in the lesioned, as compared to both the control and the grafted animals ($P < .01$). Notably, in spite of obvious individual variabilities, the abundant networks of immunoreactive fibers outgrowing from the transplanted neurons were seen to reinstate >75%–95% of the normal innervation density, particularly in the areas adjacent to the implant (e.g., the DG), with no obvious side-differences.

DISCUSSION

By using a highly selective lesioning tool in conjunction with well-established, hippocampus-dependent swim maze tasks, and neural transplantation procedures, the present study sought to examine the importance of the coeruleo-hippocampal noradrenergic projections in the regulation of aspects of spatial learning and memory. We hypothesized that, if selective ablation of noradrenergic afferents to the hippocampus was required to induce measurable learning deficits, then restoration of hippocampal noradrenergic neurotransmission by implanted neuroblasts, should be sufficient to reverse or ameliorate them, an issue which has never been addressed, so far.

Effects of the Anti-DBH-Saporin Lesion

Extending recent observations from our laboratory (Coradazzi et al., unpublished results), we found here that selective disruption of hippocampal noradrenergic innervation by direct intraparenchymal injections of the anti-DBH-saporin immunotoxin severely impaired animals' performance in the working memory version (i.e., the RAWM) of the MWM task, whereas reference memory abilities were unaffected. These data confirm previous evidence of working memory deficits associated to noradrenergic dysfunctions induced by either aging,[35,36] reversible LC inactivation[15,37] or nonselective lesioning,[14,38–40] and thus support the view of an important regulatory role played by hippocampal NA in specific aspects of cognition.

The lesioning paradigm employed here (i.e., the bilateral intrahippocampal infusion of anti-DBH-saporin), producing extensive noradrenergic neuronal and terminal fiber loss, proved to be suitable to address the specific contribution of the hippocampal noradrenergic innervation to spatial learning and memory. In a previous investigation, Steckler et al.[41] adopted a similar lesioning approach, and injected the cholinergic immunotoxin 192 IgG-saporin to selectively ablate the cholinergic septo-hippocampal innervation, thought to play a role in mediating spatial short-term memory. However, in that study, the lesion-induced deficits in an operant delayed response task were seen very modest at best, suggesting that hippocampal cholinergic innervation per se may not be as critical as the LC-derived noradrenergic inputs for the regulation of working memory abilities. It may be argued that a reduction in hippocampal cholinergic neurotransmission, possibly induced by the noradrenergic lesion, may have contributed to the cognitive deficits seen here. It has been proposed, for example, that LC-derived NA sustains acetylcholine release from septo-hippocampal terminals,[42] and thus that the disruptive effects on cognition of NA loss would occur via a reduced cholinergic neurotransmission.[15] In fact the two transmitter systems closely cooperate, mainly at hippocampal level, to regulate working memory.[42–45] If so, a concomitant loss of cholinergic and noradrenergic afferents to the hippocampus might be required to severely affect working memory. Although, from the present data, this possibility cannot completely be ruled out, it seems rather unlikely. In fact, other studies have found increased, rather than reduced, cholinergic activity in the hippocampus of patients with mild cognitive impairment/early AD and NA deficiency[11,46] or rats with neurotoxic NA depletions.[47,48] Likewise, the magnitude of the functional restoration promoted by the noradrenergic tissue transplants (see discussion below) argues against the possibility of a significant cholinergic involvement in the disruptive effects of the lesion. Future studies, entailing selective cholinergic-noradrenergic lesioning and grafting will be required to address this issue.

Effects of Transplants

In previous investigations, grafts of embryonic LC tissue implanted into the deafferented hippocampus of adult rats have been shown to become functionally integrated into the host circuitry,[49] restore fairly normal patterns of noradrenergic innervation, transmitter turnover and release,[50,51] and reinstate a noradrenergic influence onto hippocampal neurons able to suppress or retard seizure development.[52,53] Surprisingly, although an association between noradrenergic loss and cognitive disturbances has long been known, not much preclinical work has been carried out to investigate the transplant-promoted reinstatement of noradrenergic innervation and neurotransmission as a prerequisite to restore cognitive abilities in experimental animals. Thus, in the only published study addressing this issue, solid blocks of fetal LC tissue, implanted into the third ventricle of aged rats, were seen to significantly ameliorate aspects of working memory.[54] The procedure employed here, whereby fetal LC neurons were implanted in form of cell suspensions in the selectively deafferented hippocampus proved to be highly reliable, in that: (1) it allowed a much better reproducibility of the approach and monitoring of the activity and viability of grafted cells and (2) it enabled to specifically address the contribution of the hippocampal noradrenergic innervation in aspects of spatial learning and memory. Notably, the functional recovery induced by the noradrenergic-rich LC transplants in the present study appeared far more complete than that promoted by intrahippocampal grafts of cholinergic-rich basal forebrain neurons in animals with selective cholinergic lesions.[55] In fact, whereas in that study the basal forebrain tissue grafts were seen to mainly normalize reference memory, but not working memory abilities, we found here that reinnervation of the deafferented hippocampus by implanted noradrenergic-rich tissue from fetal LC was sufficient to completely reverse the lesion-induced working memory impairments, without seemingly affecting any other aspect of spatial navigation. Again, these observations suggest that the septo-hippocampal cholinergic projections are less critical than LC-derived noradrenergic afferents for sustaining working memory abilities, and thus that the two transmitter systems may be involved in the regulation of different aspects of memory performance.

CONCLUSIONS

The results of the present study provide compelling evidence substantiating an important role for the coeruleo-hippocampal noradrenergic inputs in the regulation of spatial working memory.

ABBREVIATIONS

AD Alzheimer's disease
CA cornu ammonis
DBH dopamine-beta-hydroxylase
DG dentate gyrus
DMEM Dulbecco's modified Eagle medium
KPBS potassium phosphate-buffered saline
LC locus coeruleus
MWM Morris water maze
NA noradrenaline
PBS phosphate-buffered saline

PD postnatal day
RAWM radial arm water maze
SubC subcoeruleus

ACKNOWLEDGMENTS

For expert technical assistance, we thank the staff at the facility for animal care, University of Trieste. For generous support to the study, we warmly thank the Kathleen Foreman-Casali Foundation, the Beneficentia Stiftung and the Kleiner and Bono families.

REFERENCES

1. Amaral DG, Sinnamon HM. The locus coeruleus: neurobiology of a central noradrenergic nucleus. *Progr Neurobiol.* 1977;9:147–196.
2. Aston-Jones G, Cohen JD. An integrative theory of locus coeruleus-norepinephrine function: adaptive gain and optimal performance. *Annu Rev Neurosci.* 2005;28:403–450.
3. Sara SJ. The locus coeruleus and noradrenergic modulation of cognition. *Nat Rev Neurosci.* 2009;10:211–223.
4. Sara SJ. Locus coeruleus in time with the making of memories. *Curr Opin Neurobiol.* 2015;35:87–94.
5. Marien MR, Colpaert FC, Rosenquist AC. Noradrenergic mechanisms in neurodegenerative diseases: a theory. *Brain Res Rev.* 2004;45:38–78.
6. Gannon M, Che P, Chen Y, Jiao K, Roberson ED, Wang Q. Noradrenergic dysfunction in Alzheimer's disease. *Front Neurosci.* 2015;9:220.
7. German DC, Manaye KF, White CL, Woodward DJ, McIntire DD, Smith WK, et al. Disease-specific patterns of locus coeruleus cell loss. *Ann Neurol.* 1992;32:667–676.
8. Matthews KL, Chen CPLH, Esiri MM, Keene J, Minger SL, Francis PT. Noradrenergic changes, aggressive behavior, and cognition in patients with dementia. *Biol Psychiatry.* 2002;51:407–416.
9. Zarow C, Lyness SA, Mortimer JA, Chui HC. Neuronal loss is greater in the locus coeruleus than nucleus basalis and substantia nigra in Alzheimer and Parkinson diseases. *Arch Neurol.* 2003;60:337–341.
10. Haglund M, Sjöbeck M, Englund E. Locus ceruleus degeneration is ubiquitous in Alzheimer's disease: possible implications for diagnosis and treatment. *Neuropathology.* 2006;26:528–532.
11. Grudzien A, Shaw P, Weintraub S, Bigio E, Mash DC, Mesulam MM. Locus coeruleus neurofibrillary degeneration in aging, mild cognitive impairment and early Alzheimer's disease. *Neurobiol Aging.* 2007;28:327–335.
12. Braak H, Del Tredici K Where. when, and in what form does sporadic Alzheimer's disease begin? *Curr Opin Neurol.* 2012;25:708–714.
13. Chamberlain SR, Robbins TW. Noradrenergic modulation of cognition: therapeutic implications. *J Psychopharmacol.* 2013;27:694–718.
14. Sontag TA, Hauser J, Kaunzinger I, Gerlach M, Tucha O, Lange KW. Effects of the noradrenergic neurotoxin DSP4 on spatial memory in the rat. *J Neural Transm.* 2008;115:299–303.
15. Khakpour-Taleghani B, Lashgari R, Motamedi F, Naghdi N. Effect of reversible inactivation of locus ceruleus on spatial reference and working memory. *Neuroscience.* 2009;158:1284–1291.

16. Szot P, Miguelez C, White SS, Franklin A, Sikkema C, Wilkinson CW, et al. A comprehensive analysis of the effect of DSP4 on the locus coeruleus noradrenergic system in the rat. *Neuroscience*. 2010;166:279–291.
17. Picklo MJ, Wiley RG, Lappi DA, Robertson D. Noradrenergic lesioning with an anti-dopamine-β-hydroxylase immunotoxin. *Brain Res*. 1994;666:195–200.
18. Barthelemy I, Martineau D, Ong M, Matsunami R, Ling N, Benatti L, et al. The expression of saporin, a ribosome-inactivating protein from the plant *Saponaria officinalis*, in *Escherichia coli*. *J Biol Chem*. 1993;268:6541–6548.
19. Lappi DA, Esch FS, Barbieri L, Stirpe F, Soria M. Characterization of a *Saponaria officinalis* seed ribosome-inactivating protein: immunoreactivity and sequence homologies. *Biochem Biophys Res Commun*. 1985;129:934–942.
20. Studelska DR, Brimijoin S. Partial isolation of two classes of dopamine beta-hydroxylase-containing particles undergoing rapid axonal transport in rat sciatic nerve. *J Neurochem*. 1989;53:622–631.
21. Weinshilboum RM. Serum dopamine beta-hydroxylase. *Pharmacol Rev*. 1978;30:133–166.
22. Contestabile A, Stirpe F. Ribosome-inactivating proteins from plants as agents for suicide transport and immunolesioning in the nervous system. *Eur J Neurosci*. 1993;5:1292–1301.
23. Wiley RG, Kline IV RH. Neuronal lesioning with axonally transported toxins. *J Neurosci Methods*. 2000;103:73–82.
24. Leanza G, Cataudella T, Dimauro R, Monaco S, Stanzani S. Release properties and functional integration of noradrenergic-rich tissue grafted to the denervated spinal cord of the adult rat. *Eur J Neurosci*. 1999;11:1789–1799.
25. Coradazzi M, Gulino R, Garozzo S, Leanza G. Selective lesion of the developing central noradrenergic system: short- and long-term effects and reinnervation by noradrenergic-rich tissue grafts. *J Neurochem*. 2010;114:761–771.
26. Björklund A, Stenevi U, Schmidt RH, Dunnett SB, Gage FH. Intracerebral grafting of neuronal cell suspensions. I. Introduction and general methods of preparation. *Acta Physiol Scand Suppl*. 1983;522:1–7.
27. Antonini V, Prezzavento O, Coradazzi M, Marrazzo A, Ronsisvalle S, Arena E, et al. Anti-amnesic properties of (±)-PPCC, a novel sigma receptor ligand, on cognitive dysfunction induced by selective cholinergic lesion in rats. *J Neurochem*. 2009;109:744–754.
28. Morris R. Spatial localization does not require the presence of local cues. *Learn Motivat*. 1981;12:239–260.
29. Morris R. Developments of a water-maze procedure for studying spatial learning in the rat. *J Neurosci Methods*. 1984;11:47–60.
30. Morris R, Garrud P, Rawlins J, O'Keefe J. Place navigation impaired in rats with hippocampal lesions. *Nature*. 1982;297:681–683.
31. West MJ, Slomianka L, Gundersen HJ. Unbiased stereological estimation of the total number of neurons in the subdivisions of the rat hippocampus using the optical fractionator. *Anat Rec*. 1991;231:482–497.
32. Grzanna R, Molliver ME. The locus coeruleus in the rat: an immunohistochemical delineation. *Neuroscience*. 1980;5:21–40.
33. Rasband WS, Bright DS. *NIH Image: a public domain image processing program for Macintosh. Microbeam Anal Soc J*. 4. 1995137–149.
34. Loughlin SE, Foote SL, Grzanna R. Efferent projections of nucleus locus coeruleus: morphologic subpopulations have different efferent targets. *Neuroscience*. 1986;18: 307–319.

35. Leslie FM, Loughlin SE, Sternberg DB, McGaugh JL, Young LE, Zornetzer SF. Noradrenergic changes and memory loss in aged mice. *Brain Res.* 1985;359:292–299.
36. Markowska AL, Stone WS, Ingram DK, Reynolds J, Gold PE, Conti LH, et al. Individual differences in aging: behavioral and neurobiological correlates. *Neurobiol Aging.* 1989;10:31–43.
37. Mair RD, Zhang Y, Bailey KR, Toupin MM, Mair RG. Effects of clonidine in the locus coeruleus on prefrontal- and hippocampal-dependent measures of attention and memory in the rat. *Psychopharmacology.* 2005;181:280–288.
38. Wenk G, Hughey D, Boundy V, Kim A, Walker L, Olton D. Neurotransmitters and memory: role of cholinergic, serotonergic, and noradrenergic systems. *Behav Neurosci.* 1987;101:325–332.
39. Ohno M, Yamamoto T, Kobayashi M, Watanabe S. Impairment of working memory induced by scopolamine in rats with noradrenergic DSP-4 lesions. *Eur J Pharmacol.* 1993;238:117–120.
40. Compton DM, Dietrich KL, Smith JS, Davis BK. Spatial and non-spatial earning in the rat following lesions to the nucleus locus coeruleus. *Neuroreport.* 1995;7:177–182.
41. Steckler T, Keith AB, Wiley RG, Sahgal A. Cholinergic lesions by 192 IgG-saporin and short-term recognition memory: Role of the septohippocampal projection. *Neuroscience.* 1995;66:101–114.
42. Decker MW, McGaugh JL. Effects of concurrent manipulations of cholinergic and noradrenergic function on learning and retention in mice. *Brain Res.* 1989;477:29–37.
43. Decker MW, Gallagher M. Scopolamine-disruption of radial arm maze performance: modification by noradrenergic depletion. *Brain Res.* 1987;417:59–69.
44. Ohno M, Kobayashi M, Kishi A, Watanabe S. Working memory failure by combined blockade of muscarinic and beta-adrenergic transmission in the rat hippocampus. *Neuroreport.* 1997;8:1571–1575.
45. Ohno M, Yoshimatsu A, Kobayashi M, Watanabe S. Noradrenergic DSP-4 lesions aggravate impairment of working memory produced by hippocampal muscarinic blockade in rats. *Pharmacol Biochem Behav.* 1997;57:257–261.
46. DeKosky ST, Ikonomovic MD, Styren SD, Beckett L, Wisniewski S, Bennett DA, et al. Upregulation of choline acetyltransferase activity in hippocampus and frontal cortex of elderly subjects with mild cognitive impairment. *Ann Neurol.* 2002;51:145–155.
47. Jackisch R, Gansser S, Cassel JC. Noradrenergic denervation facilitates the release of acetylcholine and serotonin in the hippocampus: towards a mechanism underlying upregulations described in MCI patients? *Exp Neurol.* 2008;213:345–353.
48. Vizi ES. Modulation of cortical release of acetylcholine by noradrenaline released from nerves arising from the rat locus coeruleus. *Neuroscience.* 1980;5:2139–2144.
49. Björklund A, Segal M, Stenevi U. Functional reinnervation of rat hippocampus by locus coeruleus implants. *Brain Res.* 1979;170:409–426.
50. Björklund A, Nornes H, Gage FH. Cell suspension grafts of noradrenergic locus coeruleus neurons in rat hippocampus and spinal cord: Reinnervation and transmitter turnover. *Neuroscience.* 1986;18:685–698.
51. Kalén P, Cenci MA, Lindvall O, Björklund A. Host brain regulation of fetal locus coeruleus neurons grafted to the hippocampus in 6-hydroxydopamine-treated rats. An intracerebral microdialysis study. *Eur J Neurosci.* 1991;3:905–918.
52. Bengzon J, Kokaia Z, Lindvall O. Specific functions of grafted locus coeruleus neurons in the kindling model of epilepsy. *Exp Neurol.* 1993;122:143–154.

53. Barry DI, Kikvadze I, Brundin P, Bolwig TG, Björklund A, Lindvall O. Grafted noradrenergic neurons suppress seizure development in kindling-induced epilepsy. *Proc Natl Acad Sci USA*. 1987;84:8712–8715.
54. Collier TJ, Gash DM, Sladek JR. Transplantation of norepinephrine neurons into aged rats improves performance of a learned task. *Brain Res*. 1988;448:77–87.
55. Leanza G, Martìnez-Serrano A, Björklund A. Amelioration of spatial navigation and short-term memory deficits by grafts of foetal basal forebrain tissue placed into the hippocampus and cortex of rats with selective cholinergic lesions. *Eur J Neurosci*. 1998;10:2353–2370.

Chapter 11

Enteric Astroglia and Noradrenergic/Purinergic Signaling

Vladimir Grubišić[1,✉] **and Vladimir Parpura**[2,✉]
[1]*Michigan State University, East Lansing, MI, United States,* [2]*University of Alabama School of Medicine, Birmingham, AL, United States*

Chapter Outline

Introduction 222
Innervation of the Gut Wall 222
Enteric Glia–Essentials 224
Enteric Glia Cells Respond to the Direct Sympathetic Input: Ca^{2+} Excitability 226
Enteric Glial Ca^{2+} Responses Regulate Gut Motility 227
Other Selected Roles of Sympathetic Innervation and Enteric Glia in the Gut 228
Sympathetic Nervous System and Enteric Glia in GI Disorders/Diseases 231
Conclusions 233
Abbreviations 234
Acknowledgment 234
References 234

ABSTRACT
Reflexive behavior of the intestine is under the direct control of the enteric nervous system, an extensive network of neurons and glial cells embedded in the gut wall. While the basic neuronal circuitry underlying local reflexes is firmly established, the active role of enteric glia, traditionally recognized as neuron supportive cells, in this process emerged only recently. Neuronal control of the gut function is an integrated system involving reflex loops at several different levels, including local/intrinsic enteric reflexes as well as those that involve an extrinsic input from sympathetic ganglia; the latter, abundantly and directly innervate both the enteric neurons and glia. We summarize the evidence for innervation of enteric glial cells and integrate current understanding of their associated roles in regulation of the gut motility in health and disease.

✉Correspondence address
E-mail: grubisic@msu.edu; vlad@uab.edu

Noradrenergic Signaling and Astroglia. DOI: http://dx.doi.org/10.1016/B978-0-12-805088-0.00011-6
© 2017 Elsevier Inc. All rights reserved.

Keywords: Enteric glial cells; enteric nervous system; autonomic nervous system; sympathetic nervous system; gut motility

INTRODUCTION

The enteric nervous system (ENS) is the largest collection of neurons and neuroglia outside the central nervous system (CNS), residing within the wall of the digestive tract and primarily regulating local gut reflexes involved in gastrointestinal (GI) motility and fluid transport. Since these GI functions can be preserved in the complete absence of the extrinsic innervation from the CNS,[1] the ENS has been recognized as an independent branch of the autonomic nervous system.[2] Normally, however, local ENS circuits work in concert with reflex loops at the level of sympathetic ganglia and the CNS (reviewed in Refs. 3,4). While it is well established that the gut reflexes are controlled by the neuronal cir cuitry, it is becoming apparent that enteric glial cells are also actively involved in this process. Recent studies depict these glial cells as local modulators affecting gut functions, including gut motility, epithelial barrier and inflammation, in health and disease (reviewed in Ref. 5). Here, we summarize the known effects of the sympathetic gut innervation of the enteric glia with contextual review of the recently discovered roles of these cells in gut function, motility in particular.

INNERVATION OF THE GUT WALL

The mammalian GI system has extensive intrinsic innervation originating from the ENS (Fig. 11.1), which is comprised of 2–1000 million neurons[6,7,8,9] and one to seven times as many glial cells[10,11] mainly organized into two networks/plexuses; there are some glial cells scattered throughout the gut wall. The myenteric plexus is primarily responsible for gut motility, while the submucosal plexus controls secretomotor functions. The ENS structurally resembles more the CNS than the peripheral ganglia. Consequently, due to its structure, size, and ability to perform functions without the CNS innervation, the ENS is often referred to as "the second brain."

Here, we only provide a brief summary of the sympathetic innervation of the gut, given that the details are described elsewhere (e.g., Ref. 3). Somata of presynaptic sympathetic neurons resides in the intermediolateral nucleus, i.e., the lateral gray column, of the spinal cord; this column exists at vertebral levels T1-L2 and mediates autonomic innervation of internal organs, including that of the gut (Fig. 11.1A). These autonomic neurons of the spinal cord give rise to preganglionic fibers, constituting thoracolumbar splanchnic nerves (visceral efferents), which synapse onto the second-order neurons located in the pre- and

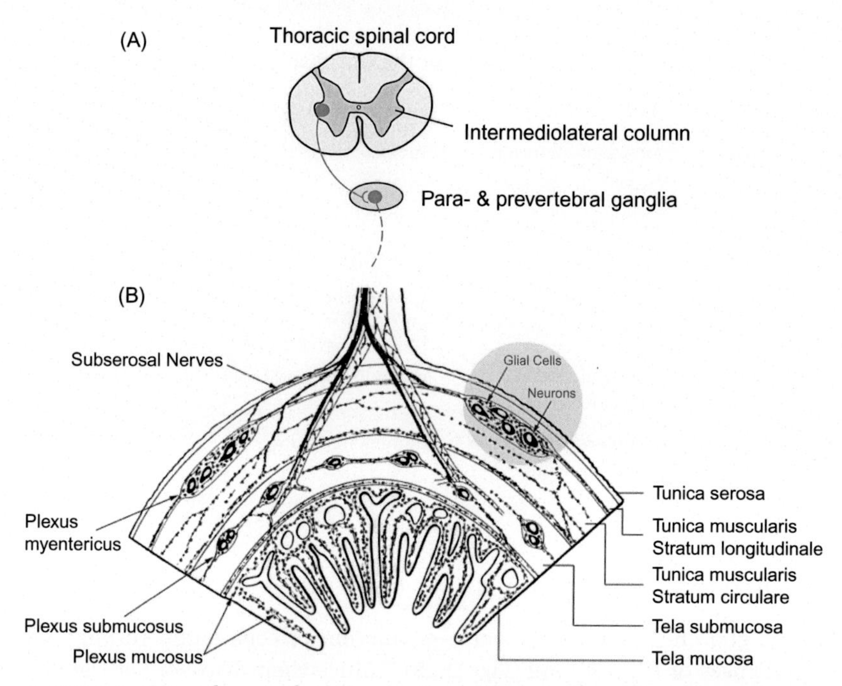

FIGURE 11.1 ENS and sympathetic innervation of the intestine. Gut physiology is under control of the extrinsic innervation, including sympathetic nerves (A), and the intrinsic ENS (B). Of note, parasympathetic innervation of the gut is provided by the vagal and pelvic nerves (not shown here). (A) Visceral efferents of the sympathetic system originate in the intermediolateral columns of the thoracolumbar spinal cord, which relays to the para- and prevertebral ganglia. In turn the ganglia send projections to densely innervate enteric plexuses, vasculature, and smooth muscle sphincter regions (B; also see Table 11.1). (B) The ENS is a network of neurons and glial cells, organized in myenteric and submucosal plexuses residing within the intestinal wall. *Figure in B obtained from Ref. 12.*

para-vertebral ganglia. In turn, these sympathetic ganglionic neurons emanate postganglionic fibers, which innervate the effector organs. In the gut, sympathetic postganglionic fibers densely innervate the four major sites: the myenteric and submucosal plexuses, blood vessels, and the smooth muscle sphincters. Nonsphincter regions of the smooth muscles in the gut wall, along with mucosa, and the lymphoid tissue are additional areas in the gut that also receive postganglionic, albeit sparser, innervation (Fig. 11.1B and Table 11.1). The functions of the sympathetic innervation are reduction of gut motility, inhibition of mucosal secretion and immunomodulation (Table 11.1). Parasympathetic innervation is provided by the vagal and pelvic nerves. Postganglionic sympathetic fibers are noradrenergic in nature, and they can corelease adenosine 5′-triphosphate (ATP). Acetylcholine is released by parasympathethic fibers and all the preganglionc fibers; ATP could be coreleased at any of the cholinergic fiber terminals as well.

TABLE 11.1 Sympathetic Innervation of the Gut Wall

Innervation	Target Location	Functional Outcome
Mayor (dense)	Muscle of sphincter regions	Sphincter contraction (inhibition of motility)
	Myenteric plexus	Inhibition of gut motility
	Submucosal plexus	Inhibition of mucosal secretion
	Blood vessels	Vasoconstriction (inhibition of secretion)
Minor (sparse)	Mucosa	Inhibition of secretion
	Muscle of nonsphincter regions	Inhibition of motility
	Lymphoid tissue (Peyer's patches)	Immunomodulation

ENTERIC GLIA—ESSENTIALS

Enteric glia encompasses a diverse population of cells found throughout the plexuses of the gut wall (Fig. 11.1B).[13] Since their discovery at the end of the 19th century,[14] our perspective of these cells changed several times. As they reside outside the CNS, they were initially classified as Schwann cells,[15] peripheral nerve-ensheathing glial cells. However, enteric glia do not ensheath nerve fibers, but rather only partially separate them. Owing to their star-shape appearance, the close proximity to enteric neurons and expression of known astroglia markers, such as glial fibrillary acidic protein (GFAP) and Ca^{2+} binding protein S100β,[16,17] enteric glial cells (EGCs) were also classified as an astrocyte sub-type and termed "intestinal astrocytes." However, there are important distinctions between the enteric glia and astrocytes of the CNS. These two cell types have different embryonic origin, whereby enteric glia derives from the neural crest, while astrocytes do so from the neuroectoderm. Additionally, aldehyde dehydrogenase 1 family member L1, an astrocyte-specific protein, is not reliably expressed in EGCs.[18] Furthermore, recent transcriptional profiling shows that EGCs actually have significant similarities not only to astrocytes, but also to oligodendrocytes and neurons of the CNS.[19] For these reasons it is important to acknowledge EGCs as a distinct glial cell type.

Proximity to enteric neurons and various nerve fibers may indicate that EGCs could be intimately interacting with the neuronal circuits. Indeed, there are direct inputs from neurons through synapse-like, synaptoid contacts (Fig. 11.2A), where axons containing presynaptic specialization and vesicles end on EGCs[20]; "synaptoid" because they lack one essential feature of synaptic contacts, the postsynaptic membrane

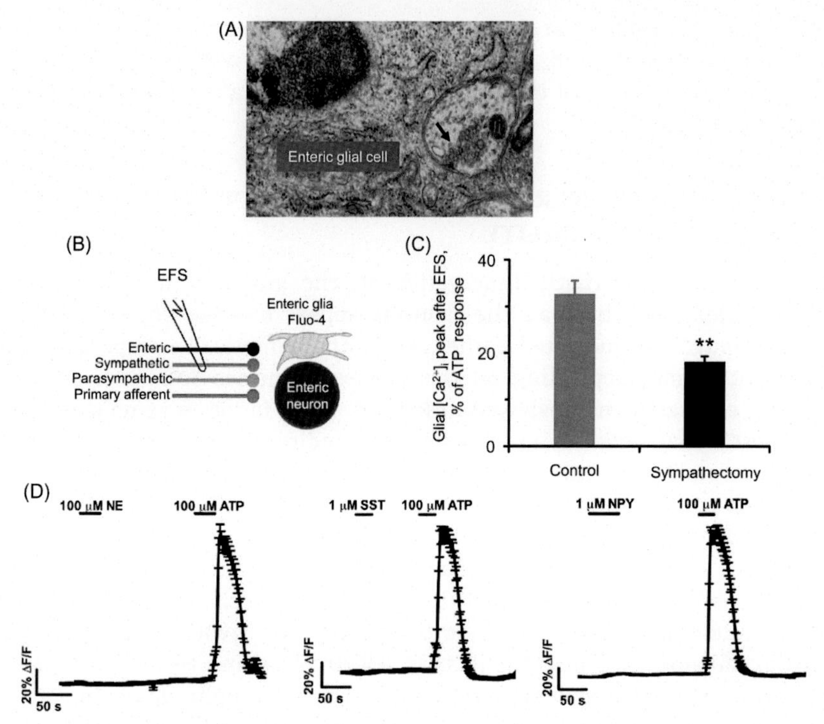

FIGURE 11.2 Enteric glia are innervated by sympathetic nervous system. (A) An electron micrograph of a synaptoid contact between an axon and an EGC of the guinea-pig ileum. An axon contains clear secretory vesicles (*arrow*; likely filled with small neurotransmitter molecules rather than being neuropeptide-laden) and electron dense "presynaptoid" membrane specialization. Note complete absence of electron dense "postsynaptoid" membrane specialization in the EGC; figure obtained from Ref. 20. (B–D) Sympathetic component of the enteric glial innervation; figures obtained from Ref. 21. (B) Experimental schematics. Electrical field stimulation (EFS) activates all nerve fibers innervating the ENS of guinea pig. Fluo-4, a Ca^{2+} indicator dye, is used to detect Ca^{2+} excitability in enteric glia. (C) Chemical sympathetctomy with 6-hydroxydopamine causes significant reduction in glial Ca^{2+} responses upon EFS, expressed scaled (%) to the response evoked by the addition of exogenous ATP; see (D). $^{**}P = .002$ (Student's *t*-test). Plot made based on the data from Ref. 21. (D) ATP, but neither norepinephrine (NE), somatostatin (SST) nor neuropeptide Y (NPY), evokes Ca^{2+} responses in enteric glia (shown as percent change of fluorescence intensity over time, $\Delta F/F$), implicating ATP as the sole active sympathetic neurotransmitter in neuron-glia signaling in the guinea pig distal colon.

specialization. It should be noted that synaptoid contacts are not an isolated phenomenon of the ENS and were also described elsewhere, most notably on pituicytes, stellate astrocyte-like glia of the neurohypophysis.[22–24] As EGCs do not exhibit active electrical currents,[25] synaptoid contacts were initially deemed nonproductive, and EGCs were considered to be passive cells with their sole purpose being the support of neuronal wellbeing. However, with the development of cytosolic Ca^{2+}

indicators, it became clear that EGCs are capable of responding to a variety of fast neurotransmitters, such as acetylcholine and ATP, by intercellular Ca^{2+} increases (reviewed in Ref. 5), thus exhibiting Ca^{2+} rather than electrical excitability.

ENTERIC GLIA CELLS RESPOND TO THE DIRECT SYMPATHETIC INPUT: Ca^{2+} EXCITABILITY

The densest sympathetic innervation in the gut wall is within the enteric plexuses where all the neurons appear to have postganglionic fibers input.[26] As indicated earlier, EGCs are numerically preponderant in the ENS and, depending on the species, comprise 50%–80% of the cellular compartment (reviewed in Ref. 27). Thus, it comes as no surprise that EGCs may receive inputs, direct or indirect, from various nerve fibers. Overall, the ENS is innervated by local enteric neurons, primary afferents (sensory information from the gut wall), and extrinsic sympathetic and parasympathetic fibers (Fig. 11.2B). It is the activity of the extrinsic sympathetic fibers that leads to the functional response in EGCs, seen as an increase in the glial Ca^{2+} levels.[21] EGCs in the gut tissue undergone chemical sympathectomy by treating it with 6-hydroxydopamine, a neurotoxin that selectively eliminates noradrenergic neurons, exhibited about 45% reduction in the amplitude of the Ca^{2+} responses to electrical stimulation of the innervating fibers (Fig. 11.2C). Subsequent application of the sympathetic (co)neurotransmitters showed that enteric glia respond only to the application of ATP, but not to that of noradrenaline/norepineprine, somatostatin, or neuropeptide Y (Fig. 11.2D). Indeed, EGCs glial cells express receptors that can sense sympathetic (co)transmitters (Table 11.2). Expression of an α_{2A}-adrenergic receptor (α_{2A}-AR) in the gut was only detected by immunohistochemistry,[28] so currently it is not clear whether enteric glial α_{2A}-AR is functional and its role is unknown. In contrast, several studies documented expression of functional purinergic receptors $P2Y_1$ and $P2Y_4$ (Table 11.2). They are both G_q-protein coupled receptors that, upon preferentially binding adenosine 5′ diphosphate (ADP) or ATP, respectively, lead to an increase in Ca^{2+} cytosolic concentration in EGCs via the phospholipase C and inositol trisphosphate-receptor signaling pathway (reviewed in Ref. 5). It appears that the receptor expression profile on EGCs is well suited for purinergic signaling carried out by postganglionic sympathetic fibers.

ATP can be extracellularly degraded by membrane-bound ecto-nucleotidases, most commonly nucleoside triphosphate diphosphohydrolase (NTPDase), otherwise expressed in the ENS.[34,35] NTPDase2 is expressed on enteric glia and primarily hydrolyzes ATP, while NTPDase3 is expressed on enteric neurons and utilizes both ATP and ADP as substrates.[36] The

TABLE 11.2 Sympathetic (Co)transmitter Receptors Expressed on Enteric Glial Cells

Neurotransmitter	Receptor	Method	References
Catecholamines	α_{2A} adrenergic	IHC	28
Nucleotides	$P2Y_1$ (ADP > ATP)	Fluo-4	29
			30
			31
	$P2Y_4$ (UTP ≥ ATP)	IHC/Fluo-4	30
		IHC	32
		ICC/Fura-2	33

ADP, adenosine 5′ diphosphate; *ATP*, adenosine 5′-triphosphate; *Fluo-4/Fura-2*, Ca^{2+} indicator dyes; *ICC*, immunocytochemistry; *IHC*, immunohistochemistry; *UTP*, uridine 5′-triphosphate. Sensitivity of $P2Y_{1/4}$ receptors to relevant natural agonists is shown parenthetically. Table is modified from a more comprehensive review on enteric glia.[5]

final NTPDase digest product, the adenosine monophosphate, can be further hydrolyzed by ecto-5′-nucleotidase to adenosine. The abundant products of extracellular ATP hydrolysis, ADP, and adenosine, can differently activate various plasma membrane receptors. However, there is a lack of evidence for functional role of nucleoside/adenosine receptors, albeit adenosine A2B receptor type was detected by immunocytochemistry in the ENS.[37,38] Taken together, in the ENS it appears that EGCs utilize both ATP and ADP as natural stimuli in the plexuses to preferentially activate their $P2Y_4$ and $P2Y_1$, respectively.

It is unclear whether the ENS utilizes additional complexity in the extracellular regulation of nucleotide/nucleoside levels as seen in the CNS (reviewed in Ref. 39). For instance, the exchange of γ-phosphates between adenine- and uracil-based nucleotides via nucleoside diphosphokinase (NDPK) could occur, which, among other species, would provide for uridine 5′ triphosphate (UTP). This nucleotide is an agonist to EGCs' $P2Y_4$ with greater than, or similar binding affinity as, that of ATP[33] and has a role in the (patho)physiology of the GI system, so that this pathway could have consequences on the gut function. Further investigation is needed to tease out this possibility. Of note, this signaling pathway could be affected by infestation with parasites that secrete a range of nucleotide-metabolizing enzymes, including NDPK.[40]

ENTERIC GLIAL Ca^{2+} RESPONSES REGULATE GUT MOTILITY

Neuronal circuity of the myenteric plexus regulates gut motility. During physiological intestinal movements myenteric glia exhibit Ca^{2+}

responses.[41] EGCs Ca^{2+} excitability can also be evoked by ATP released via unpaired connexons/hemichannels in autocrine/paracrine fashion.[42] The major constituent of these connexons in EGCs is connexin 43 (Cx43), which not only mediates EGC Ca^{2+} excitability, but in turn is required for physiological gut motility.[29] Hence, tampering with the glial Ca^{2+} signaling by pharmacological inhibition of Cx43 hemichannels or ablation of Cx43 encoding gene reduces gut smooth muscle contraction and physiological gut motility, the latter seen as an increase in colonic transit time, in otherwise healthy mice (Fig. 11.3A–C).

Be that as it may, enteric glia increased Ca^{2+} excitability could conversely/positively modulate gut motility. Indeed, a study utilizing DREDD (designer receptor exclusively activated by designer drugs) technology showed that selective activation of EGCs Ca^{2+} responses is sufficient to elicit gut smooth muscle contractions and to increase gut motility, the latter seen as a decrease in colonic transit time[43] (Fig. 11.3D–F). Taken together, EGC Ca^{2+} dynamics are sufficient and necessary to modulate the gut motility.

Based on these findings and the existence of EGCs direct innervation by sympathetic fibers coreleasing ATP, one would except that sympathetic activity would lead to an enhancement of gut motility. Quite the contrary, ever since the mid-19th century, it has been known that splanchnic nerve stimulation causes inhibition of peristalsis.[44] This could be in part due to noradrenaline, by acting on presynaptic α_{2A}-ARs[45,46] of the enteric pacemaker neurons, i.e., interstitial cells of Cajal, causing a reduction in release of acetylcholine[47] and consequently peristalsis. Additionally, direct sympathetic stimulation constricts the sphincter regions of intestinal smooth muscles, which decreases gut motility.[46]

Taken together, it appears as that the outputs of selective glial Ca^{2+} excitability (which can be caused by sympathetic stimulus) and that of the ENS upon direct sympathetic stimulation have opposing roles in regulation/modulation of gut motility. Our perspective on these seemingly disparate findings is that EGCs perhaps act as a "buffering" mechanism that prevents complete inhibition of the gut motility under the sympathetic drive. Certainly, this represents an area where further experimental work is needed.

OTHER SELECTED ROLES OF SYMPATHETIC INNERVATION AND ENTERIC GLIA IN THE GUT

Sympathetic innervation and enteric glia exhibit many other roles in intestines including regulation of epithelial functions and immune system modulation. We briefly disused these two roles, with the cautionary note that there is no demonstration of a linear signaling flow from the

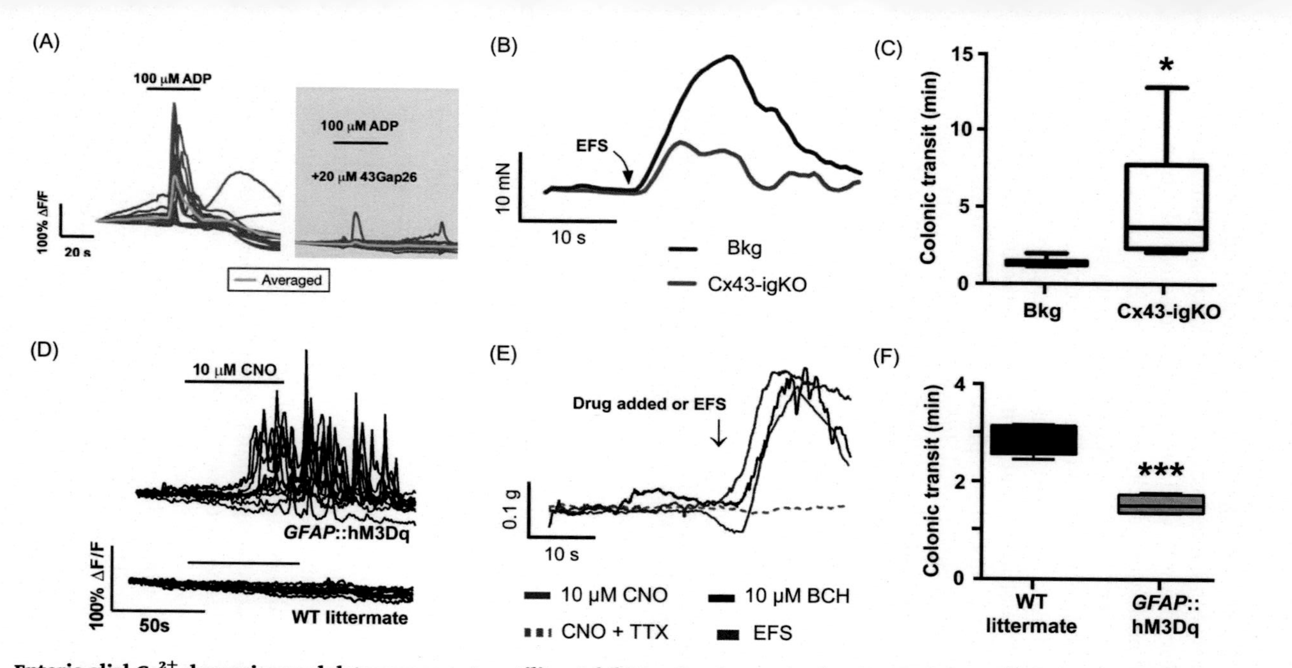

FIGURE 11.3 **Enteric glial Ca^{2+} dynamics modulate mouse gut motility.** Inhibition (A–C) or activation (D–F) of glial Ca^{2+} signaling (A, D) results in reduction or stimulation of the gut motor reflexes, respectively, as assessed by smooth muscle tension recordings (B, E); there are corresponding changes in the distal colon motility tested in vivo (C, F; respectively). (A–C) Experiments from tamoxifen-induced glia-specific knockout of connexin 43 (Cx43) mice (Cx43-igKO) and the tamoxifen-treated background, control strains (Bkg); figures obtained from Ref. 29. (A) ADP-evoked Ca^{2+} responses from wild type (WT) EGCs are blocked by the Cx43 mimetic peptide 43Gap26, an inhibitor of the Cx43 hemichannels. (B) Electrical field stimulation (EFS)-induced smooth gut muscle contractions are reduced in the Cx43-igKO mice. (C) Selective reduction of the Ca^{2+} response due to deletion of Cx43 gene in the enteric glia increases colonic transit time, which reports on a reduction in distal colon motility in vivo. (D–F) Experiments using DREDD (designer receptors exclusively activated by designer drugs) technology, i.e., *GFAP*::hM3Dq transgenic mice, where human *GFAP* promoter drives expression of an engineered G_q-coupled human M3 muscarinic receptor (hM3Dq); WT littermates were used as controls; figures obtained from Ref. 43. (D) Clozapine N-oxide (CNO), a synthetic hM3Dq agonist, elicits cytosolic Ca^{2+} increases in EGCs expressing hM3Dq, but not in controls. (E) CNO application evokes contractions of smooth muscles of gut isolated from *GFAP*::hM3Dq mice; the extent of the response is similar to those evoked by bethanechol (BCH) or EFS, which directly activate smooth muscle and enteric neurons, respectively. Note that CNO effect was blocked by tetrodotoxin (TTX) indicating that glia-specific effects are mediated via enteric neurons. (F) Selective activation of glial Ca^{2+} signaling enhances in vivo motility of the distal colon.

sympathetic innervation to EGCs that results in change in physiological output.

EGCs are necessary for promoting health of intestinal epithelium, a single cell layer that prevents bacteria and other noxious stimuli in the gut lumen from entering the gut wall. Mice with selective ablation of EGCs develop severe inflammation of the gut wall[48] owing to the breakdown of the epithelial barrier.[49] Many glia-derived factors are important for epithelial differentiation, healing, and barrier protection (reviewed in Ref. 50). Proper development of EGCs in intestinal mucosa relies on the glial interaction with gut microbiota and seems that bacterial cues regulate glial migration from the myenteric plexus to the gut lamina propria[51]; the latter is an extra-plexal glial location. Although the exact mechanism of this microbiota–glia interaction is currently unknown, it is possible that bacterial and viral components directly influence enteric glia through actions on glial Toll-like receptors (TLRs).[52] Interestingly, bacteria possess adrenergic sensors/receptors that allows them to interact with their hosts[53,54] and adrenergic signaling indeed modulates bacterial phenotype increasing their pathogenic potential.[55,56] These studies suggest that enteric glia may protect the gut epithelium from the effects of the overactive sympathetic nervous system. In support of this notion, vagal stimulation activated enteric glia to safeguard intestinal barrier function in the states of severe stress.[57]

Sympathetic nerves and enteric glia in their own rights are known to have immunomodulatory roles. Immune cells express ARs allowing them to react to stress responses through the direct sympathetic innervation of the local lymphoid tissue in the gut wall (Table 11.1) or circulating catecholamines.[58] It seems that this interaction depends on the type of the adrenergic receptor such as $\alpha_{1,2}$- and β_3-ARs, which have pro- and antiinflammatory roles, respectively.[59,60] This delineation, however, is probably not absolute as β_3-AR signaling might also have a proinflammatory component.[61] Enteric glia have a way to bidirectionally communicate with the immune cells as they respond to and release immunomodulatory molecules such as interleukins (ILs).[62] These glial cells also express mayor histocompatibility complex type II (MHC-II) molecules,[63] which, combined with the aforementioned property of EGCs to interact with microbes via TLR receptors, suggests that enteric glia could participate as antigen presenting cells. EGCs seem to have a dual role in immunomodulation. They release proinflammatory IL-6 and tumor necrosis factor alpha (TNFα)[64] and mediate inflammation-induced neuronal death through release of ATP and NF-κB (nuclear factor kappa-light-chain-enhancer of activated B cells)-induced production of nitric oxide.[65] On the other hand, enteric glia exhibit immunosuppressive properties inhibiting T cell proliferation.[66] Of note, exposure of cultured EGCs to classical proinflammatory cytokines, such as IL-1β and TNFα, induce expression of the glial cell line-derived neurotrophic factor,[67] which protects enteric glia from apoptosis[68] and has neurotrophic effect on enteric neurons.[69] In summary, these studies collectively show intricate interactions between enteric glia, immune system, and sympathetic nerves, suggesting a biological framework that allows for appropriate adjustment of the local gut immune reaction in response to foreign pathogens and current psychophysical body state.

SYMPATHETIC NERVOUS SYSTEM AND ENTERIC GLIA IN GI DISORDERS/DISEASES

As enteric glia and sympathetic innervation have several regulatory/modulatory roles in normal functioning of intestines, their dysregulation may lead to pathology. In this section we discuss some disorders/diseases to which both EGCs and sympathetic innervation contribute albeit there are no indices of them acting in a concerted manner. We suggest to the reader several review articles for more details regarding individual pathophysiological roles of either sympathetic nervous system[26,58] or EGCs.[70,71]

Unsurprisingly, sympathetic innervation and enteric glia take part in GI motility disorders. Postoperative ileus, a transient episode of impaired GI motility in patients that underwent abdominal surgery,[72] is still a significant clinical and economic burden.[73] Earlier studies showed that inhibition of the sympathetic nervous system through either pharmacological means or sympathectomy relieves the acute phase of paralytic ileus.[74,75] A recent study suggested that glial (along with other cell expressing) IL-1 receptor (IL-1R) participates in development of postoperative ileus since global IL-1R knockout mice do not develop this disorder.[76] Inhibition of IL-1R signaling, therefore, could be a novel therapeutic target in treating paralytic ileus.

Speculatively, an imbalance between sympathetic nerves and enteric glia activity might have a role in development of age-related constipation. Studies in mice showed a decrease in absolute numbers of enteric glia with aging.[77] Furthermore, Cx43, whose loss otherwise slows down gut transit, gets transcriptionally dysregulated leading to reduced protein expression in aging mice.[29] In the light of these studies functional constipation in older mice could be, at least partially, a result of sympathetic tone being less unopposed by glial activity. Since in humans the glial density actually increases with age,[10] the data on mouse gut could be translationally inapplicable to human. Certainly, this imbalance hypothesis deserves more experimental evidence from both animal models and human samples.

Nonetheless, a teeter totter between the enteric glia and sympathetic innervation activity could hypothetically take part in gut infection, diarrheal diseases, and inflammatory bowel disease (IBD). Fluid secretion is important for food digestion and host defenses, as it flushes adverse agents from the gut lumen. Daily fluid transport across mucosa is estimated to be more than the blood volume and physiologically sympathetic system prevents excessive water loss and consequent hypovolemic shock.[4] One of the important functions of the sympathetic innervation in the gut is to limit fluid loss through inhibition of mucosal secretion and vasoconstriction of local blood vessels (Table 11.1) (reviewed in Ref. 6). Elevated stress, i.e., heightened sympathetic activity, could excessively reduce fluid secretion, which in turn weakens gut defenses and is also sensed by gut microbiota as a signal to transition into more pathogenic state.[53,54] EGCs can modulate the fluidic content of fecal pellets as seen in experiments using mice with Cx43 ablation in EGCs.[29] Pellets from these mice contained less matter and higher fluid content, even though their number and total wet mass was similar to those of controls. This suggests that EGCs may play a role in management of water balance in the gut. In support, there is indirect evidence of EGCs' modulatory role of transepithelial fluid transport,[78,79] albeit the conditions (health and/or disease) under which this role is exhibited is still debatable. Recent study demonstrated that glial activity significantly contributes to the neuron depolarization-evoked electrogenic ion transport.[80] Glia-specific stimulation elicited responses that are comparable to those caused by neuron depolarization.[80] Furthermore, glia-stimulated secretory responses are mediated via neurons and also additional mechanisms that may include direct signaling between enteric glial cells and enterocytes.[80] Taken together, these findings suggest that enteric glial activation increases intestinal fluid transport, which is just the opposite from the functional outcomes of the sympathetic innervation (Table 11.1). Of course, this is another area requiring further experimental work.

Patients with IBD have altered microbiota or dysbiosios.[81] Although it is not clear if dysbiosis causes IBD or develops during the course of the disease, prolonged and heightened activity of the sympathetic system due to stressful situations, including inflammation,[82,83] favors infection and, in turn, activates the immune system. Interactions between the microbiome, epithelial barrier function, and immune system are recognized in the pathogenesis of the IBD and current treatments target dysbiosis and overactive immune system.[84] Physiological roles of enteric glia in protection of the intestinal epithelium and immunomodulation, as discussed above, are potentially new targets in IBD translational research. In support, palmitoylethanolamide reduces intestinal inflammation through peroxisome proliferator-activated receptor-α-dependent inhibition of proinflammatory NF-κB singling in EGCs.[85] Due to low toxicity, specific site of action and effective reduction of inflammation, this amide is recognized as potential pharmacological agent against ulcerative colitis, a form of IBD.[85]

CONCLUSIONS

Enteric glia receive direct sympathetic innervation and respond to sympathetic cotransmitter ATP and its extracellular "descendant" ADP with increased cytosolic Ca^{2+} concentration. There is a positive correlation between the enteric glial Ca^{2+} responses and both gut motility and epithelial secretomotor function. Enteric glia and sympathetic innervation appear to have opposing effects on both the gut motility and epithelial fluid transport suggesting a homeostatic role of glial cells in the states of stress (Fig. 11.4). It is important to keep in mind that other (co)transmitters can elicit enteric glial Ca^{2+} responses, including acetylcholine, which is the mayor neurotransmitter of the parasympathetic innervations of the gut and of cholinergic enteric neurons in the gut wall. Clearly, future experiments are warranted to explore possible differential effects that various transmitters may have on EGCs Ca^{2+} excitability and consequently on gut motility in (patho)physiology. It is possible that similar EGC-mediated modulation is applicable to the other intestinal epithelium functions, such as epithelial barrier. Furthermore, EGCs and sympathetic nerves also modulate activity of the gut immune system.

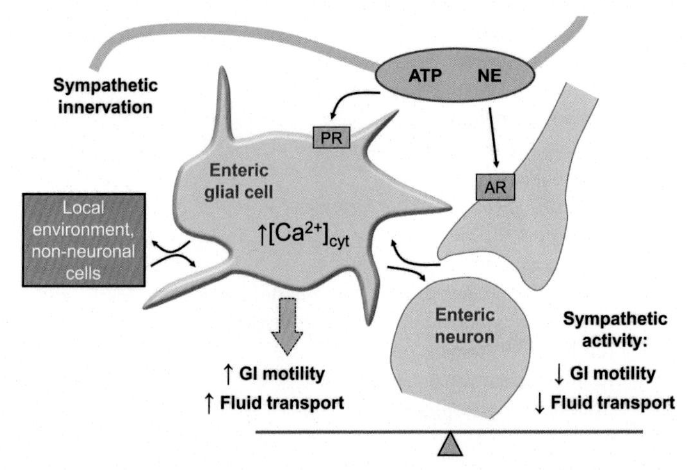

FIGURE 11.4 **Enteric glia as homeostatic regulators of basic gut functions in the states of stress.** Sympathetic varicosities release norepinephrine (NE) activating adrenergic receptors (AR) expressed on enteric neurons and consequently reducing gut motility and fluid secretion from the gut epithelium. In addition, sympathetic varicosities also corelease ATP that binds to glial purinergic receptors (PR) stimulating increase in cytosolic Ca^{2+} concentration. This activation of enteric glial cells, in turn, increases gut motility and epithelial secretion. Enteric glia perhaps acts as a "buffering" mechanism that prevents complete inhibition of the intestinal functions under the symapthetic drive. Of note, enteric glia also sense local environment and bidirectionally communicate with enteric neurons and other cells outside the ENS.

Loss of homeostasis leads to GI diseases such as functional motility disorder, postoperative ileus, infection, diarrhea, and IBD. Therefore better understanding of the (patho)physiological roles of enteric glia and the sympathetic nervous system, as well as their interactions, is necessary for development of new and more effective approaches to treat GI disorders.

ABBREVIATIONS

α_{2A}-AR	α_{2A}-adrenergic receptor
ADP	adenosine 5′-diphosphate
ATP	adenosine 5′-triphosphate
BCH	bethanechol
Fluo-4/Fura-2	Ca^{2+} indicator dyes
Cx43	connexin 43
CNO	clozapine N-oxide
CNS	central nervous system
DREDD	designer receptors exclusively activated by designer drugs
EFS	electrical field stimulation
EGCs	enteric glial cells
ENS	enteric nervous system
GFAP	glial fibrillary acidic protein
GI	gastrointestinal
hM3Dq	G_q-coupled human M3 muscarinic receptor
IBD	inflammatory bowel disease
ICC	immunocytochemistry
IHC	immunohistochemistry
ILs	interleukins
MHC	mayor histocompatibility complex
NDPK	nucleoside diphosphokinase
NE	noradrenaline/norepinephrine
NF-κB	nuclear factor kappa-light-chain-enhancer of activated B cells
NTPDase	nucleoside triphosphate diphosphohydrolase
TLRs	toll-like receptors
TNFα	tumor necrosis factor alpha
UTP	uridine 5′ triphosphate

ACKNOWLEDGMENT

Vladimir Parpura's work is supported by the National Institutes of Health (The Eunice Kennedy Shriver National Institute of Child Health and Human Development award HD078678).

REFERENCES

1. Bayliss WM, Starling EH. The movements and the innervation of the large intestine. *J Physiol.* 1900;26(1–2):107–118.

2. Langley JN. *The Autonomic Nervous System (Pt. I)*. Oxford, England: Heffer; 1921.
3. Furness JB. The enteric nervous system and neurogastroenterology. *Nat Rev Gastroenterol Hepatol.* 2012;9(5):286–294.
4. Furness JB, Callaghan BP, Rivera LR, Cho HJ. The enteric nervous system and gastrointestinal innervation: integrated local and central control. *Adv Exp Med Biol.* 2014; 817:39–71.
5. Grubisic V, Gulbransen BD. Enteric glia: the most alimentary of all glia. *J Physiol.* 2017;595(2):557–570.
6. Furness JB. *The Enteric Nervous System*. Oxford: Blackwell Publishing; 2006.
7. Gabella G. The number of neurons in the small intestine of mice, guinea-pigs and sheep. *Neuroscience.* 1987;22:737–752.
8. Furness JB, Costa M. *The Enteric Nervous System*. Edinburgh: Churchill Livingstone; 1987.
9. Karaosmanoglu T, Aygun B, Wade PR, Gershon MD. Regional differences in the number of neurons in the myenteric plexus of the guinea pig small intestine and colon: an evaluation of markers used to count neurons. *Anat Rec.* 1996;244:470–480.
10. Hoff SS, Zeller FF, Weyhern von CWHC, Wegner MM, Schemann MM, Michel KK, Rühl AA. Quantitative assessment of glial cells in the human and guinea pig enteric nervous system with an anti-Sox8/9/10 antibody. *J Comp Neurol.* 2008;509:356–371.
11. Gabella G, Trigg P. Size of neurons and glial cells in the enteric ganglia of mice, guinea-pigs, rabbits and sheep. *J Neurocytol.* 1984;13:49–71.
12. Ruhl A. Glial cells in the gut. *Neurogastroenterol Motil.* 2005;17(6):777–790.
13. Gulbransen BD, Sharkey KA. Novel functional roles for enteric glia in the gastrointestinal tract. *Nat Rev Gastroenterol Hepatol.* 2012;9(11):625–632.
14. Dogiel AS. Über den Bau der Ganglien in den Geflechten des Darmes und der Gallenblase des Menschen und der Säugetiere [German]. *Arch Anat Physiol Leipzig Anat Abt Jg.* 1899:130–158.
15. Stohr Jr. P. Synopsis of research results on the microscopic innervation of the gastrointestinal tract. *Ergeb Anat Entwicklungsgesch.* 1952;34:250–401.
16. Jessen KR, Mirsky R. Glial cells in the enteric nervous system contain glial fibrillary acidic protein. *Nature.* 1980;286(5774):736–737.
17. Ferri GL, Probert L, Cocchia D, Michetti F, Marangos PJ, Polak JM. Evidence for the presence of S-100 protein in the glial component of the human enteric nervous system. *Nature.* 1982;297(5865):409–410.
18. Boesmans W, Rocha NP, Reis HJ, Holt M, Vanden Berghe P. The astrocyte marker Aldh1L1 does not reliably label enteric glial cells. *Neurosci Lett.* 2014;566:102–105.
19. Rao M, Nelms BD, Dong L, et al. Enteric glia express proteolipid protein 1 and are a transcriptionally unique population of glia in the mammalian nervous system. *Glia.* 2015. Available from: http://dx.doi.org/10.1002/glia.22876.
20. Gabella G. Fine structure of the myenteric plexus in the guinea-pig ileum. *J Anat.* 1972;111(Pt 1):69–97.
21. Gulbransen BD, Bains JS, Sharkey KA. Enteric glia are targets of the sympathetic innervation of the myenteric plexus in the guinea pig distal colon. *J Neurosci.* 2010;30 (19):6801–6809.
22. Wittkowski W, Brinkmann H. Changes of extent of neuro-vascular contacts and number of neuro-glial synaptoid contacts in the pituitary posterior lobe of dehydrated rats. *Anat Embryol (Berl).* 1974;146(2):157–165.
23. Buijs RM, van Vulpen EH, Geffard M. Ultrastructural localization of GABA in the supraoptic nucleus and neural lobe. *Neuroscience.* 1987;20(1):347–355.

24. van Leeuwen FW, Pool CW, Sluiter AA. Enkephalin immunoreactivity in synaptoid elements on glial cells in the rat neural lobe. *Neuroscience*. 1983;8(2):229–241.
25. Hanani M, Francke M, Hartig W, Grosche J, Reichenbach A, Pannicke T. Patch-clamp study of neurons and glial cells in isolated myenteric ganglia. *Am J Physiol Gastrointest Liver Physiol*. 2000;278(4):G644–G651.
26. Lomax AE, Sharkey KA, Furness JB. The participation of the sympathetic innervation of the gastrointestinal tract in disease states. *Neurogastroenterol Motil*. 2010;22(1):7–18.
27. Gulbransen B.D. Enteric Glia. Morgan & Claypool Publishers; Colloquium Digital Library of Life Sciences; 2014.
28. Nasser Y, Ho W, Sharkey KA. Distribution of adrenergic receptors in the enteric nervous system of the guinea pig, mouse, and rat. *J Comp Neurol*. 2006;495(5):529–553.
29. McClain JL, Grubisic V, Fried D, et al. Ca2 + responses in enteric glia are mediated by connexin-43 hemichannels and modulate colonic transit in mice. *Gastroenterology*. 2014;146(2):497–507:e491.
30. Gulbransen BD, Sharkey KA. Purinergic neuron-to-glia signaling in the enteric nervous system. *Gastroenterology*. 2009;136(4):1349–1358.
31. Gulbransen BD, Bashashati M, Hirota SA, et al. Activation of neuronal P2X7 receptor-pannexin-1 mediates death of enteric neurons during colitis. *Nat Med*. 2012;18 (4):600–604.
32. Van Nassauw L, Costagliola A, Van Op den Bosch J, et al. Region-specific distribution of the P2Y4 receptor in enteric glial cells and interstitial cells of Cajal within the guinea-pig gastrointestinal tract. *Auton Neurosci*. 2006;126-127:299–306.
33. Kimball BC, Mulholland MW. Enteric glia exhibit P2U receptors that increase cytosolic calcium by a phospholipase C-dependent mechanism. *J Neurochem*. 1996;66(2):604–612.
34. Braun N, Sevigny J, Robson SC, Hammer K, Hanani M, Zimmermann H. Association of the ecto-ATPase NTPDase2 with glial cells of the peripheral nervous system. *Glia*. 2004;45(2):124–132.
35. Lavoie EG, Gulbransen BD, Martin-Satue M, Aliagas E, Sharkey KA, Sevigny J. Ectonucleotidases in the digestive system: focus on NTPDase3 localization. *Am J Physiol Gastrointest Liver Physiol*. 2011;300(4):G608–G620.
36. Kukulski F, Levesque SA, Lavoie EG, et al. Comparative hydrolysis of P2 receptor agonists by NTPDases 1, 2, 3 and 8. *Purinergic Signal*. 2005;1(2):193–204.
37. Christofi FL, Zhang H, Yu JG, et al. Differential gene expression of adenosine A1, A2a, A2b, and A3 receptors in the human enteric nervous system. *J Comp Neurol*. 2001;439 (1):46–64.
38. Vieira C, Ferreirinha F, Silva I, Duarte-Araujo M, Correia-de-Sa P. Localization and function of adenosine receptor subtypes at the longitudinal muscle--myenteric plexus of the rat ileum. *Neurochem Int*. 2011;59(7):1043–1055.
39. Lazarowski ER, Boucher RC, Harden TK. Mechanisms of release of nucleotides and integration of their action as P2X- and P2Y-receptor activating molecules. *Mol Pharmacol*. 2003;64(4):785–795.
40. Gounaris K. Nucleotidase cascades are catalyzed by secreted proteins of the parasitic nematode Trichinella spiralis. *Infect Immun*. 2002;70(9):4917–4924.
41. Broadhead MJ, Bayguinov PO, Okamoto T, Heredia DJ, Smith TK. Ca2 + transients in myenteric glial cells during the colonic migrating motor complex in the isolated murine large intestine. *J Physiol*. 2012;590(2):335–350.
42. Zhang W, Segura BJ, Lin TR, Hu Y, Mulholland MW. Intercellular calcium waves in cultured enteric glia from neonatal guinea pig. *Glia*. 2003;42(3):252–262.

43. McClain JL, Fried DE, Gulbransen BD. Agonist-evoked Ca^{2+} signaling in enteric glia drives neural programs that regulate intestinal motility in mice. *Cell Mol Gastroenterol Hepatol.* 2015;1(6):631–645.
44. Pflüger EFW. *Ueber das Hemmungs-Nervensystem für die peristaltischen Bewegungen der Gedärme.* Berlin: A. Hirschwald; 1857.
45. Hirst GD, McKirdy HC. Presynaptic inhibition at mammalian peripheral synapse? *Nature.* 1974;250(465):430–431.
46. Stebbing M, Johnson P, Vremec M, Bornstein J. Role of alpha(2)-adrenoceptors in the sympathetic inhibition of motility reflexes of guinea-pig ileum. *J Physiol.* 2001;534(Pt. 2):465–478.
47. Vizi ES, Knoll J. The effects of sympathetic nerve stimulation and guanethidine on parasympathetic neuroeffector transmission; the inhibition of acetylcholine release. *J Pharm Pharmacol.* 1971;23(12):918–925.
48. Bush TG, Savidge TC, Freeman TC, et al. Fulminant jejuno-ileitis following ablation of enteric glia in adult transgenic mice. *Cell.* 1998;93(2):189–201.
49. Savidge TC, Newman P, Pothoulakis C, et al. Enteric glia regulate intestinal barrier function and inflammation via release of S-nitrosoglutathione. *Gastroenterology.* 2007;132 (4):1344–1358.
50. Neunlist M, Van Landeghem L, Mahe MM, Derkinderen P, des Varannes SB, Rolli-Derkinderen M. The digestive neuronal-glial-epithelial unit: a new actor in gut health and disease. *Nat Rev Gastroenterol Hepatol.* 2013;10(2):90–100.
51. Kabouridis PS, Lasrado R, McCallum S, et al. Microbiota controls the homeostasis of glial cells in the gut lamina propria. *Neuron.* 2015;85(2):289–295.
52. Barajon I, Serrao G, Arnaboldi F, et al. Toll-like receptors 3, 4, and 7 are expressed in the enteric nervous system and dorsal root ganglia. *J Histochem Cytochem.* 2009;57 (11):1013–1023.
53. Sperandio V, Torres AG, Jarvis B, Nataro JP, Kaper JB. Bacteria-host communication: the language of hormones. *Proc Natl Acad Sci USA.* 2003;100(15):8951–8956.
54. Clarke MB, Hughes DT, Zhu C, Boedeker EC, Sperandio V. The QseC sensor kinase: a bacterial adrenergic receptor. *Proc Natl Acad Sci USA.* 2006;103(27):10420–10425.
55. Cogan TA, Thomas AO, Rees LE, et al. Norepinephrine increases the pathogenic potential of Campylobacter jejuni. *Gut.* 2007;56(8):1060–1065.
56. Reading NC, Rasko DA, Torres AG, Sperandio V. The two-component system QseEF and the membrane protein QseG link adrenergic and stress sensing to bacterial pathogenesis. *Proc Natl Acad Sci USA.* 2009;106(14):5889–5894.
57. Costantini TW, Bansal V, Krzyzaniak M, et al. Vagal nerve stimulation protects against burn-induced intestinal injury through activation of enteric glia cells. *Am J Physiol Gastrointest Liver Physiol.* 2010;299(6):G1308–G1318.
58. Cervi AL, Lukewich MK, Lomax AE. Neural regulation of gastrointestinal inflammation: role of the sympathetic nervous system. *Auton Neurosci.* 2014;182:83–88.
59. Vasina V, Abu-Gharbieh E, Barbara G, et al. The beta3-adrenoceptor agonist SR58611A ameliorates experimental colitis in rats. *Neurogastroenterol Motil.* 2008;20 (9):1030–1041.
60. Bai A, Lu N, Guo Y, Chen J, Liu Z. Modulation of inflammatory response via alpha2-adrenoceptor blockade in acute murine colitis. *Clin Exp Immunol.* 2009;156(2):353–362.
61. Straub RH, Stebner K, Harle P, Kees F, Falk W, Scholmerich J. Key role of the sympathetic microenvironment for the interplay of tumour necrosis factor and interleukin 6 in normal but not in inflamed mouse colon mucosa. *Gut.* 2005;54(8):1098–1106.

62. Ruhl A, Franzke S, Collins SM, Stremmel W. Interleukin-6 expression and regulation in rat enteric glial cells. *Am J Physiol Gastrointest Liver Physiol.* 2001;280(6):G1163–G1171.
63. Geboes K, Rutgeerts P, Ectors N, et al. Major histocompatibility class II expression on the small intestinal nervous system in Crohn's disease. *Gastroenterology.* 1992;103(2):439–447.
64. Bhave S., Brun P., Dewey W., Akbarali H. The role of toll-like receptor 4 in enteric glia. *FASEB J.* 2015;29:no. 1 Supplement (628.626).
65. Brown IA, McClain JL, Watson RE, Patel BA, Gulbransen BD. Enteric glia mediate neuron death in colitis through purinergic pathways that require connexin-43 and nitric oxide. *Cell Mol Gastroenterol Hepatol.* 2016;2(1):77–91.
66. Kermarrec L, Durand T, Neunlist M, Naveilhan P, Neveu I. Enteric glial cells have specific immunosuppressive properties. *J Neuroimmunol.* 2016;295-296:79–83.
67. von Boyen GB, Steinkamp M, Geerling I, et al. Proinflammatory cytokines induce neurotrophic factor expression in enteric glia: a key to the regulation of epithelial apoptosis in Crohn's disease. *Inflamm Bowel Dis.* 2006;12(5):346–354.
68. Steinkamp M, Gundel H, Schulte N, et al. GDNF protects enteric glia from apoptosis: evidence for an autocrine loop. *BMC Gastroenterol.* 2012;12:6.
69. Gougeon PY, Lourenssen S, Han TY, Nair DG, Ropeleski MJ, Blennerhassett MG. The pro-inflammatory cytokines IL-1beta and TNFalpha are neurotrophic for enteric neurons. *J Neurosci.* 2013;33(8):3339–3351.
70. Ochoa-Cortes F, Turco F, Linan-Rico A, et al. Enteric glial cells: a new frontier in neurogastroenterology and clinical target for inflammatory bowel diseases. *Inflamm Bowel Dis.* 2016;22(2):433–449.
71. Sharkey KA. Emerging roles for enteric glia in gastrointestinal disorders. *J Clin Invest.* 2015;125(3):918–925.
72. van Bree SH, Nemethova A, Cailotto C, Gomez-Pinilla PJ, Matteoli G, Boeckxstaens GE. New therapeutic strategies for postoperative ileus. *Nat Rev Gastroenterol Hepatol.* 2012;9(11):675–683.
73. Barletta JF, Senagore AJ. Reducing the burden of postoperative ileus: evaluating and implementing an evidence-based strategy. *World J Surg.* 2014;38(8):1966–1977.
74. Neely J, Catchpole B. Ileus: the restoration of alimentary-tract motility by pharmacological means. *Br J Surg.* 1971;58(1):21–28.
75. Petri G, Szenohradszky J, Porszasz-Gibiszer K. Sympatholytic treatment of "paralytic" ileus. *Surgery.* 1971;70(3):359–367.
76. Stoffels B, Hupa KJ, Snoek SA, et al. Postoperative ileus involves interleukin-1 receptor signaling in enteric glia. *Gastroenterology.* 2014;146(1):176–187:e171.
77. Stenkamp-Strahm C, Patterson S, Boren J, Gericke M, Balemba O. High-fat diet and age-dependent effects on enteric glial cell populations of mouse small intestine. *Auton Neurosci.* 2013;177(2):199–210.
78. MacEachern SJ, Patel BA, Keenan CM, et al. Inhibiting Inducible Nitric Oxide Synthase in Enteric Glia Restores Electrogenic Ion Transport in Mice With Colitis. *Gastroenterology.* 2015;149(2):445–455:e443.
79. MacEachern SJ, Patel BA, McKay DM, Sharkey KA. Nitric oxide regulation of colonic epithelial ion transport: a novel role for enteric glia in the myenteric plexus. *J Physiol.* 2011;589(Pt 13):3333–3348.
80. Grubisic V, Gulbransen BD. Enteric glial activity regulates secretomotor function in the mouse colon but does not acutely affect gut permeability. *J Physiol.* 2017. Available from: http://dx.doi.org/10.1113/JP273492 [Epub ahead of print].

81. Sha S, Xu B, Wang X, et al. The biodiversity and composition of the dominant fecal microbiota in patients with inflammatory bowel disease. *Diagn Microbiol Infect Dis.* 2013;75(3):245–251.
82. Sharkey KA, Parr EJ, Keenan CM. Immediate-early gene expression in the inferior mesenteric ganglion and colonic myenteric plexus of the guinea pig. *J Neurosci.* 1999;19(7):2755–2764.
83. Dong XX, Thacker M, Pontell L, Furness JB, Nurgali K. Effects of intestinal inflammation on specific subgroups of guinea-pig celiac ganglion neurons. *Neurosci Lett.* 2008;444(3):231–235.
84. Vindigni SM, Zisman TL, Suskind DL, Damman CJ. The intestinal microbiome, barrier function, and immune system in inflammatory bowel disease: a tripartite pathophysiological circuit with implications for new therapeutic directions. *Therap Adv Gastroenterol.* 2016;9(4):606–625.
85. Esposito G, Capoccia E, Turco F, et al. Palmitoylethanolamide improves colon inflammation through an enteric glia/toll like receptor 4-dependent PPAR-alpha activation. *Gut.* 2014;63(8):1300–1312.

Chapter 12

Noradrenaline Drives Structural Changes in Astrocytes and Brain Extracellular Space

Ang D. Sherpa[1,2,✉], **Chiye Aoki**[2,✉] **and Sabina Hrabetova**[1,✉]
[1]State University of New York Downstate Medical Center, Brooklyn, NY, United States, [2]New York University, New York, NY, United States

Chapter Outline

The Noradrenergic System—General Remarks 242
Diversity of Noradrenergic Receptor Expression Underlies Diversity of Astrocytic Responses 242
Noradrenergic System Relates to Function of Astrocytes 243
Noradrenergic System's Effects on Astrocytes In Vitro 245
Noradrenergic System's Effects on Astrocytes In Situ 245
Brain Extracellular Space 247
Noradrenergic System's Effects on Extracellular Space Structure 249
Conclusions 251
Abbreviations 252
Acknowledgments 252
References 252

ABSTRACT
Locus coeruleus neurons innervate multiple brain regions. These neurons release noradrenaline through their axonal varicosities into the extracellular space through synaptic and volume transmission during states of arousal. The extracellular space is a channel that surrounds brain cells, facilitating diffusion-mediated transport of signaling molecules, ions, and drugs. Distal astrocytic processes expressing β-adrenergic receptors are targets of noradrenaline. In this review, we discuss work in cortical tissue indicating that β-adrenergic agonist, isoproterenol, expands astrocytic processes. Isoproterenol-driven increase in volume of astrocytic processes contributes partially to decrease in the extracellular space volume from 22% to 18%. Decrease in the extracellular space volume suggests increased concentration of ions, neurotransmitters, and neuromodulators diffusing in the

✉Correspondence address
E-mail: ads420@nyu.edu; ca3@nyu.edu; sabina.hrabetova@downstate.edu

Noradrenergic Signaling and Astroglia. DOI: http://dx.doi.org/10.1016/B978-0-12-805088-0.00012-8
© 2017 Elsevier Inc. All rights reserved.

extracellular space, which, in turn, facilitates neuronal signaling during noradrenaline release in cortex.

Keywords: Noradrenergic system; volume transmission; astrocytes; diffusion; extracellular space; volume fraction; tortuosity; real-time iontophoretic (RTI) method; electron microscopy; sleep—wake cycle

THE NORADRENERGIC SYSTEM—GENERAL REMARKS

Noradrenergic signaling is implicated in learning and memory, attention, anxiety and stress, arousal, and mood[1,2] and in experience-dependent cortical plasticity.[3–6] A major source of noradrenergic signal in the central nervous system arises from a collection of brain stem neurons, called locus coeruleus (LC). LC neurons synchronously fire together in tonic and phasic modes. A single LC neuron innervates various brain regions, including the cortex, the cerebellum, the hypothalamus, and the spinal cord, through axonal branching to exert a widespread synchronous influence on these brain circuits.[7] Recent work also suggests that there are functionally and anatomically separate LC projection groups, since Chandler et al.[8] showed that LC neurons innervating subregions of the prefrontal cortex are distinct from those innervating the motor cortex.

DIVERSITY OF NORADRENERGIC RECEPTOR EXPRESSION UNDERLIES DIVERSITY OF ASTROCYTIC RESPONSES

The functional outcome of noradrenaline (NA) released from axonal varicosities of LC neurons depends on the distribution of noradrenergic receptors among brain cells within and across brain regions. All three noradrenergic receptors (ARs), α_1 (subtypes: α_{1A}, α_{1B}, and α_{1D}), α_2 (subtypes: α_{2A}, α_{2B}, α_{2C}, and α_{2D}), and β (β_1, β_2, and β_3), are metabotropic G-protein coupled receptors. In neurons, α_1-AR and β-AR are postsynaptic while α_2-ARs are both pre and postsynaptic and β_2-ARs are shown to act presynaptically.[1] In astrocytes, activation of α_1-ARs activates the enzyme, phospholipase C, that catalyzes the conversion of phosphatidylinositol 4,5-bisphosphate into two signaling molecules, inositol 1,4,5-trisphosphate (IP_3) and diacylglycerol. The released IP_3 binds to a transmembrane glycoprotein on the endoplasmic reticulum called IP_3 receptor ($InsP_3R$), which induces a conformational change in $InsP_3R$ that subsequently allows $InsP_3R$ to release calcium from the endoplasmic reticulum into the cytoplasm.[9] Aoki et al.[10] localized α_{2A}-ARs on postsynaptic membranes and in nonsynaptic locations of axons, dendritic shafts, and astrocytic processes in the monkey dorsolateral prefrontal cortex. Astrocytic α_2-ARs interact preferentially with G_i-proteins that inhibit adenylyl cyclases and 3′,5′-cyclic adenosine monophosphate (cAMP) production. Within astrocytes, inactivation of adenylyl cyclases promotes glycogenesis, but through its $\beta\gamma$ subunit, α_2-ARs also couple positively with protein kinase C and calcium to promote glycogenolysis under certain situations.[9]

Activation of β-ARs stimulates adenylyl cyclases, which, in turn, increase the formation of cAMP. Upregulation of cAMP within astrocytes activates cAMP-dependent protein kinases that phosphorylate cytoskeletal proteins, such as glial fibrillary acidic protein (GFAP) and vimentin to alter the morphology of astrocytic processes[11,12] and glycogen phosphorylase to promote glycogenolysis.[9] Through these downstream effectors, adrenergic receptors (α_1, α_2, and β) contribute to the many astrocytic functions such as glycogen metabolism, immune response regulation, release of neurotrophins, such as brain-derived neurotrophic factor and nerve growth factor, and morphological changes.[9,13,14] Astrocytic β-ARs are also implicated in brain diseases, such as Alzheimer's disease, stroke, multiple sclerosis, and human immunodeficiency virus encephalitis.[13]

Using immuno-electron microscopy, Aoki[15] found a ninefold greater encounter of β-ARs on astrocytic processes of adult cortex, compared to neuronal processes, suggesting that nonneuronal component is more responsive to NA. This notion is supported by a recent study showing a robust calcium signaling induced in neocortical astrocytes but not neurons by NA, acting predominantly through α_1-ARs.[16] Additionally, it is important to note that β-ARs expressed on astrocytes that ensheath synapses occur displaced from noradrenergic terminals.[15,17] The distribution of ARs on multiple cell targets such as neurons, glia, and microglia within the central nervous system[18,19] suggests that NA released into the surrounding extracellular space (ECS) can have multiple targets besides the synaptic junction. This is achieved through volume transmission, i.e., extrasynaptic diffusion-mediated transport of transmitters, modulators, trophic factors, and neurotransmitters, including NA, in the ECS,[14] that allows for interactions among their many targets.

NORADRENERGIC SYSTEM RELATES TO FUNCTION OF ASTROCYTES

Astrocytes are a major type of glial cells in the central nervous system and provide many additional functions. Astrocytes participate in maintaining ionic composition and pH of extracellular fluid,[20] and they accommodate a powerful uptake system for a major excitatory neurotransmitter, glutamate.[21] Astrocytes express the major Na^+-dependent glutamate transporters GLT-1 (Glutamate Transporter-1) and GLAST (Glutamate Aspartate Transporter) that remove extracellular glutamate released during synaptic transmission, thereby preventing excitotoxicity.[21] Once accumulated inside astrocytes, glutamate is converted to glutamine with the enzyme glutamine synthetase and is transported back to neurons.[22] Application of an α_1-AR agonist, phenylephrine, increases glutamate uptake in vivo, indicating that the noradrenergic signal intensifies glutamate uptake.[23] Astrocytes are involved in maintaining extracellular level of potassium ($[K^+]_{ECS}$) around 3 mM. During neuronal activity, $[K^+]_{ECS}$ increases to 10–12 mM, and during anoxic depolarization associated with ischemic conditions in brain, $[K^+]_{ECS}$ can rise to become as high as 70 mM.[24,25] Astrocytes remove excess potassium through passive "spatial buffering," facilitated by inwardly rectifying K^+ channels and coupling of astrocytes via gap junctions, and through a Na^+/K^+-ATPase pump that pumps K^+, along with water molecules, inward of astrocytes.[26,27] Astrocytic β-AR agonist, isoproterenol (ISO), activates Na^+/K^+-ATPase, to increase the removal of $[K^+]_{ECS}$.[28] Thus, the noradrenergic system enables astrocytic functions through α_1-AR and β-ARs. Astrocytes also express water channels, aquaporin 4 (AQP4) that contribute to water homeostasis and transport within brain.[29,30] While functional interplay between the AQP4 channels and the NA system remains to be tested, it has been shown that the levels of NA are increased in the medial prefrontal cortex of AQP4-knockout mice.[31] In short, through a variety of mechanisms, astrocytes modulate neuronal communication and make an impact on excitability of neuronal population, both of which may be enhanced further by the release of a number of substances from glia.

Morphology of astrocytes and their distribution in the neuropil are well suited in accomplishing these diverse functions. A small cell body of an astrocyte is surrounded by a network of astrocytic processes. Proximal astrocytic processes extending from the cell body are less numerous but thick and contain the intermediate protein, GFAP,[32,33] which undergoes phosphorylation in response to changes in cAMP level that, in turn, alter astrocyte morphology.[11,12] The proximal astrocytic processes extend into morphologically complex distal processes devoid of GFAP. In contrast to the proximal processes, distal astrocytic processes are very thin and thread-like but account for about 85% of an astrocyte's volume.[34] These fine distal astrocytic processes are interposed between neurons and their processes, often wrapping structures, such as synapses. They are also positioned at the interface between the brain and surrounding tissue, forming glia limitans that line the pia mater or the astrocytic endfeet on blood vessels.[32,35] In fact, since these distal fine processes express ARs, transporters, and channels, they are the main functional domain of astrocytes. As will be described below,[36] we have evidence indicating that these fine distal processes that are devoid of GFAP nevertheless undergo morphological changes induced by β-ARs. These changes may occur via depolymerization of F-actin,[37] which cannot be visualized readily by conventional electron microscopic methods.

Astrocytic processes need to be distributed throughout the neuropil in order to be effective in ionic and water regulation and, indeed, they are.

At the same time, astrocytic processes need to be present at specific locations within the neuropil (e.g., close to synapses) in order to remove excess glutamate from perisynaptic regions as well as to participate in the modulation of synaptic transmission. This is accomplished through a uniform distribution of astrocytes within the neuropil. It has been reported that each astrocyte occupies a separate anatomical domain, resulting in a nonoverlapping tiled layout in the neuropil.[34,38,39] Because of their complex morphology and uniform distribution, astrocytes function as diffusion barriers for signaling molecules, including neuromodulators and ions released into the ECS synaptically and extrasynaptically, thereby contributing to the structural and molecular properties of the ECS.

NORADRENERGIC SYSTEM'S EFFECTS ON ASTROCYTES IN VITRO

Noradrenergic signal can induce morphological reorganization of astrocytic processes and morphological changes of astrocytes in culture and hypothalamic slices, where ISO has been used to activate β-AR on cultured astrocytes.[11,40] Cultured astrocytes from adult rat neurohypophyses[41] and astrocytes cultured from rat cortex at postnatal days 2–3[12] alter their morphology upon activation of β-AR. Bicknell et al.[40] found that pituicytes cultured from neurohypophyses are transformed from amorphous to stellate morphology when a β-AR agonist, ISO, is added to the medium. Vardjan et al.[12] performed quantitative analysis of astrocytes cultured from cortical brain tissue during incubation with ISO. It was reported that the cross-sectional area of astrocytes is reduced while the perimeter of astrocytes is increased upon β-AR activation, and that cultured astrocytes acquire a stellate morphology. We note that Vardjan's morphometric analyses focused on the more proximal astrocytic processes containing GFAP detected by immunolabeling, and therefore the fine GFAP-negative portions of astrocytic processes that could not be detected by confocal microscopy were excluded from this analysis.

NORADRENERGIC SYSTEM'S EFFECTS ON ASTROCYTES IN SITU

Aoki[15,42] reported a rich expression of β-AR at neuronal membranes of cortical synapses neonatally, but of their dominance at astrocytes in adulthood. Based on this study, Sherpa et al.[36] speculated that noradrenergic input to the adult visual cortex is likely to change the morphology of astrocytic processes, and that this morphological change, in turn, changes the structure of ECS and impacts the diffusional transport of substances in the ECS. In contrast to previous studies in cultured astrocytes, the β-AR agonist, ISO, was applied to astrocytes embedded within the microenvironment of intact neuropil of acutely prepared adult visual cortex slices, and the analysis was focused on fine, distal astrocytic processes, previously estimated to occupy 85% of an astrocytic volume.[34] In agreement with previous studies in culture,[12] Sherpa et al.[36] reported significant changes of astrocytic morphology.

Sherpa et al.[36] found that ISO induced changes in several parameters of astrocytic morphology using electron microscopy (Fig. 12.1). First, the

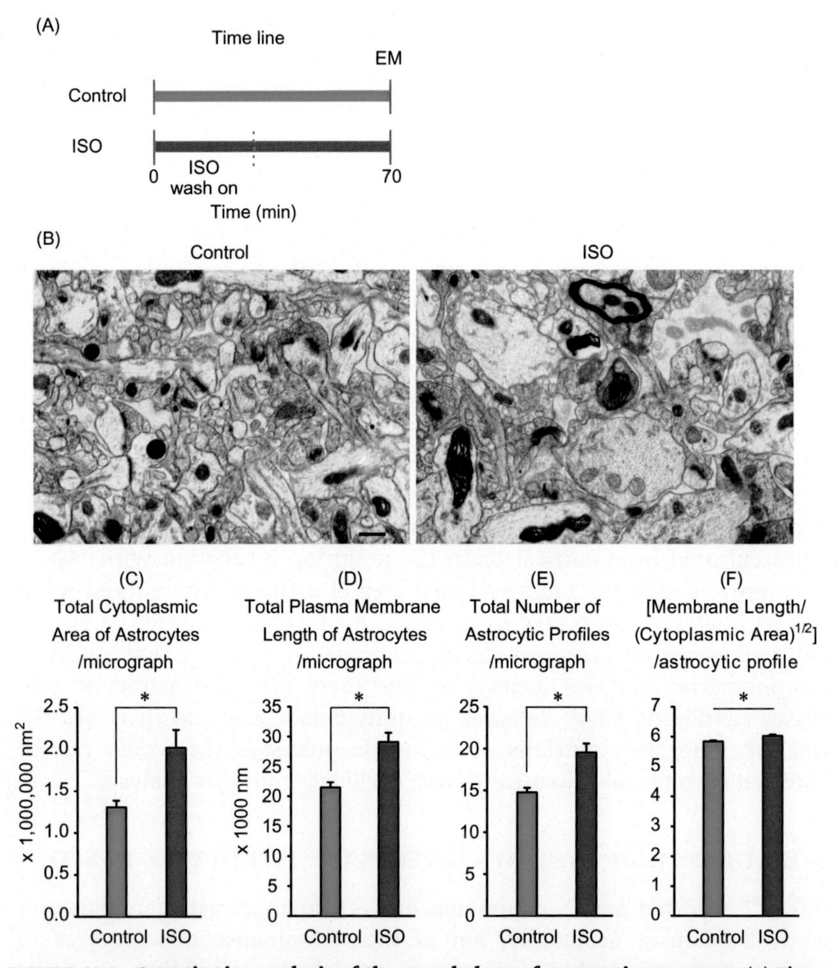

FIGURE 12.1 Quantitative analysis of the morphology of astrocytic processes. (A) Time line of electron microscopy experiment. (B) Representative electron microscopic images of neuropil of the visual cortex under control and ISO conditions. Astrocytic profiles are colored in yellow. Scale bar: 500 nm. (C–F) Summary of morphometric analyses (mean ± SEM). Control, green; ISO, red; asterisk, significant. *Reproduced from Sherpa AD, Xiao F, Joseph N, Aoki C, Hrabetova S. Activation of β-adrenergic receptors in rat visual cortex expands astrocytic processes and reduces extracellular space volume.* Synapse. *2016;70:307–316, with the permission of John Wiley & Sons, Inc.*

total cytoplasmic area of astrocytic processes was increased, suggesting expansion of distal astrocytic processes. Second, the total plasma membrane length of astrocytic processes encountered within electron micrographs was increased, suggesting new membrane synthesis. Third, the total number of astrocytic profiles encountered per unit area was increased. Fourth, the ratio of the plasma membrane length to (cytoplasmic area)$^{1/2}$ for each astrocytic profile was increased. Changes in the last two parameters suggest the formation of additional astrocytic profiles after ISO treatment.

All findings of Sherpa et al.,[36] except for the first one, agree with quantitative analyses of cultured astrocytes exposed to ISO.[12] While Sherpa et al.[36] found an increase in the total cytoplasmic area of astrocytic processes, Vardjan et al.[12] reported that the cross-sectional area of astrocytes was reduced upon β-AR activation. As alluded to above, this discrepancy may arise from the different domains of astrocytes that were studied: while Sherpa et al.[36] analyzed distal astrocytic processes, Vardjan et al.[12] analyzed the astrocytic cell body and proximal processes. It is likely that these results are, in fact, in agreement with each other. As a cultured astrocyte changes from the amorphous to stellate morphology, its cytoplasm is likely to be redistributed between the cell body and the distal processes. This would result in findings reported by Vardjan et al.[12] and Sherpa et al.[36] Taken together, study of astrocytic morphology in intact brain neuropil suggests that β-AR-induced changes in astrocytic morphology may have important functional implications by allowing for repositioning of targets of signaling molecules, such as neurotransmitter transporters, ion channels, receptors (including β-AR), and aquaporin channels within brain neuropil.

Besides β-AR, astrocytes respond with activation of α_1-ARs during the release of NA in the cortex. Ding et al.[43] observed a widespread increase in astrocytic calcium level in cortical astrocytes of awake mice from activation of α_1-ARs. Later, Paukert et al.[44] observed that activation of astrocytic α_1-AR induces a rise of intracellular calcium level simultaneously in cerebellum and visual cortex during locomotion. The NA mediated enhancement in calcium signaling in astrocytes is thought to facilitate astrocytes in detecting changes in neural activity. β-AR-mediated morphological changes in astrocytes also occur over an area of visual cortical neuropil[36] similar to the increase in cortical calcium signaling during α_1-AR activation.

BRAIN EXTRACELLULAR SPACE

The β-AR-induced expansion of astrocytic processes contributes to the diffusion properties of brain ECS. Brain ECS is a large compartment of brain tissue formed by numerous narrow spaces that surround brain cells. It occupies about 20% of the total brain tissue volume,[45,46] but the individual intercellular gaps are only about 30−60 nm wide.[47,48] The ECS is filled with ionic solution and macromolecules of the extracellular matrix (ECM), predominantly proteoglycans and glycosaminoglycans. The ECS has a fundamental role in brain function. It facilitates diffusional transport of neuroactive substances, nutrients, metabolites, and therapeutic agents, while also serving as a reservoir of ions and growth factors sequestered by the ECM. Extracellular concentration and distribution of these substances is determined by the ECS structure.

Early studies of the ECS structure employed electron microscopy, which has a power to resolve individual narrow intercellular channels. Today, the morphometric approaches are sparingly used whenever the ECS parameters need to be precisely quantified. The main drawback of morphometric approaches is that the conventional fixation procedures cause significant water redistribution, leading to distortion of the ECS structure.[47,49] It has been shown that aldehydes required for ultrastructural preservation causes greater than 10% loss of the ECS.[50]

Transport of substances in the ECS is primarily mediated by diffusion. Diffusion, in turn, can be exploited as an experimental tool to quantify macroscopic parameters of ECS structure in live tissue.[51] Diffusion-based methods that study the ECS employ small extracellular probe molecules, such as a cation, tetramethylammonium (TMA^+, MW 74), to quantify two parameters of the ECS structure: volume fraction and tortuosity. This method is called the real-time iontophoretic (RTI) method.[45] Since TMA^+ has been used mostly, it is also known as the TMA^+ method. The volume fraction (α) represents the proportion of tissue volume occupied by the ECS. The tortuosity (λ) quantifies the hindrance imposed on the diffusion process by the tissue, relative to an obstacle-free medium. Tortuosity is defined as $(D/D^*)^{1/2}$, where D is the free diffusion coefficient in a free medium and D^* is the effective diffusion coefficients in brain.[45,51] Alternatively, diffusion hindrance can be defined as diffusion permeability (θ), which is a ratio of the effective diffusion coefficient in the brain tissue and the free diffusion coefficient.[52] In isotropic healthy brain, α is about 0.2 and λ is about 1.6 (i.e., diffusion of a small extracellular probe molecule in brain tissue is slowed down about 2.5 times, relative to an obstacle-free medium).[46,51]

Light microscopic images of brain tissue may lead us to believe that brain microstructure is static. But that is far from the truth. Neuronal spines have been shown to be pruned during development and to grow or move during synapse formation and to underlie synaptic plasticity.[53] Similarly, astrocytic processes were shown to be highly mobile, and be attracted to sites of high neuronal activity.[54] Cellular elements thus appear to be in a state of constant change and rearrangement, and these processes contribute to brain plasticity essential for learning, adaptation, and survival. Since the ECS compartment is a counterpart of the cellular compartments, its structure also changes. These changes may be acute and reversible, such as during physiological conditions, when neurons are activated or osmotically challenged.[55,56] These changes may also be permanent, such as during brain trauma and in disease states. For example, the ECS structure changes significantly in many pathological states associated with cellular edema.[55,57,58] These dynamic or permanent changes in the ECS structure impact the extracellular concentration and distribution of substances diffusing in the ECS.

NORADRENERGIC SYSTEM'S EFFECTS ON EXTRACELLULAR SPACE STRUCTURE

A number of RTI studies performed in anesthetized animals, mostly rodents, isolated brain slabs such as an isolated turtle cerebellum and acute brain slices reported that the ECS occupies about 20% of the total brain volume.[56] Until recently, it was assumed that this value of ECS volume would apply to awake brain state as well. However, a recent study from Maiken Nedergaard's group[59] reported that the ECS volume dramatically decreases during an awake state. In this study, RTI measurements were carried out in the cortex of mice during sleep, awake, and under anesthesia. The ECS volume recorded in sleeping animals at midday and in animals anesthetized with a ketamine–xylazine mixture ($\alpha = 0.23$) was in good agreement with previously reported values of the ECS volume within acute brain slices.[56] However, the value of α was only about 0.14, when animals woke up in the evening. This represents a significant decrease (40%) in α during the awake state. Interestingly, no significant change in tortuosity was found between the sleep state and the awake state. Xie et al.[59] also reported that noradrenergic signaling is involved in the ECS volume changes during the sleep–wake cycle. They showed that a cocktail of adrenergic antagonists applied on the cortical surface of an awake animal increased the ECS volume from 0.14 to 0.23, indicating that blockade of adrenergic activity reverses increase in the ECS volume from that of the awake to the sleep state.

The study by Xie et al.[59] raised several questions. First, which brain compartment alters its volume in a manner reciprocal to the ECS volume changes? Second, which adrenergic receptors are responsible for the ECS volume changes? The study by Sherpa et al.[36] provided some answers to these questions (Fig. 12.2). In this study, ultrastructural analysis with electron microscopy and diffusion analysis of the ECS parameters was carried out in acutely prepared slices of the rat visual cortex with and without exposure to the βAR agonist, ISO. As already described, ultrastructural analysis found that astrocytic processes expanded during ISO application. It was also reported that α significantly decreased from

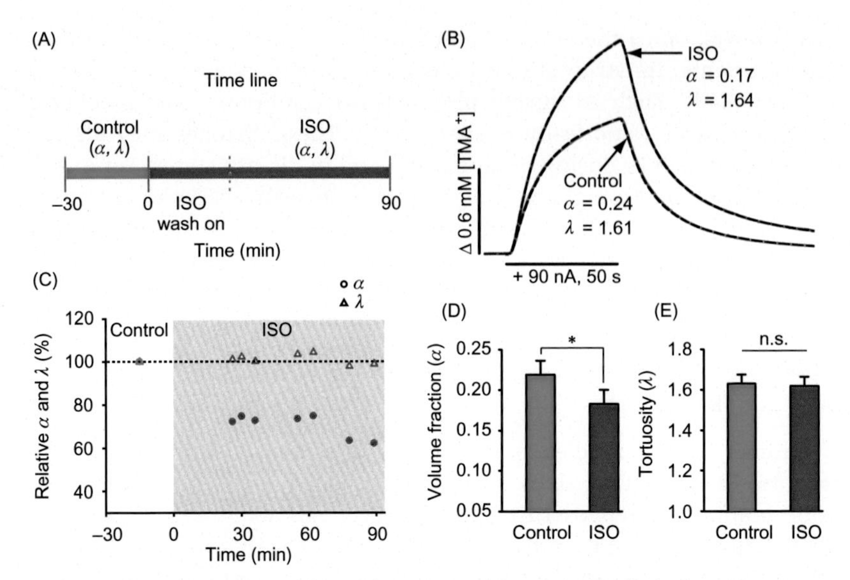

FIGURE 12.2 **Quantitative analysis of ECS parameters.** (A) Time line of RTI experiment. (B) Representative diffusion curves in the visual cortex under control (green) and ISO (red) conditions. Theoretical curves (dashed black) from the fitting procedure are superimposed. (C) Time course of α and λ values from one experiment. (D–E) Summary of α and λ values (mean $\pm$ SD). Asterisk, significant; n.s., nonsignificant. *Reproduced from Sherpa AD, Xiao F, Joseph N, Aoki C, Hrabetova S. Activation of β-adrenergic receptors in rat visual cortex expands astrocytic processes and reduces extracellular space volume.* Synapse. *2016;70:307–316, with the permission of John Wiley & Sons, Inc.*

0.22 to 0.18 in the ISO condition, while λ remained constant. The ECS volume obtained during ISO application in slices[36] is larger than during awake state in vivo.[59] This discrepancy may result from an enhancement of NA signaling from an intact LC in an awake brain whereas only a selective enhancement of NA signaling occurs in slices through βAR activation. Taken together, Sherpa et al.[36] identified astrocytes as one compartment that changes its volume in response to noradrenergic activation. Furthermore, this study identified βARs as one type of noradrenergic receptors involved in this process.

ECS structural parameters determine the spatiotemporal distribution of neuroactive substances diffusing in the ECS. As α is reduced under ISO conditions, cells and cellular processes in such a neuropil experience higher concentrations of released ions, neurotransmitters, and neuromodulators, and this enhances their actions on target sites. Our simulation model of extracellular diffusion in control conditions and in ISO conditions predicts such an effect (Fig. 12.3). Simulations show that the concentration of diffusion molecules reaches higher maximum and persists at a higher level in the ISO condition than in the control condition.

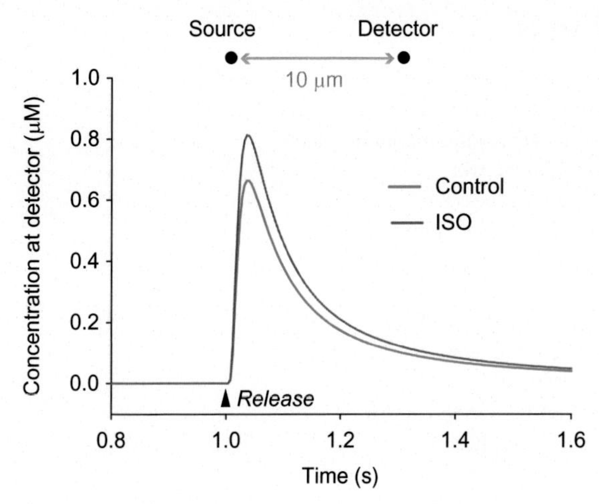

FIGURE 12.3 Simulations of diffusion in the ECS. In these simulations, informed by ECS parameters measured in the acute slices of the visual cortex in control and ISO conditions (Fig. 12.2), the same number of molecules was released (1 ms) from a point source and allowed to diffuse. Concentration profiles were recorded at a distance 10 μm from each source.

CONCLUSIONS

The noradrenergic system projects to the entire cortex and its functional outcome depends on the LC neurons innervating the cortex and diverse ARs situated among neurons, astrocytes, and microglia in the cortex. Recent work in the visual cortex reported that through volume transmission, NA activates β-ARs located on astrocytes, to expand distal astrocytic processes. Expansion of astrocytic processes accounts, partly, for the reduction in the ECS volume in cortex. Future studies may address whether this phenomenon is generalized to the other cortical regions that are innervated by the LC neurons. A reduced ECS volume will increase concentration of ions, neuromodulators, and neurotransmitters, leading to a signaling boost of these molecules on target sites. Owing to these changes in concentration of diffusing molecules in the ECS, we attribute a functional role to NA-driven expansion of astrocytic distal processes, i.e., modulation of communication among brain cells.

This review presents new important findings on how noradrenergic signal exerts its widespread effect on brain function through astrocytic morphology. It would be important in future to test for the β-AR-induced morphological changes in microglia, which also express β-ARs[19] and have received much attention regarding synaptic plasticity during the critical period for developmental plasticity.[60]

ABBREVIATIONS

AR adrenergic receptor
AQP4 aquaporin 4
cAMP 3′,5′-cyclic adenosine monophosphate
ECM extracellular matrix
ECS extracellular space
GFAP glial fibrillary acidic protein
GLAST glutamate aspartate transporter
GLT-1 glutamate transporter-1
IP_3 inositol 1,4,5-trisphosphate
ISO isoproterenol or isoprenaline
LC locus coeruleus
NA noradrenaline, norepinephrine
RTI real-time iontophoretic
TMA^+ tetramethylammonium

ACKNOWLEDGMENTS

The work was supported by NIH NINDS grants R56 NS047557 and R01 NS047557 (to S.H. and C.A.) and NIH NEI grant P30 EY013079 (to C.A.).

REFERENCES

1. Berridge CW, Waterhouse BD. The locus coeruleus-noradrenergic system: modulation of behavioral state and state-dependent cognitive processes. *Brain Res Rev.* 2003;42:33–84.
2. Sara SJ. The locus coeruleus and noradrenergic modulation of cognition. *Nat Rev Neurosci.* 2009;10:211–223.
3. Bear MF, Singer MF. Modulation of visual cortical plasticity by acetylcholine and noradrenaline. *Nature.* 1986;320:172–176.
4. Kasamatsu T, Pettigrew JD. Depletion of brain catecholamines: failure of ocular dominance shift after monocular occlusion in kittens. *Science.* 1976;194:206–209.
5. Kasamatsu T, Pettigrew JD, Ary M. Restoration of visual cortical plasticity by local microperfusion of norepinephrine. *J Comp Neurol.* 1979;185:163–181.
6. Shepard KN, Liles LC, Weinshenker D, Liu RC. Norepinephrine is necessary for experience-dependent plasticity in the developing mouse auditory cortex. *J Neurosci.* 2015;35:2432–2437.
7. Moore RY, Bloom FE. Central Catecholamine neuron systems: anatomy and physiology of the norepinephrine and epinephrine systems. *Annu Rev Neurosci.* 1979;2:113–168.
8. Chandler DJ, Gao WJ, Waterhouse BD. Heterogenous organization of the locus coeruleus projections to prefrontal and motor cortices. *Proc Natl Acad Sci USA.* 2014;111:6816–6821.
9. O'Donnell J, Zeppenfeld D, McConnell E, Pena S, Nedergaard M. Norepinephrine: a neuromodulator that boosts the function of multiple cell types to optimize CNS performance. *Neurochem Res.* 2012;37:2496–2512.
10. Aoki C, Venkatesan C, Go CG, Forman R, Kurose H. Cellular and subcellular sites for noradrenergic action in the monkey dorsolateral prefrontal cortex as revealed by the immunocytochemical localization of noradrenergic receptors and axons. *Cereb Cortex.* 1998;8:269–277.

11. Shain W, Forman DS, Madelian V, Turner JN. Morphology of astroglial cells is controlled by beta-adrenergic receptors. *J Cell Biol.* 1987;105:2307–2314.
12. Vardjan N, Kreft M, Zorec R. Dynamics of β-adrenergic/cAMP signaling and morphological changes in cultured astrocytes. *Glia.* 2014;62:566–579.
13. Laureys G, Clinckers R, Gerlo S, Spooren A, Wilczak N, Kooijman R, et al. Astrocytic β_2-adrenergic receptors: from physiology to pathology. *Prog Neurobiol.* 2010;91:189–199.
14. Fuxe K, Agnati LF, Marcoli M, Borroto-Escuela DO. Volume transmission in central dopamine and noradrenaline neurons and its astroglial targets. *Neurochem Res.* 2015;40:2600–2614.
15. Aoki C. Beta-adrenergic receptors: astrocytic localization in the adult visual cortex and their relation to catecholamine axon terminals as revealed by electron microscopic immunocytochemistry. *J Neurosci.* 1992;12:781–792.
16. Pankratov Y, Lalo U. Role for astroglial α1-adrenoreceptors in gliotransmission and control of synaptic plasticity in the neocortex. *Front Cell Neurosci.* 2015;9:1–11.
17. Aoki C, Pickel VM. C-terminal tail of beta-adrenergic receptors: immunocytochemical localization of beta-adrenergic receptors: immunocytochemical localization within astrocytes and their relation to catecholaminergic neurons in N. tractus solitarii and area postrema. *Brain Res.* 1992;571:35–49.
18. Mori K, Ozaki E, Zhang B, Yang L, Yokoyama A, Takeda I, et al. Effects of norepinephrine on rat cultured microglial cells that express alpha1, alpha2, beta1 and beta2 adrenergic receptors. *Neuropharmacology.* 2002;43:1026–1034.
19. Tanaka KF, Kashima H, Suzuki H, Ono K, Sawada M. Existence of functional beta1- and beta2-adrenergic receptors on microglia. *J Neurosci Res.* 2002;70:232–237.
20. Svichar N, Esquenazi S, Waheed A, Sly WS, Chesler M. Functional demonstration of surface carbonic anhydrase IV activity on rat astrocytes. *Glia.* 2006;53:241–247.
21. Danbolt NC. Glutamate uptake. *Prog Neurobiol.* 2001;65:1–105.
22. McBean GJ. Inhibition of the glutamate transporter and glial enzymes in rat striatum by the gliotoxin, alpha aminoadipate. *Br J Pharmacol.* 1994;113:536–540.
23. Alexander GM, Grothusen JR, Gordon SW, Schwartzman RJ. Intracerebral microdialysis study of glutamate reuptake in awake, behaving rats. *Brain Res.* 1997;766:1–10.
24. Vyskocil F, Kriz N, Bures J. Potassium-selective microelectrodes used for measuring the extracellular brain potassium during spreading depression and anoxic depolarization in rats. *Brain Res.* 1972;39:255–259.
25. Hansen AJ. Extracellular potassium concentration in juvenile and adult rat brain cortex during anoxia. *Acta Physiol Scand.* 1977;99:412–420.
26. Reichenbach A, Wolburg H. Astrocytic swelling in neuropathology. In: Kettenmann HO, Ransom BR, eds. *Neuroglia.* New York: Oxford University Press; 2005:521–531.
27. Ma B, Buckalew R, Du Y, Kiyoshi CM, Alford CC, Wang W, et al. Gap junction coupling confers isopotentiality on astrocyte syncytium. *Glia.* 2016;64:214–226.
28. Hajek I, Subbarao KV, Hertz L. Acute and chronic effects of potassium and noradrenaline on Na^+, K^+-ATPase activity in cultured mouse neurons and astrocytes. *Neurochem Int.* 1996;28:335–342.
29. Nielsen S, Nagelhus EA, Amiry-Moghaddam M, Bourque C, Agre P, Ottersen OP. Specialized membrane domains for water transport in glial cells: high-resolution immunogold cytochemistry of aquaporin-4 in rat brain. *J Neurosci.* 1997;17:171–180.
30. Yao X, Hrabetova S, Nicholson C, Manley GT. Aquaporin-4-deficient mice have increased extracellular space without tortuosity change. *J Neurosci.* 2008; 28:5460–5464.

31. Fan Y, Zhang J, Sun XL, Gao L, Zeng XN, Ding JH, et al. Sex- and region-specific alterations of basal amino acid and monoamines metabolism in the brain of aquaporin-4 knockout mice. *J Neurosci Res.* 2005;82:458–464.
32. Grosche J, Matyash V, Möller T, Verkhratsky A, Reichenbach A, Kettenmann H. Microdomains for neuron-glia interaction: parallel fiber signaling to Bergmann glial cells. *Nat Neurosci.* 1999;2:139–143.
33. Pekny M, Leveen P, Pekna M, Eliasson C, Berthold CH, Westermark B. Betsholtz. Mice lacking glial fibrillary acidic protein display astrocytes devoid of intermediate filaments but develop and reproduce normally. *EMBO J.* 1995;14:1590–1598.
34. Bushong EA, Martone ME, Jones YZ, Ellisman MH. Protoplasmic astrocytes in CA1 stratum radiatum occupy separate anatomical domains. *J Neurosci.* 2002;22:183–192.
35. Kosaka T, Hama K. Three-dimensional structure of astrocytes in the rat dentate gyrus. *J Comp Neurol.* 1986;249:242–260.
36. Sherpa AD, Xiao F, Joseph N, Aoki C, Hrabetova S. Activation of β-adrenergic receptors in rat visual cortex expands astrocytic processes and reduces extracellular space volume. *Synapse.* 2016;70:307–316.
37. Goldman JE, Abramson B. Cyclic AMP-induced shape changes of astrocytes are accompanied by rapid depolymerization of actin. *Brain Res.* 1990;528:189–196.
38. Ogata K, Kosaka T. Structural and quantitative analysis of astrocytes in the mouse hippocampus. *Neuroscience.* 2002;113:221–233.
39. Halassa MM, Fellin T, Takano H, Dong JH, Haydon PG. Synaptic islands defined by the territory of a single astrocyte. *J Neurosci.* 2007;27:6473–6477.
40. Bicknell RJ, Luckman SM, Inenaga K, Mason WT, Hatton GI. Beta-adrenergic and opioid receptors on pituicytes cultured from adult rat neurohypophysis: regulation of cell morphology. *Brain Res Bull.* 1989;22:379–388.
41. Hatton GI. Function-related plasticity in hypothalamus. *Annu Rev Neurosci.* 1997; 20:375–397.
42. Aoki C. Differential timing for the appearance of neuronal and astrocytic beta-adrenergic receptors in the developing rat visual cortex as revealed by light and electron-microscopic immunohistochemistry. *Vis Neurosci.* 1997;14:1129–1142.
43. Ding F, O'Donnell J, Thrane AS, Zeppenfeld D, Kang H, Xie L, et al. Alpha1-adrenergic receptors mediate coordinated Ca^{2+} signaling of cortical astrocytes in awake, behaving mice. *Cell Calcium.* 2013;54:387–394.
44. Paukert M, Agarwal A, Cha J, Doze VA, Kang JU, Bergles DE. Norepinephrine controls astroglial responsiveness to local circuit activity. *Neuron.* 2014;82:1263–1270.
45. Nicholson C, Phillips JM. Ion diffusion modified by tortuosity and volume fraction in the extracellular microenvironment of the rat cerebellum. *J Physiol.* 1981;321:225–257.
46. Nicholson C, Sykova E. Extracellular space structure revealed by diffusion analysis. *Trends Neurosci.* 1998;21:207–215.
47. Thorne RG, Nicholson C. In vivo diffusion analysis with quantum dots and dextrans predicts the width of brain extracellular space. *Proc Natl Acad Sci USA.* 2006;103: 5567–5572.
48. Xiao F, Nicholson C, Hrabe J, Hrabetova S. Diffusion of flexible random-coil dextran polymers measured in anisotropic brain extracellular space by integrative optical imaging. *Biophys J.* 2008;95:1382–1392.
49. Kinney JP, Spacek J, Bartol TM, Bajaj CL, Harris KM, Sejnowski TJ. Extracellular sheets and tunnels modulate glutamate diffusion in hippocampal neuropil. *J Comp Neurol.* 2013;521:448–464.

50. Korogod N, Petersen CCH, Knott GW. Ultrastructural analysis of adult mouse neocortex comparing aldehyde perfusion with cryo fixation. *eLife*. 2015;4:e05793.
51. Nicholson C. Diffusion and related transport mechanism in brain tissue. *Rep Prog Phys*. 2001;64:815–884.
52. Hrabe J, Hrabetova S, Segeth K. A model of effective diffusion and tortuosity in the extracellular space of the brain. *Biophys J*. 2004;87:1606–1617.
53. Bhatt DH, Zhang S, Gan WB. Dendritic spine dynamics. *Annu Rev Physiol*. 2009; 71:261–282.
54. Bernardinelli Y, Randall J, Janett E, Nikonenko I, Konig S, Jones EV, et al. Activity-dependent structural plasticity of perisynaptic astrocytic domains promotes excitatory synapse stability. *Curr Biol*. 2014;24:1679–1688.
55. Sykova E, Mazel T, Vargova L, Vorisek I, Prokopova-Kubinova S. Extracellular space diffusion and pathological states. In: Agnati LF, Fuxe K, Nicholson C, Sykova E, eds. *Progress in Brain Research*. Vol 125. Amsterdam: Elsevier Sciences; 2000:155–178.
56. Sykova E, Nicholson C. Diffusion in brain extracellular space. *Physiol Rev*. 2008; 88:1277–1340.
57. Hrabetova S, Chen KC, Masri D, Nicholson C. Water compartmentalization and spread of ischemic injury in thick-slice ischemia model. *J Cereb Blood Flow Metab*. 2002; 22:80–88.
58. Hrabetova S, Hrabe J, Nicholson C. Dead-space microdomains hinder extracellular diffusion in rat neocortex during ischemia. *J Neurosci*. 2003;23:8351–8359.
59. Xie L, Kang H, Xu Q, Chen MJ, Liao Y, Thiyagarajan M, et al. Sleep drives metabolic clearance from the adult brain. *Science*. 2013;342:373–377.
60. Tremblay M-E, Lowery RL, Majewska AK. Microglial interactions with synapses are modulated by visual experience. *PLoS Biol*. 2010;8:e1000527.

Chapter 13

Signaling Pathway of β-Adrenergic Receptor in Astrocytes and its Relevance to Brain Edema

Baoman Li[1], Dan Song[1], Ting Du[1], Alexei Verkhratsky[2,3,4,5,6,7,✉] and Liang Peng[1,✉]

[1]China Medical University, Shenyang, P. R. China, [2]University of Manchester, Manchester, United Kingdom, [3]IKERBASQUE, Basque Foundation for Science, Bilbao, Spain, [4]University of the Basque Country UPV/EHU and CIBERNED, Leioa, Spain, [5]University of Nizhny Novgorod, Nizhny Novgorod, Russia, [6]University of Ljubljana, Ljubljana, Slovenia, [7]Celica Biomedical, Ljubljana, Slovenia

Chapter Outline

Introduction 258
β_1-Adrenergic Receptor 259
Extracellular Ions During Ischemia and/or Reperfusion 262
MAPK/$ERK_{1/2}$ Signaling Pathway During Ischemia and/or Reperfusion 263
Effect of β_1-Adrenergic Receptor Antagonist on Brain Edema During Ischemia/Reperfusion 264
Conclusions 266
List of Abbreviations 267
Acknowledgments 268
References 268

ABSTRACT

Astroglia express β_1- and β_2-adrenergic receptors. Stimulation of astrocytic β_1-adreneergic receptors induces extracellular regulated kinase 1 and 2 ($ERK_{1/2}$) phosphorylation via protein kinase A, G_s/G_i switching, Ca^{2+} release from intracellular stores, metalloproteinase-catalyzed release of growth factor, and transactivation of epidermal growth factor receptor; while stimulation of β_2-adrenergic receptors induces $ERK_{1/2}$ phosphorylation by β-arrestin-mediated Src activation, without the involvement of epidermal growth factor receptor activation. Brain edema

✉Correspondence address
E-mail: hkkid08@yahoo.com

Noradrenergic Signaling and Astroglia. DOI: http://dx.doi.org/10.1016/B978-0-12-805088-0.00013-X
© 2017 Elsevier Inc. All rights reserved.

after 3 h of focal ischemia followed by 8 h reperfusion can be prevented by antagonists of β_1-adrenergic receptor and inhibitors of the associated signaling pathway, whereas inhibition of β_2-adrenergic cascade has no effect. In astrocytes in primary cultures, stimulation of β_1-adrenergic receptor increases the activity of both Na,K-ATPase and Na–K–Cl cotransporter NKCC1. Here we discuss mechanisms underlying the effects of β_1-adrenergic receptor activation on brain edema, with particular emphasis on the signaling pathway of β_1-adrenergic receptor, extracellular ions and mitogen-activated protein kinase/$ERK_{1/2}$ cascade during ischemia and reperfusion periods.

Keywords: Astrocyte; brain edema; β_1-adrenergic receptor; NKCC1; Na,K-ATPase; EGF receptor; MAPK/ERK pathway

INTRODUCTION

Brain edema accompanies many neurological disorders, such as brain ischemia, hemorrhages, traumatic brain injury, hyperammonemia, and brain tumors. According to the underlying mechanisms brain edema may be classified into cytotoxic edema (cell swelling) that emerges in minutes after brain injury, ionic edema, an extracellular edema that occurs immediately after cellular edema with blood–brain barrier (BBB) still intact, while water can only influx to extracellular space from plasma, and vasogenic edema that occurs hours after the injury when BBB is compromised and induces extravasation of plasma proteins.[1] Cytotoxic edema can occur in all the cell types in the brain. Astrocytes are especially important in cytotoxic edema since their contribution to brain volume homeostasis.

Astrocytes express multiple proteins contributing to the development of the brain edema, such as Na,K-ATPase, $Na^+/K^+/Cl^-$ cotransporter (NKCC1), and water channels of aquaporin 4 (AQP4) type. The Na,K-ATPase translocates 3 Na^+ outside of the cell and 2 K^+ into the cytosol at the expense of hydrolysis of 1 ATP molecule. In contrast the NKCC1 cotransports 1 Na^+, 1 K^+, and 2 Cl^- into the cytosol; this transporter is driven by the transmembrane ionic gradients generated by the ATP-dependent pumps. Finally AQP4 transports water across the plasmalemma by diffusion. In astrocytes but not in neurons, Na,K-ATPase can be stimulated by β-adrenergic receptor (β-AR) activation.[2] Based on the clinical symptoms, the Lund concept was introduced at Lund University Hospital, Sweden in order to control brain volume and cerebral perfusion in severe head injury.[3,4] In this protocol, β_1-AR antagonist metoprolol and α_2-AR agonist, clonidine or dexmedetomidine, are used together to decrease systemic blood pressure. Since these two receptors are highly expressed in astrocytes, we hypothesized that these drugs may affect astroglia. Subsequently, we found that β_1-AR antagonist betaxolol reduced brain edema after 3 h of medial cerebral artery occlusion (MCAO) followed by 8 h reperfusion,[5] whereas α_2-AR agonist dexmedetomidine-induced release of heparin-binding epidermal growth factor (HB-EGF) from cultured astrocytes.[6,7] The later had protective effect on neurons treated with H_2O_2.[8] In this paper, we will focus on the mechanisms of β_1-AR inhibition on brain edema.

β_1-ADRENERGIC RECEPTOR

There are three subtypes of β-ARs, β_1, β_2, and β_3, all of which are G_s protein-coupled receptors. Both β_1- and β_2-ARs are abundantly expressed in mammalian brain, perhaps with a ratio of 2:1,[9,10] whereas the density of β_3-AR expression in rodent cerebral cortex amounts to only 3% of that in the brown adipose tissue.[11] Astrocytes, both in primary cultures and freshly isolated from brain tissue of mice expressing fluorescent marker under control of an astroglia-specific promoter (glial fibrillary acidic protein) and sorted by fluorescence-activated cell sorting express β_1- and β_2-ARs.[12] In addition to cyclic adenosine monophosphate/protein kinase A (PKA) pathway, stimulation of β_1-AR also induces a significant increase of phosphorylation of extracellular regulated kinase 1 and 2 ($ERK_{1/2}$) in astrocytes in primary cultures via transactivation of epidermal growth factor receptor (EGFR).[13]

Stimulation of EGFR by growth factors activates two associated intracellular signal pathways, Ras/Raf/mitogen-activated protein kinase (MAPK)/$ERK_{1/2}$ and PI3K/AKT. However, EGFR can also be transactivated by stimulation of a $G_{i/o}$ or G_q proteins, by an increase of intracellular Ca^{2+} or by other stimuli,[6,14–16] such as reactive oxygen species (ROS).[17] Activation of a $G_{i/o}$ or G_q proteins leads to metalloproteinase (MMP)-catalyzed shedding of an agonist of the EGFR, for example HB-EGF, which in turn stimulates EGFRs. Transactivation of EGFR in astrocytes has been demonstrated in signaling pathways linked to various receptors,[7,13,18] to an increased concentration of ammonium and to high extracellular concentration of K^+.[19,20] The functional significance of EGFR transactivation, however, may depend on the time and/or level of the stimulation and downstream signaling, i.e., the MAPK/$ERK_{1/2}$ and/or PI3K/AKT pathways. The $ERK_{1/2}$ is a serine/threonine kinase that acts as a critical component of the MAPK signal transduction pathway. We have found that this signaling pathway mediates astrocytic responses to various stimuli including α_2-AR agonist, dexmeditomidine,[7] β_1- and β_2-ARs agonist, isoproterenol,[10] vasopressin,[21] antidepressant fluoxetine,[18] 5-HT,[22] high extracellular K^+,[19] and ammonium.[20]

The signaling pathways of β_1- and β_2-AR-induced $ERK_{1/2}$ phosphorylation are quite complicated (Fig. 13.1).[13] The agonist of β_1/β_2-ARs isoproterenol at nanomolar concentration stimulates $ERK_{1/2}$ phosphorylation via activation of β_2-ARs. This $ERK_{1/2}$ phosphorylation is induced by β-arrestin-mediated Src activation, without activation of EGFR. At micromolar concentrations, however, isoproterenol acts on β_1-ARs stimulating $ERK_{1/2}$

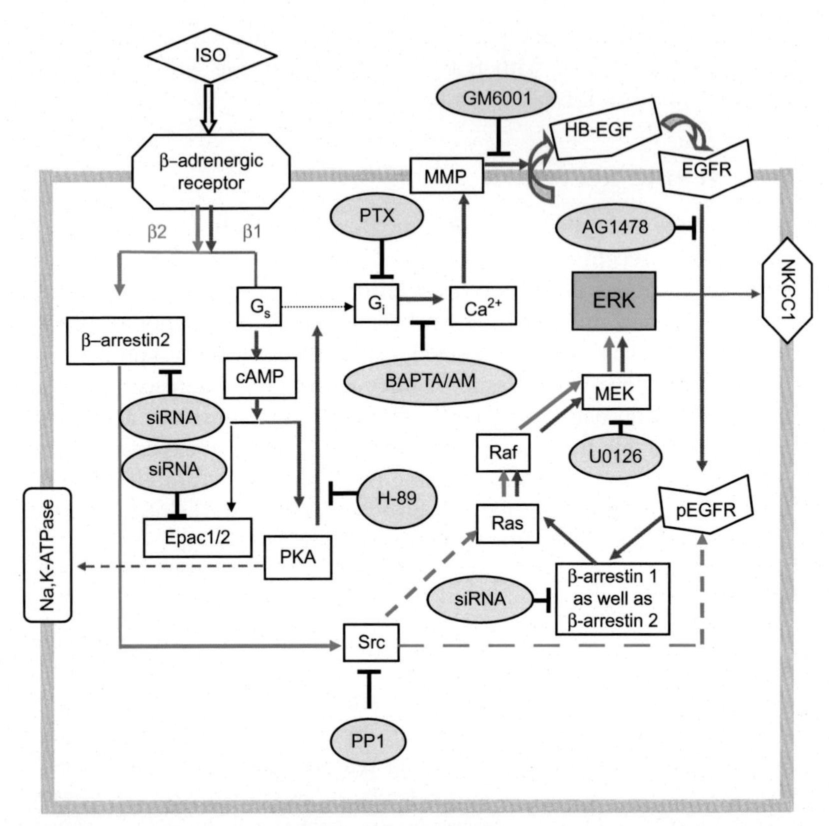

FIGURE 13.1 Schematic illustration of stimulation of ERK phosphorylation by β-adrenergic receptors in astrocytes. Isoproterenol (ISO) binds to β-ARs. At high concentrations (≥1 μM), the activation of the receptors induces β_1-AR (red arrows), PKA-dependent "G_s/G_i switching," which in turn induces an enhancement of intracellular Ca^{2+} concentration by Ca^{2+} release from intracellular stores. The latter activates Zn-dependent MMPs and leads to shedding of growth factor(s). The released EGF ligand stimulates autophosphorylation of EGFR at Y1173 site in the same and adjacent cells. The downstream target of EGFR, ERK (shown in blue) is phosphorylated via the Ras/Raf/MEK pathway, contingent upon recruitment of β-arrestins 1 and 2. The ERK phosphorylation by ISO at high concentration can be inhibited by H89, an inhibitor of PKA, by PTX, an inhibitor of G_i protein, by BAPTA/AM, an intracellular Ca^{2+} chelator, by GM6001, an inhibitor of Zn-dependent MMP, by AG1478, an inhibitor of the receptor-tyrosine kinase of the EGFR, by siRNA against β-arrestin 1 and less completely by siRNA against β-arrestin 2, and by U0126, a MEK inhibitor (all inhibitors shown in yellow). In contrast, at low concentration (≤100 nM) β_2-AR (green arrows) activation of the receptors activates Src via recruitment of β-arrestin 2. Src in turn stimulates ERK phosphorylation and phosphorylates EGFR at Y845 and Y1045 sites without involvement of the receptor-tyrosine kinase. $ERK_{1/2}$ phosphorylation is secondary to MEK activation, which may be induced by direct activation of Raf or Ras by Src, or by Src-mediated phosphorylation of the EGFR. The ERK phosphorylation by ISO at low concentration can be inhibited by siRNA against β-arrestin 2, by PP1, a Src inhibitor, and by U0126, a MEK inhibitor. ERK activates NKCC1, and PKA stimulates Na,K-ATPase indirectly. *Modified from Du T, Li B, Li H, Li M, Hertz L, Peng L. Signaling pathways of isoproterenol-induced ERK1/2 phosphorylation in primary cultures of astrocytes are concentration-dependent. J Neurochem. 2010;115:1007–1023* [13].

phosphorylation via PKA and G_s/G_i switching, Ca^{2+} release from intracellular stores, MMP-catalyzed release of growth factor, and transactivation of EGFR, with no involvement of Src.

Stimulation of β-ARs increases Na,K-ATPase activity in astrocytes, but not in neurons.[2] Similar link is also observed in lung alveolar cells[23,24] and in cardiomyocytes.[25,26] The PKA seems to be involved in short-term regulation of pump activity. In alveolar epithelial cells, isoproterenol increases Na,K-ATPase activity by rapidly (in 15 min) recruiting Na,K-ATPase proteins from intracellular pools to the cell membrane.[24] As the α subunit of Na,K-ATPase cannot be phosphorylated by PKA in situ, the major PKA substrate is phospholemman (PLM), a member of the FXYD family of proteins that interacts with Na,K-ATPase. It has been reported that PLM inhibits Na,K-ATPase activity by decreasing Na^+ affinity of the intracellular binding site, and this inhibitory effect is relieved by PKA activation.[26] In contrast, $MAPK/ERK_{1/2}$ signaling pathway is responsible for long-term regulation of Na,K-ATPase activity (in days) by increasing its gene expression.[27]

The NKCC1, a cotransporter of Na^+, K^+, $2Cl^-$, and water,[28,29] in adult brain is present mainly or even exclusively in astrocytes.[30,31] Since NKCC1 is a secondary transporter it is driven by the ion gradients established by the Na,K-ATPase.[32,33] The NKCC1 plays a major role in importing ions and water into brain parenchyma during stroke,[34] and is also crucial for another essential component of stroke-induced edema, the cytotoxic cellular swelling, which contributes to the creation of the abnormal concentrations of ions and increases the water content in the extracellular space that further facilitate ions and water transport across the BBB into the brain.[35] The NKCC1 has six transmembrane domains with both N and C terminals located intracellularly. The NKCC1 has three phosphorylation sites, Thr^{184}, Thr^{189}, and Thr^{262}. Phosphorylation of Thr^{189} is necessary for NKCC1 activity.[36] We have found that NKCC1 activity in astrocytes is increased by β_1-AR stimulation.[5] Most cells shrink during exposure to a hypertonic medium with 100 mM sucrose and NKCC1 activity is important for a subsequent regulatory volume increase.[32] The stimulation of NKCC1-dependent regulatory volume increase in astrocytes by β-AR agonist, isoproterenol, is inhibited by β_1-AR antagonist, betaxolol, but not by β_2-AR antagonist, ICI118551.[5] Inhibitors in β_1-adrenergic signaling pathway in astrocytes also abolish the effect of isoproterenol. These include PKA inhibitor H89, G_i inhibitor, pertussis toxin (PTX), MMP inhibitor GM6001, EGFR inhibitor AG1478, and $ERK_{1/2}$ inhibitor U0126.[5] In various tissues isoproterenol also induces NKCC1 phosphoryation.[37–39] Although PKA is involved in NKCC1 phosphorylation by β-adrenergic stimulation, it is not due to the direct phosphorylation on the transporter.[40] The site of isoproterenol-induced phosphorylation is found in the NH_2-terminal cytosolic tail of NKCC1,[41] which does not contain the sole PKA consensus site in this cotransporter.[40] This is in agreement with the signaling pathway of β_1-AR stimulation of NKCC1 in astrocytes.[5]

We believe that distinct effects of β_1- and β_2-ARs on NKCC1 result from the difference in their intracellular signaling pathways, although both receptors stimulate $ERK_{1/2}$ phosphorylation in astrocytes. NKCC1 is a secondary transporter and it is driven by the ion gradients established by the Na/K-ATPase. $ERK_{1/2}$ activates NKCC1 but has no acute effect on Na, K-ATPase activity. The activation of β_1- but not β_2-ARs stimulates $ERK_{1/2}$ phosphorylation via PKA and triggers Ca^{2+} release from intracellular stores. The activity of PKA is probably necessary for the simultaneous stimulation of Na,K-ATPase. At the same time, however, PKA is not involved in β_2-AR-induced $ERK_{1/2}$ phosphorylation.

EXTRACELLULAR IONS DURING ISCHEMIA AND/OR REPERFUSION

During ischemia, there is a large increase in extracellular K^+ concentration concomitant with a decrease of Ca^{2+} and Na^+ concentrations in cerebral ischemia induced by inflation of a pneumatic cervical cuff and in focal ischemia induced by MCAO.[42] This most likely reflects the Na,K-ATPase insufficiency because of the ATP depletion. In ischemia, extracellular K^+ concentrations can rise up to 80 mM and extracellular Ca^{2+} concentrations fall to 0.1 mM. These disturbances in ionic content can be recovered after reperfusion; this recovery takes only a few minutes for K^+ and Na^+, but lasts substantially longer for Ca^{2+}.[42]

An increase in extracellular K^+ concentrations stimulates Na,K-ATPase activity in astrocytes.[43,44] In cultured cerebral astrocytes, K_m value of Na, K-ATPase for K^+ is 1.9 mM and hence increase in $[K^+]_o$ by 5–12 mM may significantly increase the activity of the pump.[45] The signaling pathway of K^+, as well as ammonia,[20] is similar with nanomolar concentrations of ouabain or endogenous ouabain-like compounds. Low concentrations of ouabain act on α subunit of Na,K-ATPase and induce the interaction of Src and EGFR, that subsequently leads to EGFR transactivation.[46] The stimulation of EGFR leads to activation of two pathways. One of these is activation of phospholipase C (PLC), releasing inositol trisphosphate (IP_3) and 1,2-diacylglycerol from membrane-bound phosphatidylinositide-4,5-biphosphate. The second pathway is represented by Raf/MAPK/$ERK_{1/2}$ cascade. Alternatively, Na/K-ATPase tethers PLC and IP_3 receptor into a Ca^{2+}-regulatory complex and induces Ca^{2+} release from intracellular Ca^{2+} store.[47] Minor increases in extracellular K^+ stimulate K^+ uptake, which can be inhibited by an inhibitor of IP_3 receptor, xestospongine.[48] In cultured astrocytes, this stimulation of K^+ uptake by Na,K-ATPase is abolished when Ca^{2+} is removed from the incubation medium.[5]

When extracellular K^+ concentration increases to 15 mM or higher, NKCC1 activity is also stimulated. This leads to an influx of Na^+, K^+, and

Cl^- together with osmotically obliged water. The cell swelling induced by an increase in $[K^+]_o$ depends on $ERK_{1/2}$ phosphorylation.[19] Elevation of extracellular K^+ concentration above 15 mM depolarizes the cell membrane and thereby leads to Ca^{2+} entry through voltage-dependent L-channels. An increase in intracellular Ca^{2+} concentration leads to a Src-dependent MMP-catalyzed release of EGF ligand from its membrane-bound precursor. The released EGF ligand activates EGFR and its associated MAPK/$ERK_{1/2}$ signaling pathway.[19] Obviously, at extracellular K^+ concentrations above 15 mM or higher NKCC1 activity depends on extracellular Ca^{2+}.

MAPK/$ERK_{1/2}$ SIGNALING PATHWAY DURING ISCHEMIA AND/OR REPERFUSION

In MCAO model, EGFR phosphorylation was found at 3, 7, and 14 days after 1 h ischemia.[49] The PI3K/AKT pathway is believed to mediate cell protection,[50] whereas the functions of MAPK/$ERK_{1/2}$ are more complicated. Transient, moderate $ERK_{1/2}$ phosphorylation promotes cell protection,[50] whereas sustained, extensive $ERK_{1/2}$ phosphorylation may induce cell injury.[51] In the brain ischemia cell damage occurs because of ATP depletion that leads to a loss of ionic homeostasis, membrane depolarization, increase in intracellular Ca^{2+}, decrease in intracellular pH and uncontrolled release of glutamate. In reperfusion, cells experience extensive oxidative damage because restoration of oxygen results in generation of large amount of ROS.[50] Understanding signaling pathways in ischemia and reperfusion could provide detailed information about mechanisms underlying cell protection or cell injury. Since NKCC1 is phosphorylated and stimulated by $ERK_{1/2}$, it is also applied to brain edema.

Conceptually, the PI3K/AKT and Raf/MAPK/$ERK_{1/2}$ represent two parallel signaling pathways. Interactions between PI3K/AKT and Raf/MAPK/$ERK_{1/2}$ occur at several stages and could be either positive or negative. Recently, we have found that EGFR is phosphorylated and transactivated in both core and penumbra areas of the infarction during both ischemia and reperfusion periods in the MCAO model (Fig. 13.2).[52] However, ischemia does not induce $ERK_{1/2}$ phosphorylation, which develops during 5, 15, 30, or 60 min of reperfusion after 2 h ischemia. In contrast the activity of PI3K/AKT pathway is significantly increased. Thus strongly stimulated AKT during ischemia associates and phosphorylates the inhibitory phosphorylation site Ser^{259} of Raf, whereas the stimulatory phosphorylation site Ser^{338} of Raf is not phosphorylated in ischemia. This inhibition of Raf leads to inactivation of its downstream signal MAPK/$ERK_{1/2}$.[52]

During 60 min reperfusion after 2 h ischemia, there is a significant increase in $ERK_{1/2}$ phosphorylation but no change in AKT phosphorylation. The inhibition of AKT phosphorylation cannot be prevented by $ERK_{1/2}$ inhibitor. Instead, both inhibitors of ROS and phosphatase, and

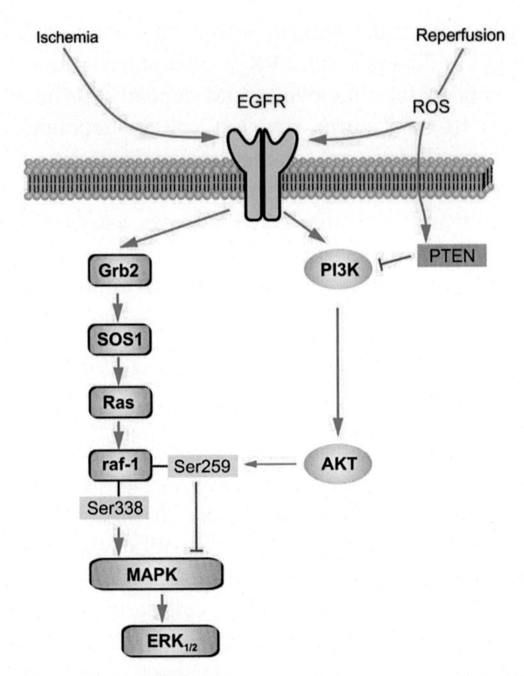

FIGURE 13.2 **Diagram of crosstalk between Raf/MAPK/ERK and PI3K/AKT signal pathways during brain ischemia and reperfusion.** Ischemia induces EGFR transactivation and phosphorylation at Y1173, Y845, and Y1045. The activation of EGFR, in turn, significantly stimulates PI3K/AKT signal pathway. Subsequently, AKT phosphorylates Raf-1 at its inhibitory phosphorylation site Ser259 and inhibits Raf-1 activity. The inhibition of Raf-1 leads to inactivation of its downstream signal MAPK/ERK. During reperfusion, ROS stimulates both EGFR and PTEN. The later, in turn, inhibits PI3K/AKT signal pathway and restores the activity of Raf-1/MAPK/$ERK_{1/2}$ signal pathway, which is responsible for reperfusion-induced cell damage. *EGFR*, epidermal growth factor receptor; *ROS*, reactive oxygen species; *PTEN*, phosphatase and tensin homolog; *SOS1*, son of sevenless 1; *AKT*, protein kinase B; *MAPK*, mitogen-activated protein kinase; *ERK*, extracellular signal-regulated kinase. *Zhou J, Du T, Li B, Rong Y, Verkhratsky A, Peng L. Crosstalk between MAPK/ERK and PI3K/AKT signal pathways during brain ischemia/reperfusion. ASN Neuro. 2015;7* [52].

tensin homolog (PTEN) not only abolishes $ERK_{1/2}$ phosphorylation, but also restores AKT phosphorylation during reperfusion, suggesting that the ROS-dependent increase in PTEN activity in reperfusion period relieves $ERK_{1/2}$ from inhibition of AKT.[52]

EFFECT OF β_1-ADRENERGIC RECEPTOR ANTAGONIST ON BRAIN EDEMA DURING ISCHEMIA/REPERFUSION

Administration of β_1-AR antagonist before the onset of brain ischemia has neuroprotective effects against transient focal ischemia.[53] The same group reported that administration β_1-AR antagonist 30 min after the

TABLE 13.1 Brain Water Content in MCAO Model After 3 h Ischemia and 8 h Reperfusion in the Right Hemisphere Under Control Conditions (only Saline Injected Intra-Ventricularly Before the Occlusion) and After Injection of a β_1- or β_2-AR Subtype-Specific Antagonists, Dissolved in Saline[5]

	Left Hemisphere	Right Hemisphere
Saline	77.97 ± 0.17 ($n = 8$)	81.28 ± 0.34 ($n = 8$)*
Betaxolol	78.02 ± 0.17 ($n = 7$)	78.44 ± 0.58 ($n = 7$)
ICI118551	77.78 ± 0.32 ($n = 7$)	81.08 ± 0.18 ($n = 7$)*

In rats with MCAO in the right hemisphere drugs were added 15 min before the occlusion. Water content was calculated as [(wet weight − dry weight)/wet weight] × 100%. In control animals (saline) an increase in the ipsilateral hemisphere was significant ($P < 0.05$), as marked with *. In the presence of β_1-AR antagonist betaxolol no significant effect was seen, but in the presence of the β_2-AR antagonist ICI118551 cell water increased significantly ($P < 0.05$) in the ipsilateral hemisphere as marked with*.

onset of 2 h of focal ischemia decreased infarct size and improved neurological deficit score after 7 days.[54]

Our group studied effects of β-adrenergic stimulation on brain edema by intra-ventricular injection of β_1- or β_2-AR antagonist or specific inhibitors of β_1- or β_2-AR signaling pathway immediately before 3 h of focal ischemia followed by 8 h reperfusion.[5] There is no brain edema in ischemia period, whereas water content is significantly increased in the lesioned hemisphere in the reperfusion period (Table 13.1). The edema can be suppressed by the inhibitor of NKCC1 or by β_1-AR antagonist, but not by β_2-AR antagonist. All the specific inhibitors of β_1-signaling pathway also abolish the edema (Table 13.2). These include H89, inhibitor of PKA, PTX, inhibitor of G_i protein, GM6001, inhibitor of MMP, AG1478, inhibitor of EGFR and U0126, inhibitor of $ERK_{1/2}$. In contrast, inhibitor of Src, PP1 in β_1-adrenergic signaling pathway has no effect.

The effect of β-AR stimulation probably also depends on the specific pathological context. Vardjan et al.[55] have reported recently that β-AR activation reduces hypotonicity-induced astrocyte swelling in culture and in vivo, and attenuates trauma-induced astrocyte swelling during spinal cord injury. Hypotonicity would stimulate NKCC1 to transfer Na^+, K^+, $2Cl^-$, and water across the cell membrane in the direction opposite to that induced by hypertonicity. Hypertonicity by adding 100 mM sucrose induces cell shrinkage that, in turn, activates NKCC1 and triggers regulatory volume increase. We found that β_1-AR activation enhanced the rate of regulatory volume increase in astrocytes.[56] In contrast, hypotonicity induces cell swelling that, in turn activates NKCC1 and triggers regulatory volume decrease, i.e., reduces cell swelling.

TABLE 13.2 Brain Water Content in MCAO Model After 3 h Ischemia and 8 h Reperfusion in the Right Hemisphere Under Control Conditions (Intra-Cerebral Saline Only) and After Injection of Inhibitors of Either the β_1- or the β_2-AR Pathway in Astrocytes[5]

	Left Hemisphere	Right Hemisphere
Saline	77.97 ± 0.17 ($n = 8$)	81.28 ± 0.34 ($n = 8$)*
H89	77.00 ± 0.42 ($n = 3$)	77.19 ± 0.09 ($n = 3$)
PTX	77.19 ± 0.11 ($n = 4$)	77.51 ± 0.26 ($n = 4$)
GM6001	77.08 ± 0.11 ($n = 4$)	77.15 ± 0.13 ($n = 4$)
AG1478	77.14 ± 0.11 ($n = 3$)	77.27 ± 0.04 ($n = 3$)
U0126	77.39 ± 0.10 ($n = 4$)	78.22 ± 0.67 ($n = 4$)
PP1	77.52 ± 0.26 ($n = 5$)	80.04 ± 0.33 ($n = 5$)*

In rats with MCAO in the right hemisphere drugs were added 15 min before the occlusion. Water content was calculated as [(wet weight − dry weight)/wet weight] × 100%. In control animals an increase in the ipsilateral hemisphere was significant ($P < 0.05$), as marked with*. This was also the case after treatment with PP1, an inhibitor of Src, an intermediate in β_2-AR signaling, but not after administration of any of the other inhibitors, which interrupt β_1-, but not β_2-AR signaling as shown and discussed in Fig. 13.1. Most, but not all inhibitors used to delineate the signaling pathways shown in that figure were tested in this table.

CONCLUSIONS

Neural cells exposed to the period of ischemia inevitably die without reperfusion. In clinical settings, only a small part of patients experience complete reperfusion, with majority having either partial reperfusion or no reperfusion at all (see Ref. 57 and references therein). It is well known that during reperfusion cells experience extensive oxidative damage. In this paper, we suggest that reperfusion also plays a key role in development of the brain edema. Although brain edema depends on ionic edema (water influx), as well as vasogenic edema (extravasation of plasma protein that occur during reperfusion),[1] cell swelling induced by Na,K-ATPase failure cannot be excluded in ischemia period. Stimulation of β_1-ARs significantly increases NKCC1 activity. Exposure to β_1-AR antagonist could prevent brain edema completely, in spite of high extracellular K^+ concentrations in ischemia period. Several factors may be responsible for this phenomenon:

1. The function of NKCC1 is dependent on Na,K-ATPase activity. The failure of Na,K-ATPase because of ATP deficit in ischemia period not only induces substantial increase in extracellular K^+ concentrations but also obliterates the driving force for NKCC1.
2. In cultured cells the increase in NKCC1 activity is dependent on extracellular Ca^{2+}. NKCC1 could not be stimulated by high extracellular K^+

concentrations at low extracellular Ca^{2+} concentrations in ischemia period. Nevertheless, the situation is more complicated in situ because there is overload of intracellular Ca^{2+} in ischemia period since Ca^{2+} influx from extracellular space to intracellular fluid. When cultured cells are incubated in Ca^{2+}-free medium, it prevents an increase in intracellular Ca^{2+}.

3. NKCC1 is phosphorylated and stimulated by $ERK_{1/2}$ activation. In ischemia period, Raf/MAPK/$ERK_{1/2}$ signaling pathway is inhibited by AKT. Only in reperfusion period Raf/MAPK/$ERK_{1/2}$ is released from AKT inhibition. However, high extracellular K^+ concentrations last only for a few minutes during reperfusion, whereas brain edema will not occur until 8 h after reperfusion. The information about crosstalk between Raf/MAPK/$ERK_{1/2}$ and PI3K/AKT signal pathways during ischemia/reperfusion is obtained from brain in vivo. Detailed study about cell localization of these signals is required. It will be interesting to analyze NKCC1 phosphorylation that is necessary for the activity of cotransporter during ischemia and reperfusion periods.

The effect of β_1-AR antagonist on brain edema suggests that β_1-AR may be activated during ischemia and reperfusion periods. Again, Na,K-ATPase and NKCC1 are being activated only during reperfusion periods. Similar with high extracellular K^+ concentrations, there is a significant amount of adrenaline released in ischemia but it lasts longer than K^+ elevations during reperfusion.[58] During 20 min of ischemia, an 18-fold increase in hippocampal adrenaline concentration was detected in the perfusate. The levels remained elevated during early period of reperfusion and gradually returned to normal after 40 min.[58] Activation of β_1-AR stimulates Na,K-ATPase and NKCC1 in astrocytes, which could subsequently induce cytotoxic edema, ionic edema, and vasogenic edema. NKCC1 is also expressed in BBB.[34,59] It would be of interest to characterize signaling pathways of β_1- and β_2-ARs in cerebral endothelial cells since relevant signaling pathways may differ between different cells and tissues.

LIST OF ABBREVIATIONS

AQP4 aquaporin 4
AR adrenergic receptor
BBB blood–brain barrier
EGFR epidermal growth factor receptor
$ERK_{1/2}$ extracellular regulated kinase 1 and 2
HB-EGF heparin-binding epidermal growth factor
IP_3 inositol trisphosphate
MCAO medial cerebral artery occlusion
MMP metalloproteinase
NKCC1 $Na^+/K^+/Cl^-$ cotransporter

PKA protein kinase A
PLC phospholipase C complex
PLM phospholemman
PTEN phosphatase and tensin homolog
PTX pertussis toxin
ROS reactive oxygen species

ACKNOWLEDGMENTS

This study was supported by Grants No. 31171036 to L.P., No. 31000479 to B.L., No. 31400925 to D.S., and No. 31300883 to T.D. from the National Natural Science Foundation of China. A.V. was supported in part by the Federal Target Program "Research and development in priority areas of the development of the scientific and technological complex of Russia for 2014–2020" of the Ministry of Education and Science of Russia, contract 14.581.21.0016 (Project ID RFMEFI58115X0016).

REFERENCES

1. Stokum JA, Gerzanich V, Simard JM. Molecular pathophysiology of cerebral edema. *J Cereb Blood Flow Metab*. 2016;36:513–538.
2. Hajek I, Subbarao KV, Hertz L. Acute and chronic effects of potassium and noradrenaline on Na^+,K^+-ATPase activity in cultured mouse neurons and astrocytes. *Neurochem Int*. 1996;28:335–342.
3. Grande PO. New haemodynamic aspects on treatment of posttraumatic brain oedema. *Swedish Society of Anaesthesia and Intensive Care*. 1992;6:41–46.
4. Asgeirsson B, Grände PO, Nordström CH. A new therapy of post-trauma brain oedema based on haemodynamic principles for brain volume regulation. *Intensive Care Med*. 1994;20:260–267.
5. Song D, Xu J, Du T, Yan E, Hertz L, Walz W, et al. Inhibition of brain swelling after ischemia-reperfusion by β-adrenergic antagonists: correlation with increased K^+ and decreased Ca^{2+} concentrations in extracellular fluid. *Biomed Res Int*. 2014;2014:873590.
6. Peng L, Yu AC, Fung KY, Prévot V, Hertz L. Alpha-adrenergic stimulation of ERK phosphorylation in astrocytes is α_2-specific and may be mediated by transactivation. *Brain Res*. 2003;978:65–71.
7. Li B, Du T, Li H, Gu L, Zhang H, Huang J, et al. Signalling pathways for transactivation by dexmedetomidine of epidermal growth factor receptors in astrocytes and its paracrine effect on neurons. *Br J Pharmacol*. 2008;154:191–203.
8. Zhang M., Shan X., Gu L., Hertz L., Peng L. Dexmedetomidine causes neuroprotection via astrocytic α2-adrenergic receptor stimulation and HB-EGF release. *J Anesthesiol Clin Sci*. doi:10.7243/2049-9752-2-6.
9. Nahorski SR. Heterogeneity of cerebral β-adrenoceptor binding sites in various vertebrate species. *Eur J Pharmacol*. 1978;51:199–209.
10. Sastre M, Guimón J, García-Sevilla JA. Relationships between β- and α_2-adrenoceptors and G coupling proteins in the human brain: effects of age and suicide. *Brain Res*. 2001;898:242–255.
11. Summers RJ, Papaioannou M, Harris S, Evans BA. Expression of β_3-adrenoceptor mRNA in rat brain. *Br J Pharmacol*. 1995;116:2547–2548.

12. Hertz L, Lovatt D, Goldman SA, Nedergaard M. Adrenoceptors in brain: cellular gene expression and effects on astrocytic metabolism and $[Ca^{2+}]i$. *Neurochem Int.* 2010;57:411–420.
13. Du T, Li B, Li H, Li M, Hertz L, Peng L. Signaling pathways of isoproterenol-induced $ERK_{1/2}$ phosphorylation in primary cultures of astrocytes are concentration-dependent. *J Neurochem.* 2010;115:1007–1023.
14. Zwick E, Daub H, Aoki N, Yamaguchi-Aoki Y, Tinhofer I, Maly K, et al. Critical role of calcium-dependent epidermal growth factor receptor transactivation in PC12 cell membrane depolarization and bradykinin signaling. *J Biol Chem.* 1997;272:24767–24770.
15. Pierce KL, Luttrell LM, Lefkowitz RJ. New mechanisms in heptahelical receptor signaling to mitogen activated protein kinase cascades. *Oncogene.* 2001;20:1532–1539.
16. Peng L, Du T, Xu J, Song D, Li B, Zhang M, et al. Adrenergic and V1-ergic agonists/antagonists affecting recovery from brain trauma in the Lund Project Act on astrocytes. *Curr Signal Transduction Therapy.* 2012;7:43–55.
17. Zhuang S, Schnellmann RG. H_2O_2-induced transactivation of EGF receptor requires Src and mediates $ERK_{1/2}$, but not Akt, activation in renal cells. *Am J Physiol Renal Physiol.* 2004;286:F858–F865.
18. Li B, Zhang S, Zhang H, Nu W, Cai L, Hertz L, et al. Fluoxetine-mediated 5-HT_{2B} receptor stimulation in astrocytes causes EGF receptor transactivation and ERK phosphorylation. *Psychopharmacology (Berl).* 2008;201:443–458.
19. Cai L, Du T, Song D, Li B, Hertz L, Peng L. Astrocyte ERK phosphorylation precedes K^+-induced swelling but follows hypotonicity-induced swelling. *Neuropathology.* 2011;31:250–264.
20. Dai H, Song D, Xu J, Li B, Hertz L, Peng L. Ammonia-induced Na,K-ATPase/ouabain-mediated EGF receptor transactivation, MAPK/ERK and PI3K/AKT signaling and ROS formation cause astrocyte swelling. *Neurochem Int.* 2013;63:610–625.
21. Du T, Song D, Li H, Li B, Cai L, Hertz L, et al. Stimulation by vasopressin of ERK phosphorylation and vector-driven water flux in astrocytes is transactivation-dependent. *Eur J Pharmacol.* 2008;587:73–77.
22. Li B, Zhang S, Li M, Hertz L, Peng L. Serotonin increases $ERK_{1/2}$ phosphorylation in astrocytes by stimulation of 5-HT_{2B} and 5-HT_{2C} receptors. *Neurochem Int.* 2010;57:432–439.
23. Suzuki S, Zuege D, Berthiaume Y. Sodium-independent modulation of Na^+-K^+-ATPase activity by β-adrenergic agonist in alveolar type II cells. *Am J Physiol.* 1995;268:L983–L990.
24. Bertorello AM, Ridge KM, Chibalin AV, Katz AI, Sznajder JI. Isoproterenol increases Na^+-K^+-ATPase activity by membrane insertion of alpha-subunits in lung alveolar cells. *Am J Physiol.* 1999;276:L20–L27.
25. Osadchaia LM, Mugula Zh, Stefanov VE. The beta-adrenergic regulation of the Na, K-ATPase activity in the sarcolemma of the heart muscle. *Nauchnye Doki Vyss Shkoly Biol Nauki.* 1990;6:138–1347.
26. Despa S, Tucker AL, Bers DM. Phospholemman-mediated activation of Na/K-ATPase limits $[Na]_i$ and inotropic state during β-adrenergic stimulation in mouse ventricular myocytes. *Circulation.* 2008;117:1849–1855.
27. Pesce L, Guerrero C, Comellas A, Ridge KM, Sznajder JI. β-Agonists regulate Na,K-ATPase via novel MAPK/ERK and rapamycin-sensitive pathways. *FEBS Lett.* 2000;486:310–314.

28. Hamann S, Herrera-Perez JJ, Zeuthen T, Alvarez-Leefmans FJ. Cotransport of water by the $Na^+-K^+-2Cl^-$ cotransporter NKCC1 in mammalian epithelial cells. *J Physiol.* 2010;588:4089–4101.
29. Zeuthen T, Macaulay N. Cotransport of water by $Na^+-K^+-2Cl^-$ cotransporters expressed in Xenopus oocytes: NKCC1 versus NKCC2. *J Physiol.* 2012;590:1139–1154.
30. Kanaka C, Ohno K, Okabe A, Kuriyama K, Itoh T, Fukuda A, et al. The differential expression patterns of messenger RNAs encoding K-Cl cotransporters (KCC1,2) and Na-K-2Cl cotransporter (NKCC1) in the rat nervous system. *Neuroscience.* 2001;104:933–946.
31. Mikawa S, Wang C, Shu F, Wang T, Fukuda A, Sato K. Developmental changes in KCC1, KCC2 and NKCC1 mRNAs in the rat cerebellum. *Brain Res Dev Brain Res.* 2002;136:93–100.
32. Pedersen SF, O'Donnell ME, Anderson SE, Cala PM. Physiology and pathophysiology of Na^+/H^+ exchange and $Na^+-K^+-2Cl^-$ cotransport in the heart, brain, and blood. *Am J Physiol Regul Integr Comp Physiol.* 2006;291:R1–R25.
33. Hertz L. Bioenergetics of cerebral ischemia: a cellular perspective. *Neuropharmacology.* 2008;55:289–309.
34. Yuen N, Lam TI, Wallace BK, Klug NR, Anderson SE, O'Donnell ME. Ischemic factor-induced increases in cerebral microvascular endothelial cell Na/H exchange activity and abundance: evidence for involvement of $ERK_{1/2}$ MAP kinase. *Am J Physiol Cell Physiol.* 2014;306:C931–C942.
35. Kimelberg HK. Current concepts of brain edema. Review of laboratory investigations. *J Neurosurg.* 1995;83:1051–1059.
36. Darman RB, Forbush B. A regulatory locus of phosphorylation in the N terminus of the Na-K-Cl cotransporter, NKCC1. *J Biol Chem.* 2002;277:37542–37550.
37. Flemmer AW, Gimenez I, Dowd BF, Darman RB, Forbush B. Activation of the Na-K-Cl cotransporter NKCC1 detected with a phospho-specific antibody. *J Biol Chem.* 2002;277:37551–37558.
38. Paulais M, Turner RJ. β-Adrenergic upregulation of the Na^+-K^+-$2Cl^-$ cotransporter in rat parotid acinar cells. *J Clin Invest.* 1992;89:1142–1147.
39. Tanimura A, Kurihara K, Reshkin SJ, Turner RJ. Involvement of direct phosphorylation in the regulation of the rat parotid Na^+-K^+-$2Cl^-$ cotransporter. *J Biol Chem.* 1995; 270:25252–25258.
40. Kurihara K, Nakanishi N, Moore-Hoon ML, Turner RJ. Phosphorylation of the salivary $Na^{(+)}$-$K^{(+)}$-$2Cl^{(-)}$ cotransporter. *Am J Physiol Cell Physiol.* 2002;282:C817–C823.
41. Kurihara K, Moore-Hoon ML, Saitoh M, Turner RJ. Characterization of the phosphorylation event resulting in upregulation of the salivary Na^+-K^+-$2Cl^-$ cotransporter by beta-adrenergic stimulation. *Am J Physiol Cell Physiol.* 1999;277:C1184–C1193.
42. Hansen AJ, Nedergaard M. Brain ion homeostasis in cerebral ischemia. *Neurochem Pathol.* 1988;9:195–209.
43. Henn FA, Haljamäe H, Hamberger A. Glial cell function: active control of extracellular K^+ concentration. *Brain Res.* 1972;43:437–443.
44. Grisar T, Frere JM, Franck G. Effect of K^+ ions on kinetic properties of the (Na^+, K^+)-ATPase (EC 3.6.1.3) of bulk isolated glial cells, perikarya and synaptosomes from rabbit brain cortex. *Brain Res.* 1979;165:87–103.
45. Hájek V, Meugnier H, Bes M, Brun Y, Fiedler F, Chmela Z, et al. *Staphylococcus saprophyticus* subsp. bovis subsp. nov., isolated from bovine nostrils. *Int J Syst Bacteriol.* 1996;46:792–796.

46. Ferrari P, Ferrandi M, Valentini G, Bianchi G. Rostafuroxin: an ouabain antagonist that corrects renal and vascular Na^{+}-K^{+}-ATPase alterations in ouabain and adducin-dependent hypertension. *Am J Physiol Regul Integr Comp Physiol.* 2006;290: R529–R535.
47. Yuan Z, Cai T, Tian J, Ivanov AV, Giovannucci DR, Xie Z. Na/K-ATPase tethers phospholipase C and IP3 receptor into a calcium-regulatory complex. *Mol Biol Cell.* 2005;16:4034–4045.
48. Xu J, Song D, Xue Z, Gu L, Hertz L, Peng L. Requirement of glycogenolysis for uptake of increased extracellular K^{+} in astrocytes: potential implications for K^{+} homeostasis and glycogen usage in brain. *Neurochem Res.* 2013;38:472–485.
49. Yang Q, Wang EY, Huang XJ, Qu WS, Zhang L, Xu JZ, et al. Blocking epidermal growth factor receptor attenuates reactive astrogliosis through inhibiting cell cycle progression and protects against ischemic brain injury in rats. *J Neurochem.* 2011;119:644–653.
50. Murphy E, Steenbergen C. Mechanisms underlying acute protection from cardiac ischemia-reperfusion injury. *Physiol Rev.* 2008;88:581–609.
51. Martin P, Pognonec P. ERK and cell death: cadmium toxicity, sustained ERK activation and cell death. *FEBS J.* 2010;277:39–46.
52. Zhou J, Du T, Li B, Rong Y, Verkhratsky A, Peng L. Crosstalk between MAPK/ERK and PI3K/AKT signal pathways during brain ischemia/reperfusion. *ASN Neuro.* 2015:7.
53. Goyagi T, Kimura T, Nishikawa T, Tobe Y, Masaki Y. β-Adrenoreceptor antagonists attenuate brain injury after transient focal ischemia in rats. *Anesth Analg.* 2006; 103:658–663.
54. Goyagi T, Horiguchi T, Nishikawa T, Tobe Y. Post-treatment with selective β_1 adrenoceptor antagonists provides neuroprotection against transient focal ischemia in rats. *Brain Res.* 2010;1343:213–217.
55. Vardjan N, Horvat A, Anderson JE, Yu D, Croom D, Zeng X, et al. Adrenergic activation attenuates astrocyte swelling induced by hypotonicity and neurotrauma. *Glia.* 2016;64:1034–1049.
56. Song D, Xu J, Hertz L, Peng L. Regulatory volume increase in astrocytes exposed to hypertonic medium requires β1-adrenergic Na(+) /K(+)-ATPase stimulation and glycogenolysis. *J Neurosci Res.* 2015;93:130–139.
57. Zhao H, Steinberg GK, Sapolsky RM. General versus specific actions of mild-moderate hypothermia in attenuating cerebral ischemic damage. *J Cereb Blood Flow Metab.* 2007;27:1879–1894.
58. Mordecai YT, Globus MY, Busto R, Dietrich WD, Martinez E, Valdés I, et al. Direct evidence for acute and massive norepinephrine release in the hippocampus during transient ischemia. *J Cereb Blood Flow Metab.* 1989;9:892–896.
59. Abbruscato TJ, Lopez SP, Roder K, Paulson JR. Regulation of blood-brain barrier Na, K,2Cl-cotransporter through phosphorylation during in vitro stroke conditions and nicotine exposure. *J Pharmacol Exp Ther.* 2004;310:459–468.

Chapter 14

Noradrenaline, Astroglia, and Neuroinflammation

José L.M. Madrigal[1,2,✉]

[1]Universidad Complutense de Madrid, Madrid, Spain, [2]Instituto de Investigación Neuroquímica and Instituto de Investigación Sanitaria, Hospital 12 de Octubre, Madrid, Spain

Chapter Outline

Introduction 274

Astrocytes and Neuroinflammation 274

Noradrenaline Depletion in Neurodegenerative Diseases 275

Astrocyte Activation in Neurodegenerative Diseases 276

Noradrenaline Regulation of Astrogliosis 277

Noradrenaline Regulation of Astroglial Chemokines 280

Conclusions 282

Abbreviations 282

References 283

ABSTRACT

Among the different functions of astrocytes, their contribution to the neuroinflammatory response is a relevant one. This chapter reviews the inflammatory response in the central nervous system and the causes and effects of astroglial activation, paying special attention to the resulting neurological damage and how this facilitates the progression of certain neurodegenerative diseases. Based on this, it is also discussed how those agents able to control the toxic actions of astrocytes are of great interest from a therapeutic point of view. Noradrenaline is one of these agents. Besides its role as a neurotransmitter, it can also regulate other actions within the central nervous system. The noradrenaline deficit in diverse neurodegenerative diseases is discussed here, as well as the mechanisms through which noradrenaline can prevent those transformations of astrocytes that lead to the damage and loss of healthy neurons.

Keywords: Noradrenaline; astrocyte; inflammation; cytokine; chemokine; neurodegeneration

✉Correspondence address
E-mail: jlmmadrigal@med.ucm.es

Noradrenergic Signaling and Astroglia. DOI: http://dx.doi.org/10.1016/B978-0-12-805088-0.00014-1
© 2017 Elsevier Inc. All rights reserved.

INTRODUCTION

Astrocytes and Neuroinflammation

The injury to a tissue in the body stimulates a series of changes aimed at restoring the homeostasis existing before the onset of the alteration. The term inflammation refers to all the processes involved in this response. An adequate inflammatory response should remove the disrupting agent and help recover from the damage, it may have caused. In the process, some unavoidable incidental damage may occur to nearby healthy cells. While this self-inflicted harm may look as a drawback, it does not constitute a major inconvenience when the affected tissue can regenerate and fully recover its function afterwards. In fact, this last part of the whole process is a constitutive portion of the inflammatory response. However, when the regenerative capacity of the affected tissue is limited or the time required for it to fully recover is too long, the collateral damage of healthy cells does constitute a threat. This is the case of the central nervous system (CNS), and this is commonly accepted to be the reason for the existence of additional defensive elements that reduce the risk of injuries such as a reinforced bone structure (skull and vertebrae), additional lining layers (meninges) and selective barriers that prevent the penetration of potential harmful substances present in the blood (blood–brain barrier, BBB) or cerebrospinal fluid (blood-cerebrospinal fluid barrier).

These defenses, while preventing the attack of external dangers, also hinder the establishment of communications between the CNS and the periphery. An example of this is the limited entrance of cells from the immune system due to the resistance caused by the BBB. For this reason, alternative defense mechanisms exist in the CNS.

The main cell types responsible for the inflammatory response within the CNS are microglia and astrocytes. Microglia continuously analyze their surroundings using their processes and respond to alterations by switching to an "activated" state characterized by morphological and functional modifications that allow them to migrate, proliferate, phagocyte and to secrete cytokines or other toxic substances in order to eliminate the homeostasis-altering agent they detected. Similarly, astrocytes can also react to different kinds of stimuli generating a response known as astrogliosis. This response can exists in different degrees and is characterized by changes in the cell morphology as well as in the expression of different genes. In cases of severe damage, the astrogliosis can reach its maximal degree, facilitating the formation of new tissue composed mainly by astrocytes known as glial scar, a mechanism necessary in order to preserve, as much as possible, affected structures and functions in the CNS. However, this anomalous accumulation of cells may obstruct the extension and connections of axons.

The changes resulting from astrogliosis have proven to facilitate the recovery, at least to some extent, from different kinds of injuries such as ischemia[1] or brain injury.[2] However, an insufficient control of microglia or astrocytes activation can lead to an excessive neuroinflammatory response. This consists of a response too long, too intense, or both, causing either way the degeneration of healthy cells including neurons and other cell types. This phenomenon is responsible for the neuronal damage existing in different brain pathologies including neurotrauma, ischemia, infections, autoimmune diseases, genetic diseases, and neurodegenerative diseases.[3]

Among the different causes of an uncontrolled neuroinflammatory response, one of the most relevant ones seems to be the loss of termination signals that reduce the proliferation of glial cells as well as their production of cytokines and toxic agents.[4] These termination signals include antiinflammatory agents and trophic factors that contribute to the recovery of function in the damaged tissue. Noradrenaline (NA) has proven to be one of the most important antiinflammatory mediators.

NORADRENALINE DEPLETION IN NEURODEGENERATIVE DISEASES

The discovery of a characteristic loss of noradrenergic neurons from the locus coeruleus (LC) in Parkinson's disease[5–7] was one of the first steps in a series of studies that allowed identifying NA as a potent antiinflammatory agent in the CNS with a key role in the progression of neurodegenerative diseases. Later, the existence of damage in the LC was also observed in many other neurodegenerative diseases. Among these Alzheimer's disease (AD) is one of the most studied ones probably due to its high prevalence in modern societies as well as to its predicted increase because of extended life expectancy. In the early 1980s, different research groups presented data confirming the reduced density of LC noradrenergic neurons in AD.[8–11] Subsequent studies that correlated the amount of noradrenergic neuron loss to the degree of disease progression,[12] reinforced the relevance of this observation.

Following these initial discoveries, alterations in NA production or signaling have been detected to occur in diverse neurodegenerative diseases. Studies performed in Huntington's disease patients revealed a reduction in the length of the LC projections and in the number of its neurons. Additionally, these changes were associated with increased dementia and severity of motor impairments.[13] Furthermore, the measurement of NA concentration in samples obtained from amyotrophic lateral sclerosis (ALS) patients and control subjects detected lower concentrations in different segments of the spinal cord.[14]

Based on previous reports describing altered noradrenergic signaling in posttraumatic stress disorder (PTSD), the number of neurons in the right LC was also counted in brain samples from war veterans who suffered PTSD. The authors of this study observed a substantially lower neuronal count in PTSD patients.[15]

While the involvement of noradrenergic alterations with multiple sclerosis has been suspected for a relatively long time, clear evidence of reduced NA levels was not available until Dr. Feinstein's research group analyzed it using human brain samples.[16] In the same study, the increased astrocyte activation and loss of NA observed were also found using a mouse multiple sclerosis model such as experimental autoimmune encephalomyelitis, confirming this way the consistency of the results.

ASTROCYTE ACTIVATION IN NEURODEGENERATIVE DISEASES

Besides the current knowledge about the existence of NA-related alterations in different neurodegenerative diseases, its precise consequences as well as the remaining questions about whether this is a cause or an effect of the disease, remain still unresolved.

On the other side, the contribution of astrocytes to most of these neurodegenerative diseases is widely acknowledged. Due to different stimuli, astrocytes experience a series of changes in their morphology and functioning commonly known as astrocytosis, astrogliosis or reactive astrogliosis. This causes, among other changes, astrocyte proliferation, the formation of glial scars that can help contain the damage in CNS tissue and the release of signaling molecules. However, when the reactive astrogliosis is not properly controlled, it can lead to alterations in the tissue structure and function as well as to direct toxicity for nearby cells which do not constitute a threat.[17]

This way, astrogliosis can be used as a marker that helps detect the existence of CNS diseases. In the case of AD, activated astrocytes are commonly found surrounding amyloid beta plaques.[18,19] The location of astrocytes and their ability to clear amyloid beta deposits[20,21] helps prevent further damage caused by amyloid beta.

However, the continuous production of amyloid beta maintains the activation of astrocytes and their direct or indirect toxicity (via activation of microglia or other cell types), which causes the damage of neighboring neurons.[22,23] In a similar fashion, astrocytes can also accumulate amyloid beta derived from neurons.[24] This process, like many others involved in neuroinflammation, could be beneficial since it would help reduce the accumulation of amyloid beta and prevent the formation of plaques, but when the amount of amyloid beta accumulated inside these cells exceeds their capacity, they undergo lysis dispersing the substances contained in their cytoplasm, including of course amyloid beta, the one responsible for the whole process and potentiating this way its detrimental effects.

In fact the neuronal toxicity observed for amyloid beta in neuronal cultures has been proven to be exacerbated by the mere presence of even a reduced number of astrocytes.[25]

Besides amyloid beta, another characteristic alteration defining AD is the accumulation of hyperphosphorylated tau. This alteration causes the disruption of axonal transport and ultimately, the accumulation of neurofibrillary tangles and the degeneration of the affected neurons.[26] While both factors, amyloid beta and tau neurofibrillary tangles, are known to be two of the main agents responsible for the progression of AD, in the last years different studies using autopsy samples[27] or animal models[28] have proven that the alteration of tau filaments has a more clear accountability for the subsequent development of AD than amyloid beta does. Therefore, it can be concluded that the detection of neurofibrillary tangles of hyperphosphorylated tau protein constitutes a more reliable predictor for the future development of AD. A genetic modification performed in mice that caused the expression of human tau protein in astrocytes allowed studying the contribution of astroglial tau to neurodegeneration. This way, the transgenic mice presented an accumulation of tau pathology that was age-dependent and was combined with mild BBB disruption, induction of low-molecular-weight heat shock proteins and focal neuronal degenerations.[29]

ALS is a progressive neurodegenerative disease that affects motor neurons in the brain cortex, brainstem, and spinal cord ultimately causing muscle paralysis and death. The mechanisms responsible for the neuronal degeneration, characteristic of this disease, have not been discovered yet. Some studies indicate that mutations in the *SOD1* gene, which encodes Cu/Zn superoxide dismutase-1, an antioxidant enzyme, may be related to the progression of ALS.[30,31] Interestingly, astrocytes have also proven to play a key role in the neuronal degeneration in ALS. The coculture of human adult primary sporadic ALS astrocytes with human embryonic stem-cell-derived motor neurons allowed observing the toxic effect of ALS astrocytes on motor neurons.[32]

Parkinson's disease is another neurodegenerative disease where astrocytes constitute one of the main sources of neurotoxic agents responsible for the loss of dopaminergic neurons, characteristic of this disease. Evidence for this fact comes from different studies that reported an elevation in the density of astrocytes as well as an increase in the expression of astrocyte specific glial fibrillary acidic protein in human samples from Parkinson's disease patients[33] and from animal models.[34]

NORADRENALINE REGULATION OF ASTROGLIOSIS

According to the information provided above, NA deficiency and astrocyte activation seem to be two features associated with multiple neurodegenerative conditions. Therefore, it is fairly reasonable to consider that, at least in some of these cases, there may be a connection between both facts. This is supported by different studies providing data on anti-inflammatory actions of NA. The apparently dominant role of microglia over astrocytes as executing agents of the neuroinflammatory response could be the reason explaining why a larger number of these studies focus on microglia. However, a good number of them are focused on astrocytes. In fact, some of the earliest studies describing NA antiinflammatory actions were those showing its ability to inhibit the expression of class II major histocompatibility complex (MHCII) antigens in cultured mouse astrocytes,[35] being the ability of astrocytes to produce MHCII in vivo as was later confirmed by Zeinstra et al.[36] The reduction of this effect by antagonist of β-adrenergic receptors (β-ARs) such as propranolol or its achievement also by the cyclic adenosine monophosphate (cAMP) analog dibutyryl-cAMP or by the phosphodiesterase inhibitor dipyridamole[37] allowed the authors to confirm that this effect of NA depends on its activation of β_2-ARs on astrocytes.

One of the main reasons why activated microglia or astrocytes may be lethal for neurons is their production of different substances with cytotoxic properties. Such substances are probably destined to eliminate the dangerous cells or microorganisms that these glial cells consider could be causing the alteration of their environment that they detected. Among these cytotoxic agents, one of the most relevant ones, due to its potency and production by different cell types, is nitric oxide (NO).[38,39] NO is produced by different isoforms of nitric oxide synthases (NOS), which convert L-arginine to L-citrulline and NO. Three isoforms of NOS have been characterized; constitutive (cNOS or NOS1), endothelial (eNOS or NOS3), and inducible (iNOS or NOS2). However, the inducible type seems to be the most relevant one for the neuroinflammatory response due to its ability to produce larger amounts of NO than the other isoforms of NOS.[40] Such elevated concentrations of NO can lead to oxidative stress, altered energy metabolism, and DNA damage among other alterations that would eventually result toxic for the cells exposed to it.[41]

In vitro analysis of NA direct effect on rat astrocytes showed how this molecule inhibits the activation of NOS2 stimulated by lipopolysaccharide (LPS). Further analysis performed using β-AR agonists or antagonists and dibutyryl-cAMP confirmed that, also in this case, NA acts trough the activation of β_2-ARs.[42] The involvement of cAMP and protein kinase A (PKA) was analyzed in detail by a different research group that used different cAMP-elevating agents such as forskolin or cAMP-mimicking ones such as 8-bromo-cAMP or (Sp)-cAMP. The treatment of cultured primary astrocytes with these substances reduced the production of NO and the expression of NOS2 that were induced in these cells by LPS or cytokines.[43] In agreement with this, compounds that decrease cAMP and PKA activity such as H-89 and (Rp)-cAMP, stimulated NO production and

NOS2 expression in cultured astrocytes. The observation of a lack of effect of PKA on NOS2 activity led the authors to propose that PKA may modulate the synthesis of NOS2 but not its activity.[43] Indirectly, NA also inhibits NOS2 through the upregulation of the cDNA clone DST11, a complement C5a receptor isoform expressed by glia and neurons.[44]

Besides the inhibition of NOS2 expression, the activation of β_2-ARs in astrocytes stimulates the production of a number of trophic factors some of which may be responsible for the increase of neuritic growth observed in neurons exposed to the culture media obtained from NA-treated astrocytes.[45] This treatment also elevates the mRNA and protein concentration of peroxisome proliferator-activated receptor gamma,[46] a transcription factor known to control diverse physiological functions and is particularly interesting due to the reduction of neuroinflammation that can be achieved by its activation.[47]

The activation of nuclear factor kappa B, another transcription factor whose nuclear translocation is required for the expression of NOS2, various cytokines, and many other proteins,[48] is also regulated by NA. The treatment of cultured astrocytes with NA increases the expression and synthesis of the inhibitory I kappa B alpha and activates its promoter being these modifications prevented by β-AR antagonism.[49]

Similar to NO, interleukin (IL)-1β is another mediator known to contribute to the development of the inflammatory response in the CNS. This fact is evidenced by the observation of elevated levels of IL-1β in tissue samples obtained from subjects who had suffered neuroinflammatory processes such as brain injury or stroke.[50] Based on this, it is of interest to analyze if NA antiinflammatory actions include the regulation of astrocytic production of this cytokine. One study focused on this matter and observed that the elevation of intracellular cAMP in human astrocytes reduced the accumulation of IL-1β mRNA induced by LPS treatment. This alteration of IL-1β mRNA did not seem to be due to a modification of its stability.[51]

IL-β mechanisms for propagation of the neuroinflammatory response include the induction of other cytokines such as IL-6. Like IL-1β, IL-6 is another cytokine expressed by different cell types within the CNS, being astrocytes one of the main ones.[52] IL-6 actions comprise a large number of processes. In fact, while neuroinflammation and the immune response are among the best studied ones,[53] further analysis of IL-6 functions and the development of mice lacking astrocyte production of IL-6 or its receptor IL-6R, showed that this cytokine plays important roles on normal brain physiology and neurogenesis.[54] Therefore, given the antiinflammatory and neuroprotective actions of NA, the induction of IL-6 mRNA and protein observed in primary rat astrocytes cultures treated with NA,[55] could indicate that some of NA effects are mediated by IL-6. The induction of IL-6 by NA is mediated mainly by β_2-ARs although α_1-ARs also contribute. Interestingly, NA effect on IL-6 expression in astrocytes can also be induced by another neuroactive substance such as the neuropeptide vasoactive intestinal peptide (VIP), being the effect of NA and VIP synergized with IL-1β and tumor necrosis factor α (TNFα).[56]

TNFα expression and synthesis in astrocytes is also regulated by NA, observing an inhibition in basal conditions and also when these cells are activated by LPS. This consistency of effects does not occur in the case of IL-1β. In the presence of LPS, NA reduces IL-1β expression, but in the absence of other stimuli, NA elevates IL-1β mRNA concentration in primary astrocytes.[57] This indicates that the changes produced in astrocytes after their detection of a threatening agent may also include some changes in the way they respond to NA.

NORADRENALINE REGULATION OF ASTROGLIAL CHEMOKINES

While astrocytic production of pro-inflammatory cytokines and NO may have a significant impact on their environment, microglia effects usually have a greater relevance given the capacity of these cells to produce larger amounts of those agents. However, this is not the case for CCL2, a chemokine also known as monocyte chemoattractant protein 1 (MCP-1). CCL2 is largely expressed by astrocytes, being these cells one of the main sources of it within the CNS.[58] The best known role of CCL2 is to attract those cells which express the specific receptor CCR2 to the sites where CCL2 accumulates.[59] Because of this, CCL2 can be used as an indicator that allows for the detection of inflammatory processes. However, CCL2 is now known to participate in many other processes where cell migration is involved such as neurogenesis or angiogenesis but also in others that allow neurotransmission and neuroprotection.[60] In addition to many studies describing the neurotoxic actions of CCL2,[61] it has also been shown that CCL2 protects neurons from different injuries including those mediated by *N*-methyl-D-aspartate (NMDA), β-amyloid,[62] endotoxemia,[63] methylmercury,[64] or HIV-tat protein.[65] NA has a strong effect on CCL2 in cultured astrocytes; it elevates its mRNA, activates its promoter and increases its release to the culture medium. These effects seem to be mediated through the activation of β_2-ARs and help protect neurons against excitotoxic injuries such as NMDA mediated injuries or the exposure of neurons to an in vitro model of stroke consisting of the deprivation of oxygen and glucose.[66] In addition, the elevation of brain NA levels in mice by the administration of the NA precursor L-threo-3,4-dihydroxyphenylserine or desipramine, an inhibitor of NA transporters (inhibition of NA uptake), caused the elevation of CCL2 production by astrocytes. Furthermore, when the animals were exposed to restraint stress as a way to physiologically elevate brain production of NA, only a moderate increase of CCL2 was observed. This was due to the inhibitory effect on CCL2 production of the glucocorticoids that are also produced during the stress response; this was confirmed because the blockade of glucocorticoids production with the selective inhibitor metyrapone led to a large increase in CCL2 production in stressed animals.[67]

Interestingly, when cultured astrocytes were treated with inhibitors of NA transporters such as desipramine or atomoxetine, an increase in the expression and synthesis of CCL2 could also be observed. Since noradrenergic neuronal terminals were not present in the conditions used, it could be deduced that a different mechanism, independent of the alteration of NA levels, is also able to modulate the production of this chemokine by astrocytes. In fact the activation of α_2-ARs with the specific agonist clonidine, caused a reduction of CCL2 production while the opposite effect could be detected using the specific α_2-AR blocker yohimbine, this treatment was able to potentiate the production of CCL2 induced by the activation of β-ARs with isoproterenol.[68] These observations indicate that α_2-ARs, and not only the β type of AR, can also regulate the production of CCL2 in astrocytes (Fig. 14.1).

CX3CL1, also known as Fractalkine, is another chemokine for which, as for CCL2, various roles have been attributed besides those strictly related to its chemoattractant function. Of particular interest are its antiinflammatory and neuroprotective effects that have been discovered during the last years.[69] These seem to be mediated primarily through the activation of

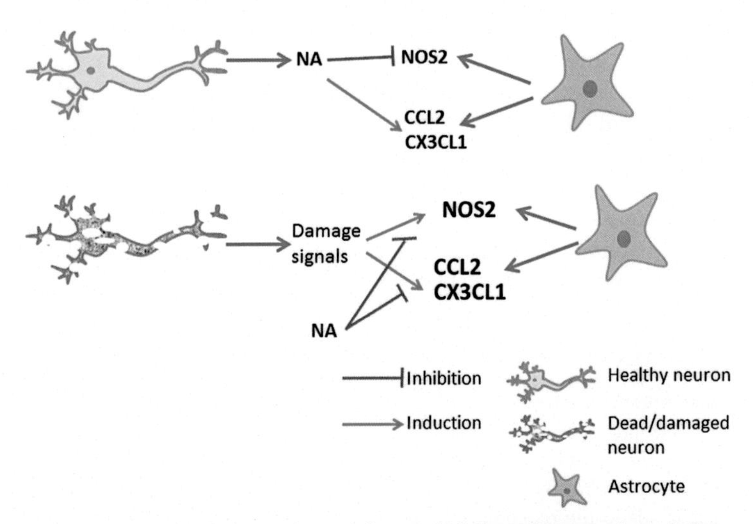

FIGURE 14.1 Noradrenaline regulation of astroglial chemokines and pro-inflammatory mediators. In normal conditions healthy neurons maintain adequate levels of noradrenaline (NA), which inhibits astrocytic production of certain chemokines. When neurons are damaged, their reduced production of NA and the signs of damage stimulate the neuroinflammatory response and the chemokine levels are excessively elevated. In these conditions, NA presents an opposite effect inhibiting the production of chemokines CCL2 and CX3CL1.

CX3CL1's specific receptor CX3CR1, which is mainly expressed by microglia although it can also act directly on neurons and protect them from excitotoxic stimuli.[70] The activation of CX3CR1 inhibits microglial activation, and since neurons are the main producers of CX3CL1 in brain, it has been proposed that CX3CL1 is used by neurons to communicate with microglia in certain situations and prevent an excessive neuroinflammation that could be detrimental for neurons.[71] While not so significant as neurons, astrocytes also constitute a relevant source of CX3CL1.[72,73] Given the antiinflammatory effects of this chemokine, the induction caused by NA[57] on astrocytes suggests that CX3CL1 could also contribute to the neuroprotective actions of NA.

As it was observed for IL-1β, in the presence of an inflammatory stimulus such as LPS, NA inhibited the production of CX3CL1, CCL2, and other chemokines. The existence of this diversity of effects may be in agreement with the existing discrepancies about whether the augmented signaling through CCL2 or CX3CL1 observed in certain neurodegenerative diseases[69,74,75] contributes to the loss of function or is part of a complex defensive and restoring response.

CONCLUSIONS

Based on NA well known neuroprotective effects, and on the existence of both neuroprotective and neurotoxic actions described for chemokines CCL2 and CX3CL1, it can be hypothesized that NA may help to maintain the adequate concentrations of certain cytokines and chemokines necessary to maintain homeostasis. This consists of a moderate induction of these mediators in basal conditions and a restriction of an exaggerated production of them in inflammatory processes. According to this, the loss of noradrenergic neurons observed in AD and other neurodegenerative diseases would lead to an uncontrolled production of different agents involved in the orchestration of the neuroinflammatory response, contributing to the progression of the disease.

ABBREVIATIONS

AD Alzheimer's disease
ALS amyotrophic lateral sclerosis
ARs adrenergic receptors
BBB blood–brain barrier
cAMP cyclic adenosine monophosphate
CNS central nervous system
CX3CL1 fractalkine
IL interleukin
LC locus coeruleus
LPS lipopolysaccharide
MCP-1 monocyte chemoattractant protein 1

MHCII class II major histocompatibility complex
NA noradrenaline
NMDA *N*-methyl-D-aspartate
NO nitric oxide
NOS nitric oxide synthase
PKA protein kinase A
PTSD posttraumatic stress disorder
TNFα tumor necrosis factor α
VIP vasoactive intestinal peptide

REFERENCES

1. Li L, Lundkvist A, Andersson D, Wilhelmsson U, Nagai N, Pardo AC, et al. Protective role of reactive astrocytes in brain ischemia. *J Cereb Blood Flow Metab.* 2008;28:468–481.
2. Okada S, Nakamura M, Katoh H, Miyao T, Shimazaki T, Ishii K, et al. Conditional ablation of Stat3 or Socs3 discloses a dual role for reactive astrocytes after spinal cord injury. *Nat Med.* 2006;12:829–834.
3. Pekny M, Nilsson M. Astrocyte activation and reactive gliosis. *Glia.* 2005;50:427–434.
4. Serhan CN, Chiang N, Van Dyke TE. Resolving inflammation: dual anti-inflammatory and pro-resolution lipid mediators. *Nat Rev Immunol.* 2008;8:349–361.
5. Beheim-Schwarzbach D. Cell changes in the nucleus coeruleus in Parkinson syndrome. *J Nerv Ment Dis.* 1952;116:619–632.
6. Forno LS, Alvord Jr. EC. The pathology of Parkinsonism. Part I. Some new observations and correlations. *Contemp Neurol Ser.* 1971;8:119–130.
7. Roveti G. Inclusion bodies in substantia nigra cells and locus caeruleus in cases without Parkinson's syndrome, with a contribution to senile changes in these regions. *Monatsschr Psychiatr Neurol.* 1956;132:347–363.
8. Bondareff W, Mountjoy CQ, Roth M. Selective loss of neurones of origin of adrenergic projection to cerebral cortex (nucleus locus coeruleus) in senile dementia. *Lancet.* 1981;1:783–784.
9. Mann DM, Yates PO. Pathological basis for neurotransmitter changes in Parkinson's disease. *Neuropathol Appl Neurobiol.* 1983;9:3–19.
10. Perry EK, Tomlinson BE, Blessed G, Perry RH, Cross AJ, Crow TJ. Neuropathological and biochemical observations on the noradrenergic system in Alzheimer's disease. *J Neurol Sci.* 1981;51:279–287.
11. Tomlinson BE, Irving D, Blessed G. Cell loss in the locus coeruleus in senile dementia of Alzheimer type. *J Neurol Sci.* 1981;49:419–428.
12. Bondareff W, Mountjoy CQ, Roth M, Rossor MN, Iversen LL, Reynolds GP, et al. Neuronal degeneration in locus ceruleus and cortical correlates of Alzheimer disease. *Alzheimer Dis Assoc Disord.* 1987;1:256–262.
13. Zweig RM, Ross CA, Hedreen JC, Peyser C, Cardillo JE, Folstein SE, et al. Locus coeruleus involvement in Huntington's disease. *Arch Neurol.* 1992;49:152–156.
14. Sofic E, Riederer P, Gsell W, Gavranovic M, Schmidtke A, Jellinger K. Biogenic amines and metabolites in spinal cord of patients with Parkinson's disease and amyotrophic lateral sclerosis. *J Neural Transm Park Dis Dement Sect.* 1991;3:133–142.
15. Bracha HS, Garcia-Rill E, Mrak RE, Skinner R. Postmortem locus coeruleus neuron count in three American veterans with probable or possible war-related PTSD. *J Neuropsychiatry Clin Neurosci.* 2005;17:503–509.

16. Polak PE, Kalinin S, Feinstein DL. Locus coeruleus damage and noradrenaline reductions in multiple sclerosis and experimental autoimmune encephalomyelitis. *Brain.* 2011;134:665–677.
17. Sofroniew MV, Bush TG, Blumauer N, Lawrence K, Mucke L, Johnson MH. Genetically-targeted and conditionally-regulated ablation of astroglial cells in the central, enteric and peripheral nervous systems in adult transgenic mice. *Brain Res.* 1999;835:91–95.
18. Canning DR, McKeon RJ, DeWitt DA, Perry G, Wujek JR, Frederickson RC, et al. beta-Amyloid of Alzheimer's disease induces reactive gliosis that inhibits axonal outgrowth. *Exp Neurol.* 1993;124:289–298.
19. Murphy Jr. GM, Ellis WG, Lee YL, Stultz KE, Shrivastava R, Tinklenberg JR, et al. Astrocytic gliosis in the amygdala in Down's syndrome and Alzheimer's disease. *Prog Brain Res.* 1992;94:475–483.
20. Lv J, Ma S, Zhang X, Zheng L, Ma Y, Zhao X, et al. Quantitative proteomics reveals that PEA15 regulates astroglial Abeta phagocytosis in an Alzheimer's disease mouse model. *J Proteomics.* 2014;110:45–58.
21. Wyss-Coray T, Loike JD, Brionne TC, Lu E, Anankov R, Yan F, et al. Adult mouse astrocytes degrade amyloid-beta in vitro and in situ. *Nat Med.* 2003;9:453–457.
22. Garcia-Matas S, de VN, Aznar AO, Marimon JM, Adell A, Planas AM, et al. In vitro and in vivo activation of astrocytes by amyloid-beta is potentiated by pro-oxidant agents. *J Alzheimers Dis.* 2010;20:229–245.
23. White JA, Manelli AM, Holmberg KH, Van Eldik LJ, LaDu MJ. Differential effects of oligomeric and fibrillar amyloid-beta 1-42 on astrocyte-mediated inflammation. *Neurobiol Dis.* 2005;18:459–465.
24. Nagele RG, Wegiel J, Venkataraman V, Imaki H, Wang KC, Wegiel J. Contribution of glial cells to the development of amyloid plaques in Alzheimer's disease. *Neurobiol Aging.* 2004;25:663–674.
25. Garwood CJ, Pooler AM, Atherton J, Hanger DP, Noble W. Astrocytes are important mediators of Abeta-induced neurotoxicity and tau phosphorylation in primary culture. *Cell Death Dis.* 2011;2:e167.
26. Iqbal K, Alonso AC, Chen S, Chohan MO, El-Akkad E, Gong CX, et al. Tau pathology in Alzheimer disease and other tauopathies. *Biochim Biophys Acta.* 1739;2005:198–210.
27. Murray ME, Lowe VJ, Graff-Radford NR, Liesinger AM, Cannon A, Przybelski SA, et al. Clinicopathologic and 11C-Pittsburgh compound B implications of Thal amyloid phase across the Alzheimer's disease spectrum. *Brain.* 2015;138:1370–1381.
28. Lonskaya I, Hebron M, Chen W, Schachter J, Moussa C. Tau deletion impairs intracellular beta-amyloid-42 clearance and leads to more extracellular plaque deposition in gene transfer models. *Mol Neurodegener.* 2014;9:46.
29. Forman MS, Lal D, Zhang B, Dabir DV, Swanson E, Lee VM, et al. Transgenic mouse model of tau pathology in astrocytes leading to nervous system degeneration. *J Neurosci.* 2005;25:3539–3550.
30. Patel SA, Maragakis NJ. Amyotrophic lateral sclerosis: pathogenesis, differential diagnoses, and potential interventions. *J Spinal Cord Med.* 2002;25:262–273.
31. Rosen DR, Siddique T, Patterson D, Figlewicz DA, Sapp P, Hentati A, et al. Mutations in Cu/Zn superoxide dismutase gene are associated with familial amyotrophic lateral sclerosis. *Nature.* 1993;362:59–62.
32. Re DB, Le V, Yu C, Amoroso MW, Politi KA, et al. Necroptosis drives motor neuron death in models of both sporadic and familial ALS. *Neuron.* 2014;81:1001–1008.

33. Forno LS, DeLanney LE, Irwin I, Di MD, Langston JW. Astrocytes and Parkinson's disease. *Prog Brain Res.* 1992;94:429–436.
34. Teismann P, Schulz JB. Cellular pathology of Parkinson's disease: astrocytes, microglia and inflammation. *Cell Tissue Res.* 2004;318:149–161.
35. Frohman EM, Vayuvegula B, van den Noort S, Gupta S. Norepinephrine inhibits gamma-interferon-induced MHC class II (Ia) antigen expression on cultured brain astrocytes. *J Neuroimmunol.* 1988;17:89–101.
36. Zeinstra E, Wilczak N, Streefland C, De KJ. Astrocytes in chronic active multiple sclerosis plaques express MHC class II molecules. *Neuroreport.* 2000;11:89–91.
37. Frohman EM, Vayuvegula B, Gupta S, van den Noort S. Norepinephrine inhibits gamma-interferon-induced major histocompatibility class II (Ia) antigen expression on cultured astrocytes via beta-2-adrenergic signal transduction mechanisms. *Proc Natl Acad Sci USA.* 1988;85:1292–1296.
38. Brown GC, Vilalta A. How microglia kill neurons. *Brain Res.* 1628;2015:288–297.
39. Dawson VL, Dawson TM, Bartley DA, Uhl GR, Snyder SH. Mechanisms of nitric oxide-mediated neurotoxicity in primary brain cultures. *J Neurosci.* 1993;13:2651–2661.
40. Vincent VA, Tilders FJ, Van Dam AM. Production, regulation and role of nitric oxide in glial cells. *Mediators Inflamm.* 1998;7:239–255.
41. Murphy MP. Nitric oxide and cell death. *Biochim Biophys Acta.* 1999;1411:401–414.
42. Feinstein DL, Galea E, Reis DJ. Norepinephrine suppresses inducible nitric oxide synthase activity in rat astroglial cultures. *J Neurochem.* 1993;60:1945–1948.
43. Pahan K, Namboodiri AM, Sheikh FG, Smith BT, Singh I. Increasing cAMP attenuates induction of inducible nitric-oxide synthase in rat primary astrocytes. *J Biol Chem.* 1997;272:7786–7791.
44. Gavrilyuk V, Kalinin S, Hilbush BS, Middlecamp A, McGuire S, Pelligrino D, et al. Identification of complement 5a-like receptor (C5L2) from astrocytes: characterization of anti-inflammatory properties. *J Neurochem.* 2005;92:1140–1149.
45. Day JS, O'Neill E, Cawley C, Aretz NK, Kilroy D, Gibney SM, et al. Noradrenaline acting on astrocytic beta(2)-adrenoceptors induces neurite outgrowth in primary cortical neurons. *Neuropharmacology.* 2014;77:234–248.
46. Klotz L, Sastre M, Kreutz A, Gavrilyuk V, Klockgether T, Feinstein DL, et al. Noradrenaline induces expression of peroxisome proliferator activated receptor gamma (PPARgamma) in murine primary astrocytes and neurons. *J Neurochem.* 2003;86:907–916.
47. Mrak RE, Landreth GE. PPARgamma, neuroinflammation, and disease. *J Neuroinflammation.* 2004;1:5.
48. Gilmore TD. Introduction to NF-kappaB: players, pathways, perspectives. *Oncogene.* 2006;25:6680–6684.
49. Gavrilyuk V, Dello RC, Heneka MT, Pelligrino D, Weinberg G, Feinstein DL. Norepinephrine increases I kappa B alpha expression in astrocytes. *J Biol Chem.* 2002;277:29662–29668.
50. Shaftel SS, Griffin WS, O'Banion MK. The role of interleukin-1 in neuroinflammation and Alzheimer disease: an evolving perspective. *J Neuroinflammation.* 2008;5:7.
51. Willis SA, Nisen PD. Inhibition of lipopolysaccharide-induced IL-1 beta transcription by cyclic adenosine monophosphate in human astrocytic cells. *J Immunol.* 1995;154:1399–1406.
52. Van Wagoner NJ, Oh JW, Repovic P, Benveniste EN. Interleukin-6 (IL-6) production by astrocytes: autocrine regulation by IL-6 and the soluble IL-6 receptor. *J Neurosci.* 1999;19:5236–5244.

53. Rincon M. Interleukin-6: from an inflammatory marker to a target for inflammatory diseases. *Trends Immunol.* 2012;33:571–577.
54. Quintana A, Erta M, Ferrer B, Comes G, Giralt M, Hidalgo J. Astrocyte-specific deficiency of interleukin-6 and its receptor reveal specific roles in survival, body weight and behavior. *Brain Behav Immun.* 2013;27:162–173.
55. Norris JG, Benveniste EN. Interleukin-6 production by astrocytes: induction by the neurotransmitter norepinephrine. *J Neuroimmunol.* 1993;45:137–145.
56. Maimone D, Cioni C, Rosa S, Macchia G, Aloisi F, Annunziata P. Norepinephrine and vasoactive intestinal peptide induce IL-6 secretion by astrocytes: synergism with IL-1 beta and TNF alpha. *J Neuroimmunol.* 1993;47:73–81.
57. Hinojosa AE, Caso JR, Garcia-Bueno B, Leza JC, Madrigal JL. Dual effects of noradrenaline on astroglial production of chemokines and pro-inflammatory mediators. *J Neuroinflammation.* 2013;10:81.
58. Glabinski AR, Balasingam V, Tani M, Kunkel SL, Strieter RM, Yong VW, et al. Chemokine monocyte chemoattractant protein-1 is expressed by astrocytes after mechanical injury to the brain. *J Immunol.* 1996;156:4363–4368.
59. Conductier G, Blondeau N, Guyon A, Nahon JL, Rovere C. The role of monocyte chemoattractant protein MCP1/CCL2 in neuroinflammatory diseases. *J Neuroimmunol.* 2010;224:93–100.
60. Semple BD, Kossmann T, Morganti-Kossmann MC. Role of chemokines in CNS health and pathology: a focus on the CCL2/CCR2 and CXCL8/CXCR2 networks. *J Cereb Blood Flow Metab.* 2010;30:459–473.
61. Bose S, Cho J. Role of chemokine CCL2 and its receptor CCR2 in neurodegenerative diseases. *Arch Pharm Res.* 2013;36:1039–1050.
62. Bruno V, Copani A, Besong G, Scoto G, Nicoletti F. Neuroprotective activity of chemokines against N-methyl-D-aspartate or beta-amyloid-induced toxicity in culture. *Eur J Pharmacol.* 2000;399:117–121.
63. Zisman DA, Kunkel SL, Strieter RM, Tsai WC, Bucknell K, Wilkowski J, et al. MCP-1 protects mice in lethal endotoxemia. *J Clin Invest.* 1997;99:2832–2836.
64. Godefroy D, Gosselin RD, Yasutake A, Fujimura M, Combadiere C, Maury-Brachet R, et al. The chemokine CCL2 protects against methylmercury neurotoxicity. *Toxicol Sci.* 2012;125:209–218.
65. Eugenin EA, D'Aversa TG, Lopez L, Calderon TM, Berman JW. MCP-1 (CCL2) protects human neurons and astrocytes from NMDA or HIV-tat-induced apoptosis. *J Neurochem.* 2003;85:1299–1311.
66. Madrigal JL, Leza JC, Polak P, Kalinin S, Feinstein DL. Astrocyte-derived MCP-1 mediates neuroprotective effects of noradrenaline. *J Neurosci.* 2009;29:263–267.
67. Madrigal JL, Garcia-Bueno B, Hinojosa AE, Polak P, Feinstein DL, Leza JC. Regulation of MCP-1 production in brain by stress and noradrenaline-modulating drugs. *J Neurochem.* 2010;113:543–551.
68. Hinojosa AE, Garcia-Bueno B, Leza JC, Madrigal JL. Regulation of CCL2/MCP-1 production in astrocytes by desipramine and atomoxetine: involvement of alpha2 adrenergic receptors. *Brain Res Bull.* 2011;86:326–333.
69. Lauro C, Catalano M, Trettel F, Limatola C. Fractalkine in the nervous system: neuroprotective or neurotoxic molecule? *Ann NY Acad Sci.* 2015;1351:141–148.
70. Limatola C, Lauro C, Catalano M, Ciotti MT, Bertollini C, Di AS, et al. Chemokine CX3CL1 protects rat hippocampal neurons against glutamate-mediated excitotoxicity. *J Neuroimmunol.* 2005;166:19–28.

71. Sheridan GK, Murphy KJ. Neuron-glia crosstalk in health and disease: fractalkine and CX3CR1 take centre stage. *Open Biol.* 2013;3:130181.
72. Sheng WS, Hu S, Ni HT, Rock RB, Peterson PK. WIN55,212-2 inhibits production of CX3CL1 by human astrocytes: involvement of p38 MAP kinase. *J Neuroimmune Pharmacol.* 2009;4:244–248.
73. Tei N, Tanaka J, Sugimoto K, Nishihara T, Nishioka R, Takahashi H, et al. Expression of MCP-1 and fractalkine on endothelial cells and astrocytes may contribute to the invasion and migration of brain macrophages in ischemic rat brain lesions. *J Neurosci Res.* 2013;91:681–693.
74. Mocchetti I, Campbell LA, Harry GJ, Avdoshina V. When human immunodeficiency virus meets chemokines and microglia: neuroprotection or neurodegeneration? *J Neuroimmune Pharmacol.* 2013;8:118–131.
75. Re DB, Przedborski S. Fractalkine: moving from chemotaxis to neuroprotection. *Nat Neurosci.* 2006;9:859–861.

Chapter 15

Astrocytic β_2-Adrenergic Receptors and Multiple Sclerosis

Jacques De Keyser[1,2,3,✉]
[1]University Hospital Brussels, Brussels, Belgium, [2]Vrije Universiteit Brussel (VUB), Brussels, Belgium, [3]University Medical Center Groningen, University of Groningen, Groningen, The Netherlands

Chapter Outline

Introduction 290
Downregulation of Astrocytic β_2-Adrenergic Receptors in Multiple Sclerosis 291
β_2-Adrenergic Receptors in Multiple Sclerosis and Progressive Multifocal Leukoencephalopathy 292
Underlying Mechanism of Astrocytic β_2-Adrenergic Receptor Downregulation 293
Pathophysiological Role in Focal Inflammatory Lesions 294
Pathophysiological Role in Axonal Degeneration 294
Pathophysiological Role in Both Axonal Degeneration and Oligodendrogliopathy 295
Abbreviations 296
Acknowledgments 296
References 296

ABSTRACT
A downregulation of astrocytic β_2-adrenergic receptors might play a role in the pathophysiology of multiple sclerosis. It may facilitate autoimmune processes, leading to the characteristic focal inflammatory demyelinating lesions, by transforming astrocytes into facultative antigen presenting cells and reducing double negative T cells, which are considered to act as regulatory T cells having the capacity of inducing antigen-specific immune tolerance. Downregulation of astrocytic β_2-adrenergic receptors may contribute to progressive axonal degeneration by allowing expression of inducible nitric oxide oxidase leading to increased levels of nitric oxide, reducing glycogenolysis leading to impaired formation of lactate as axonal energy source, reducing trophic support, and facilitating astrocytic expression of endothelin-1 causing cerebral hypoperfusion.

✉Correspondence address
E-mail: jacques.dekeyser@uzbrussel.be

Noradrenergic Signaling and Astroglia. DOI: http://dx.doi.org/10.1016/B978-0-12-805088-0.00015-3
© 2017 Elsevier Inc. All rights reserved.

Keywords: Multiple sclerosis; β_2-adrenergic receptors; inflammation; axonal degeneration; pathophysiology

INTRODUCTION

Despite intensive research, the cause of multiple sclerosis (MS) remains unknown and pathophysiologic mechanisms are not well understood. The disease is mostly diagnosed in young adults with a female preponderance.

Most commonly patients start with a relapsing–remitting disease course, which is characterized by focal inflammatory demyelinating lesions mainly affecting white matter of the central nervous system (CNS).[1] Relapses are attacks of new or worsening MS symptoms caused by inflammatory active focal lesions. T cells against myelin peptides are believed to play a central role in the pathogenesis of relapses.[2] If not treated with high dose steroids, symptoms usually last for some weeks and either resolve completely or cause residual disability. A number of immunomodulatory drugs have been developed that are partially effective in reducing new focal lesions on magnetic resonance imaging (MRI) of the brain and the frequency of relapses.[3] Although there is a popular belief that relapsing remitting MS is an autoimmune disease against myelin proteins, it remains unclear whether the inflammatory reactions in the CNS are primary or secondary. Antimyelin T cells are part of the repertoire of each individual and clonal expansion of these antimyelin T cells (indicating that they have been activated in the periphery, and hence have the ability to penetrate into the CNS), not only occurs in MS patients but also in healthy individuals and individuals with a variety of other disorders.[4,5] If the immune aggressors are present and can become active in all individuals, why do only a few people develop focal MS lesions? In addition, inflammatory infiltrates are not always present in areas of active demyelination,[6] and some demyelinating lesions appear to be caused by a primary oligodendrogliopathy.[7,8]

After some years, many patients with a relapsing–remitting course switch to a secondary progressive form in which there is a downhill course that may still be accompanied by a few relapses. A minority of patients suffers from primary progressive MS and experience progression of disability from onset, which often manifests as a chronic progressive myelopathy.[1] There is evidence that secondary and primary progressive MS are age-dependent, and may therefore be regarded as essentially similar conditions.[9,10] Worsening in progressive MS is not obviously influenced by relapses,[11] and immunomodulatory drugs that reduce relapses and new focal lesions on brain MRI have no effect on progressive MS. There is currently no therapy that can slow down the progressive degeneration of axons, which is the primary determinant of long-term disability in MS.

Several findings suggest that progressive axonal degeneration in MS is caused by impairment of axonal energy-dependent processes.[12]

Here we discuss the possible pathophysiological role of astrocytic β_2-adrenergic receptor (β_2-AR) downregulation in MS.

DOWNREGULATION OF ASTROCYTIC β_2-ADRENERGIC RECEPTORS IN MULTIPLE SCLEROSIS

In 1997, our research team was asked whether we could find applications for the use of [^{18}F]fluorocarazolol, a nonselective β_2-AR antagonist passing the blood–brain barrier, and therefore suitable for positron emission tomography (PET) imaging and quantification of β_2-ARs in the brain.[13]

Mantyh et al. had examined the expression of ARs in normal, crushed, and transected optic nerves of the rabbit and rat, and in normal and damaged human optic nerves.[14] In the normal rabbit, rat, and human optic nerves, only α_1 and β_2-ARs were observed, and these were present in low to moderate densities. Using combined immunohistochemistry and autoradiography they found that the majority of β_2-ARs were expressed by astrocytes. In damaged optic nerves, associated by hypertrophy and proliferation of astrocytes, β_2-ARs were significantly upregulated. Hypertrophy and proliferation of astrocytes occurs in focal demyelinating MS lesions resulting in a dense astrogliotic scar. Hence, the name "multiple sclerosis" referring to the pathological finding of multiple gliotic scars (plaques) disseminated throughout the CNS.[1] We started to investigate [^{18}F]fluorocarazolol for the quantification of astrogliotic lesions in the brain of individuals with MS. After local ethical committee approval we included a 33-year-old female patient who had a 10-year history of relapsing–remitting MS, and multiple plaques throughout the white matter on brain MRI. To our surprise, none of the MS plaques were labeled with [^{18}F]fluorocarazolol, whereas gray matter containing neurons expressing β_2-ARs, such as cerebral cortex, was clearly positive. Unfortunately, we could not proceed with the study because the ligand turned out to be positive in the AMES test (a test for mutagenicity).

Confronted with this unexpected finding we used immunohistochemistry to assess the presence of β_2-ARs in postmortem brain tissues of controls, patients with MS and infarctions, and in spinal cord from patients with amyotrophic lateral sclerosis. β_2-ARs were prominently expressed in reactive astrocytes surrounding cerebral infarctions and in the lateral corticospinal tracts in amyotrophic lateral sclerosis. However, in MS brain samples, β_2-ARs could not be detected in either astrocytes in normal-appearing white matter or in astrogliotic plaques, whereas they were normally present on neurons in the cerebral cortex.[15] The downregulation of β_2-ARs in MS plaques and white matter was confirmed using quantitative autoradiography with [^{3}H]dihydroalprenolol.[16] In contrast to rodents, human white matter astrocytes do not express β_1-ARs.[16,17] β_3-ARs have been identified on mouse and chick astrocytes.[18] In contrast, β_3-AR mRNA has not been detected in human cerebellum and cerebral cortex.[19]

β_2-ADRENERGIC RECEPTORS IN MULTIPLE SCLEROSIS AND PROGRESSIVE MULTIFOCAL LEUKOENCEPHALOPATHY

Here, we present in more detail an experiment that has not been published previously. Using immunohistochemistry we studied astrocytic β_2-ARs on postmortem slices of patients with MS and progressive multifocal leukoencephalopathy (PML), a demyelinating opportunistic infection of oligodendrocytes by the John Cunningham (JC) virus.[20] PML occurs predominantly in immunocompromised individuals, and is a rare but feared complication of certain immunomodulatory treatments used in MS, such as natalizumab.[21] The brain specimens used in this experiment were obtained from patients with Acquired Immune Deficiency Syndrome. A typical finding in PML lesions is the presence of astrocytes with enlarged hyperchromatic nuclei.[22]

Many commercially available antibodies against G-protein coupled receptors, including those for β_2-ARs, lack selectivity.[23] After having tested several commercially available antibodies, we selected a chicken–antihuman β_2-AR antibody from GenWay Biotechnology (San Diego, CA, USA). To reduce nonspecific binding to a minimum, the antihuman β_2-AR antibody was preincubated overnight at 4°C on 20-μm brain sections from β_2-AR knockout C57Bl/6J mice (generated by Brian Kobilka, Stanford University, Stanford, CA, USA and kindly donated by Lutz Hein, Freiburg, Germany). Thereafter, antibody solutions were removed and stored at 4°C until use. Staining of β_2-AR knockout brain tissue with this purified chicken–antihuman β_2-AR antibody showed no immunoreactivity indicating that the purified antibody was specific. Specificity of secondary antibody immunoreactivity was controlled by the incubation of tissue sections in 5% goat serum instead of primary antibodies. Monoclonal Cy3-labeled mouse-antihuman glial fibrillary acidic protein (GFAP) was purchased from Sigma Aldrich (Saint Louis, MO, USA). Alexa-fluor 488-goat-antichicken-IgG was purchased from Molecular probes (Leiden, The Netherlands). Before the addition of the first and secondary antibody solution, sections were incubated for 30 min at room temperature in normal goat serum, to suppress nonspecific antibody binding. Sections were incubated with both purified chicken–antihuman β_2-AR (1 μg for each section) and Cy3-labeled mouse-antihuman GFAP (1/100) in phosphate-buffered saline (PBS) overnight at 4°C. Thereafter, sections were incubated with the secondary antibody solution: Alexa-fluor 488 goat–antichicken-IgG (FITC-conjugated) (1/100) in PBS for 120 min at room temperature. Between all steps the sections were rinsed thoroughly with PBS. Sections

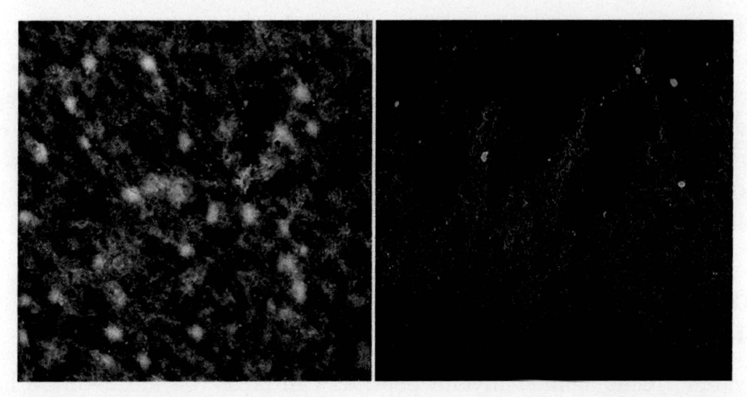

FIGURE 15.1 **Astrocytic β_2-adrenergic receptors in progressive multifocal leukoencephalopathy and multiple sclerosis lesions.** Double-staining for glial fibrillary acidic protein (GFAP) as marker for astrocytes (red) and β_2-adrenergic receptors (green) in the white matter of a progressive multifocal leukoencephalopathy (PML) lesion (left) and an astrogliotic plaque of multiple sclerosis (MS, right).

were treated with a solution of Sudan Black B (Sigma-Aldrich) in 70% ethanol for 5 min to block lipofuscin fluorescence. Excess of Sudan black B was removed by rinsing sections quickly in 70% ethanol and in distillated water. Sections were embedded in Prolong Gold antifading fluorescent mounting medium (Invitrogen, Oregon, USA). After staining, fluorescence serial images were taken using confocal scanning laser microscopy.

From the MS brain samples, we identified eight chronic demyelinated astrogliotic plaques, and from the PML cases, we selected two demyelinated lesions, characterized by the presence of swollen hypertrophic astrocytes. In PML, hypertrophic astrocytes were strongly immunopositive for the β_2-AR antibody, whereas astrocytes in the MS plaques did not stain (Fig. 15.1). These additional experiments confirm a downregulation of astrocytic β_2-ARs in MS and their upregulation in other disorders where astrocytes are activated.

UNDERLYING MECHANISM OF ASTROCYTIC β_2-ADRENERGIC RECEPTOR DOWNREGULATION

Why astrocytes in MS are deficient in β_2-ARs is unknown. A similar deficiency in astrocytic β_2-ARs has been observed in the brains of dogs with canine distemper virus (CDV) encephalitis, but not in control dogs with other cerebral inflammatory conditions.[24] CDV is a morbillivirus that preferentially infects astrocytes and causes a chronic inflammatory demyelinating disease of the CNS that mimics MS, including apoptosis of oligodendrocytes and axonal degeneration in normal-appearing white matter.[25] The virus was also found to dramatically reduce the number of

β_2-ARs in C-6 rat glioma cells.[26] The concept that morbilliviruses might play a role in MS pathogenesis originates from epidemiological and serological studies on measles virus and CDV.[27–29] Attempts to identify morbillivirus antigen in MS brain specimens have not been conclusive.[30] However, in dogs with chronic demyelinating encephalitis, CDV itself is usually no longer detectable in astrocytes,[31] suggesting a "hit-and-run" mechanism.[32]

PATHOPHYSIOLOGICAL ROLE IN FOCAL INFLAMMATORY LESIONS

Activated T cells against myelin peptides can cross the blood–brain barrier and enter the CNS searching for their cognate antigens. As they are activated, they secrete interferon γ, which at the transcriptional level can activate major histocompatibility complex (MHC) class II promoters to produce MHC class II molecules necessary for presenting the myelin antigen.[33] Studies on cultured astrocytes have shown that NA, via activation of β_2-ARs and the cyclic adenosine monophosphate (cAMP) pathway, tightly suppresses interferon γ-induced expression of MHC class II and B-7 costimulatory molecules on astrocytes, which are necessary to mount an autoimmune attack against myelin.[34,35] Thus, in a presence of interferon γ, β_2-AR-deficient astrocytes may convert from an immunosuppressive phenotype into a phenotype that expresses the necessary attributes to function as facultative antigen presenting cell.[33] This might provide an explanation for the findings that, despite that antimyelin T cells are present and sometimes activated in all individuals, only a few individuals develop inflammatory MS lesions.

Downregulation of astrocytic β_2-ARs can further contribute to focal inflammatory processes through a number of other pathways. In rat astrocytes the β-AR agonist isoproterenol suppressed LPS-induced tumor necrosis factor (TNF) and interleukin (IL)-6 promoter activities, mRNA accumulations, and protein levels.[36] β_2-ARs were found to be potent regulators of astrocytic TNF α-activated genes, both in vitro and in vivo, and to modulate the molecular network involved in the homeostasis of inflammatory cells in the CNS by increasing the number of double negative (DN) T cells that lack the expression of both CD4 and CD8 T cell coreceptors.[37] These DN T cells are thought to act as regulatory T cells having the capacity of inducing antigen-specific immune tolerance, and may therefore counteract autoimmune processes.[38,39]

PATHOPHYSIOLOGICAL ROLE IN AXONAL DEGENERATION

Downregulation of astrocytic β_2-ARs may contribute to progressive axonal degeneration by different mechanisms.[40] NA and other agents leading to an increase of intracellular cAMP inhibit the expression of inducible nitric oxide oxidase (iNOS; enzyme catalyzing the production of NO from arginine) in astrocytes.[41,42] Enhanced levels of NO can compete with oxygen for the binding domain on cytochrome C oxidase (complex IV in the mitochondrial respiratory chain), thereby reducing the electron flow and synthesis of ATP. Enhanced density of mitochondria and of cytochrome C oxidase in MS lesions coincides with enhanced mitochondrial oxidative stress and might represent a compensation mechanism to overcome NO occupancy.[43,44]

Another possible consequence of astrocytic β_2-AR deficiency is an impaired astrocytic glycogenolysis. Astrocytes are the main reservoir of glycogen in the CNS,[45–47] and activation of astrocytic β_2-ARs by NA stimulates glycogenolysis, leading to the formation of lactate. Ex vivo studies in rodent optic nerve suggested a physiological role for glycogen-derived lactate as an energy source for axons.[47,48] Using in vivo microdialysis experiments in mouse cerebellar white matter, Laureys et al. found indeed evidence for a continuous axonal lactate uptake and glial–axonal metabolic coupling of glutamate/lactate exchange. However, physiological lactate production was not influenced by either activation or blocking of β_2-ARs.[49] Further studies are required to solve these contradictory findings.

PATHOPHYSIOLOGICAL ROLE IN BOTH AXONAL DEGENERATION AND OLIGODENDROGLIOPATHY

A mechanism that might be involved in both oligodendrocyte apoptosis and axonal degeneration is a reduced trophic support by astrocytes. NA through cAMP stimulates in astrocytes the production of a number of trophic factors, including brain-derived growth factor (BDNF) and nerve growth factor (NGF).[50] NGF promotes the survival of axons and oligodendrocytes, and facilitates migration and proliferation of oligodendrocyte precursors to sites of myelin damage. BDNF also acts as survival factor for oligodendrocytes.[51,52] No information is available on the expression of these growth factors in astrocytes in MS.

Reactive astrocytes in MS plaques express high levels of endothelin (ET)-1, which is a powerful vasoconstrictor acting on ET_A receptors on arteriolar smooth muscle cells,[53] and is likely responsible for the reduced cerebral blood flow observed in patients with MS.[54] Chronic cerebral hypoperfusion has been shown to induce mitochondrial dysfunction, myelin breakdown, apoptosis of oligodendrocytes, and axonal degeneration.[55–58] As studies found that the β_2-AR agonist olodaterol reduced ET-1 expression in human lung fibroblasts,[59] it would be of interest to investigate whether the deficiency of astrocytic β_2-ARs plays a role in the enhanced ET-1 expression in astrocytes in MS plaques.

ABBREVIATIONS

AR adrenergic receptor
BDNF brain-derived growth factor
cAMP cyclic adenosine monophosphate
CDV canine distemper virus
CNS central nervous system
DNT cells double negative T cells
ET endothelin
GFAP glial fibrillary acid protein
IL interleukin
iNOS inducible nitric oxide oxidase
JC virus John Cunningham virus
MHC major histocompatibility complex
MRI magnetic resonance imaging
MS multiple sclerosis
NA noradrenaline
NGF nerve growth factor
NO nitric oxide
PET positron emission tomography
PML progressive multifocal leukoencephalopathy
TNF tumor necrosis factor

ACKNOWLEDGMENTS

This work was supported by the Charcot foundation and FWO Belgium. Nadine Wilczak performed the immunohistochemical studies in the MS and PML brain samples.

REFERENCES

1. Compston A, Coles A. Multiple sclerosis. *Lancet*. 2008;372(9648):1502–1517.
2. Constantinescu CS, Gran B. The essential role of T cells in multiple sclerosis: a reappraisal. *Biomed J*. 2014;37(2):34–40.
3. Milo R. Effectiveness of multiple sclerosis treatment with current immunomodulatory drugs. *Expert Opin Pharmacother*. 2015;16(5):659–673.
4. Banwell B, Bar-Or A, Cheung R, et al. Abnormal T-cell reactivities in childhood inflammatory demyelinating disease and type 1 diabetes. *Ann Neurol*. 2008;63(1):98–111.
5. Hellings N, Gelin G, Medaer R, et al. Longitudinal study of antimyelin T-cell reactivity in relapsing–remitting multiple sclerosis: association with clinical and MRI activity. *J Neuroimmunol*. 2002;126(1–2):143–160.
6. Barnett MH, Prineas JW. Relapsing and remitting multiple sclerosis: pathology of the newly forming lesion. *Ann Neurol*. 2004;55(4):458–468.
7. Lucchinetti C, Bruck W, Parisi J, Scheithauer B, Rodriguez M, Lassmann H. Heterogeneity of multiple sclerosis lesions: implications for the pathogenesis of demyelination. *Ann Neurol*. 2000;47(6):707–717.
8. Nakahara J, Maeda M, Aiso S, Suzuki N. Current concepts in multiple sclerosis: autoimmunity versus oligodendrogliopathy. *Clin Rev Allergy Immunol*. 2012;42(1):26–34.

9. Confavreux C, Vukusic S, Moreau T, Adeleine P. Relapses and progression of disability in multiple sclerosis. *N Engl J Med*. 2000;343(20):1430–1438.
10. Koch M, Mostert J, Heersema D, De Keyser J. Progression in multiple sclerosis: further evidence of an age dependent process. *J Neurol Sci*. 2007;255(1–2):35–41.
11. Confavreux C, Vukusic S. Natural history of multiple sclerosis: a unifying concept. *Brain*. 2006;129(Pt 3):606–616.
12. Mahad DH, Trapp BD, Lassmann H. Pathological mechanisms in progressive multiple sclerosis. *Lancet Neurol*. 2015;14(2):183–193.
13. Doze P, Van Waarde A, Elsinga PH, Van-Loenen Weemaes AM, Willemsen AT, Vaalburg W. Validation of S-1′-[18F]fluorocarazolol for in vivo imaging and quantification of cerebral beta-adrenoceptors. *Eur J Pharmacol*. 1998;353(2–3):215–226.
14. Mantyh PW, Rogers SD, Allen CJ, et al. Beta 2-adrenergic receptors are expressed by glia in vivo in the normal and injured central nervous system in the rat, rabbit, and human. *J Neurosci*. 1995;15(1 Pt 1):152–164.
15. De Keyser J, Wilczak N, Leta R, Streetland C. Astrocytes in multiple sclerosis lack beta-2 adrenergic receptors. *Neurology*. 1999;53(8):1628–1633.
16. Zeinstra E, Wilczak N, De Keyser J. [3*H*]dihydroalprenolol binding to beta adrenergic receptors in multiple sclerosis brain. *Neurosci Lett*. 2000;289(1):75–77.
17. Laureys G, Clinckers R, Gerlo S, et al. Astrocytic beta(2)-adrenergic receptors: from physiology to pathology. *Prog Neurobiol*. 2010;91(3):189–199.
18. Catus SL, Gibbs ME, Sato M, Summers RJ, Hutchinson DS. Role of beta-adrenoceptors in glucose uptake in astrocytes using beta-adrenoceptor knockout mice. *Br J Pharmacol*. 2011;162(8):1700–1715.
19. Berkowitz DE, Nardone NA, Smiley RM, et al. Distribution of beta 3-adrenoceptor mRNA in human tissues. *Eur J Pharmacol*. 1995;289(2):223–228.
20. Aksamit Jr. AJ. Progressive multifocal leukoencephalopathy. *Continuum*. 2012; 18(6 Infectious Disease):1374–1391.
21. Warnke C, Olsson T, Hartung HP. PML: The dark side of immunotherapy in multiple sclerosis. *Trends Pharmacol Sci*. 2015;36(12):799–801.
22. Kleinschmidt-DeMasters BK, Tyler KL. Progressive multifocal leukoencephalopathy complicating treatment with natalizumab and interferon beta-1a for multiple sclerosis. *N Engl J Med*. 2005;353(4):369–374.
23. Hamdani N, van der Velden J. Lack of specificity of antibodies directed against human beta-adrenergic receptors. *Naunyn-Schmiedeberg's Arch Pharmacol*. 2009;379(4):403–407.
24. De Keyser J, Wilczak N, Walter JH, Zurbriggen A. Disappearance of beta2-adrenergic receptors on astrocytes in canine distemper encephalitis: possible implications for the pathogenesis of multiple sclerosis. *Neuroreport*. 2001;12(2):191–194.
25. Seehusen F, Baumgartner W. Axonal pathology and loss precede demyelination and accompany chronic lesions in a spontaneously occurring animal model of multiple sclerosis. *Brain Pathol*. 2010;20(2):551–559.
26. Koschel K, Muenzel P. Persistent paramyxovirus infections and behaviour of beta-adrenergic receptors in C-6 rat glioma cells. *J Gen Virol*. 1980;47(2):513–517.
27. Giraudon P, Bernard A. Chronic viral infections of the central nervous system: Aspects specific to multiple sclerosis. *Revue Neurologique*. 2009;165(10):789–795.
28. Cook SD, Blumberg B, Dowling PC. Potential role of paramyxoviruses in multiple sclerosis. *Neurol Clin*. 1986;4(1):303–319.
29. Sips GJ, Chesik D, Glazenburg L, Wilschut J, De Keyser J, Wilczak N. Involvement of morbilliviruses in the pathogenesis of demyelinating disease. *Rev Med Virol*. 2007;17(4):223–244.

30. Geeraedts F, Wilczak N, van Binnendijk R, De Keyser J. Search for morbillivirus proteins in multiple sclerosis brain tissue. *Neuroreport*. 2004;15(1):27–32.
31. Cook SD, Rohowsky-Kochan C, Bansil S, Dowling PC. Evidence for multiple sclerosis as an infectious disease. *Acta Neurol Scand Suppl*. 1995;161:34–42.
32. Scarisbrick IA, Rodriguez M. Hit–hit and hit–run: viruses in the playing field of multiple sclerosis. *Curr Neurol Neurosci Rep*. 2003;3(3):265–271.
33. De Keyser J, Laureys G, Demol F, Wilczak N, Mostert J, Clinckers R. Astrocytes as potential targets to suppress inflammatory demyelinating lesions in multiple sclerosis. *Neurochem Int*. 2010;57(4):446–450.
34. Frohman EM, Vayuvegula B, Gupta S, van den Noort S. Norepinephrine inhibits gamma-interferon-induced major histocompatibility class II (Ia) antigen expression on cultured astrocytes via beta-2-adrenergic signal transduction mechanisms. *Proc Natl Acad Sci USA*. 1988;85(4):1292–1296.
35. Zeinstra EM, Wilczak N, Wilschut JC, et al. 5HT4 agonists inhibit interferon-gamma-induced MHC class II and B7 costimulatory molecules expression on cultured astrocytes. *J Neuroimmunol*. 2006;179(1–2):191–195.
36. Nakamura A, Johns EJ, Imaizumi A, Abe T, Kohsaka T. Regulation of tumour necrosis factor and interleukin-6 gene transcription by beta2-adrenoceptor in the rat astrocytes. *J Neuroimmunol*. 1998;88(1–2):144–153.
37. Laureys G, Gerlo S, Spooren A, Demol F, De Keyser J, Aerts JL. beta(2)-adrenergic agonists modulate TNF-alpha induced astrocytic inflammatory gene expression and brain inflammatory cell populations. *J Neuroinflammation*. 2014;11:21.
38. Hillhouse EE, Lesage S. A comprehensive review of the phenotype and function of antigen-specific immunoregulatory double negative T cells. *J Autoimmunity*. 2013;40:58–65.
39. Juvet SC, Zhang L. Double negative regulatory T cells in transplantation and autoimmunity: recent progress and future directions. *J Mol Cell Biol*. 2012;4(1):48–58.
40. Cambron M, D'Haeseleer M, Laureys G, Clinckers R, Debruyne J, De Keyser J. White-matter astrocytes, axonal energy metabolism, and axonal degeneration in multiple sclerosis. *J Cereb Blood Flow Metab*. 2012;32(3):413–424.
41. Feinstein DL. Suppression of astroglial nitric oxide synthase expression by norepinephrine results from decreased NOS-2 promoter activity. *J Neurochem*. 1998;70(4):1484–1496.
42. Feinstein DL, Galea E, Reis DJ. Norepinephrine suppresses inducible nitric oxide synthase activity in rat astroglial cultures. *J Neurochem*. 1993;60(5):1945–1948.
43. Mahad DJ, Ziabreva I, Campbell G, et al. Mitochondrial changes within axons in multiple sclerosis. *Brain*. 2009;132(Pt 5):1161–1174.
44. Witte ME, Bo L, Rodenburg RJ, et al. Enhanced number and activity of mitochondria in multiple sclerosis lesions. *J Pathol*. 2009;219(2):193–204.
45. Cataldo AM, Broadwell RD. Cytochemical identification of cerebral glycogen and glucose-6-phosphatase activity under normal and experimental conditions. II. Choroid plexus and ependymal epithelia, endothelia and pericytes. *J Neurocytol*. 1986;15(4): 511–524.
46. Fillenz M, Lowry JP, Boutelle MG, Fray AE. The role of astrocytes and noradrenaline in neuronal glucose metabolism. *Acta Physiol Scand*. 1999;167(4):275–284.
47. Wender R, Brown AM, Fern R, Swanson RA, Farrell K, Ransom BR. Astrocytic glycogen influences axon function and survival during glucose deprivation in central white matter. *J Neurosci*. 2000;20(18):6804–6810.

48. Tekkok SB, Brown AM, Westenbroek R, Pellerin L, Ransom BR. Transfer of glycogen-derived lactate from astrocytes to axons via specific monocarboxylate transporters supports mouse optic nerve activity. *J Neurosci Res.* 2005;81(5):644–652.
49. Laureys G, Valentino M, Demol F, et al. beta(2)-adrenergic receptors protect axons during energetic stress but do not influence basal glio-axonal lactate shuttling in mouse white matter. *Neuroscience.* 2014;277:367–374.
50. Zafra F, Lindholm D, Castren E, Hartikka J, Thoenen H. Regulation of brain-derived neurotrophic factor and nerve growth factor mRNA in primary cultures of hippocampal neurons and astrocytes. *J Neurosci.* 1992;12(12):4793–4799.
51. Althaus HH, Kloppner S, Schmidt-Schultz T, Schwartz P. Nerve growth factor induces proliferation and enhances fiber regeneration in oligodendrocytes isolated from adult pig brain. *Neurosci Lett.* 1992;135(2):219–223.
52. Koda M, Murakami M, Ino H, et al. Brain-derived neurotrophic factor suppresses delayed apoptosis of oligodendrocytes after spinal cord injury in rats. *J Neurotrauma.* 2002;19(6):777–785.
53. Zhang WW, Badonic T, Hoog A, et al. Structural and vasoactive factors influencing intracerebral arterioles in cases of vascular dementia and other cerebrovascular disease: a review. Immunohistochemical studies on expression of collagens, basal lamina components and endothelin-1. *Dementia.* 1994;5(3–4):153–162.
54. D'Haeseleer M, Beelen R, Fierens Y, et al. Cerebral hypoperfusion in multiple sclerosis is reversible and mediated by endothelin-1. *Proc Natl Acad Sci USA.* 2013;110(14):5654–5658.
55. Aliev G. Oxidative stress induced-metabolic imbalance, mitochondrial failure, and cellular hypoperfusion as primary pathogenetic factors for the development of Alzheimer disease which can be used as a alternate and successful drug treatment strategy: past, present and future. *CNS Neurol Dis Drug Targets.* 2011;10(2):147–148.
56. Aliev G, Smith MA, Obrenovich ME, de la Torre JC, Perry G. Role of vascular hypoperfusion-induced oxidative stress and mitochondria failure in the pathogenesis of Azheimer disease. *Neurotox Res.* 2003;5(7):491–504.
57. Tomimoto H, Ihara M, Wakita H, et al. Chronic cerebral hypoperfusion induces white matter lesions and loss of oligodendroglia with DNA fragmentation in the rat. *Acta Neuropathol.* 2003;106(6):527–534.
58. Farkas E, Donka G, de Vos RA, Mihaly A, Bari F, Luiten PG. Experimental cerebral hypoperfusion induces white matter injury and microglial activation in the rat brain. *Acta Neuropathol.* 2004;108(1):57–64.
59. Ahmedat AS, Warnken M, Juergens UR, Paul Pieper M, Racke K. beta(2)-adrenoceptors and muscarinic receptors mediate opposing effects on endothelin-1 expression in human lung fibroblasts. *Eur J Pharmacol.* 2012;691(1-3):218–224.

Chapter 16

Potentiation of β-Amyloid-Induced Cortical Inflammation by Noradrenaline and Noradrenergic Depletion: Implications for Alzheimer's Disease

Douglas L. Feinstein[1,2,✉] **and Michael T. Heneka**[3,4]

[1]*University Illinois at Chicago, Chicago, IL, United States,* [2]*Jesse Brown VA Medical Center, Chicago, IL, United States,* [3]*University of Bonn, Bonn, Germany,* [4]*German Center for Neurodegenerative Disease (DZNE), Bonn, Germany*

Chapter Outline

Introduction: The Locus Coeruleus and Noradrenaline Function in Alzheimer's Disease 302
Neuroinflammation in Alzheimer's Disease 302
Antiinflammatory Actions of Noradrenaline 303
Locus Coeruleus Damage in Alzheimer's Disease 305
Locus Coeruleus Damage in Mouse Models of Alzheimer's Disease 305
Clinical Trials to Modulate Noradrenaline Levels in Alzheimer's Disease Patients 306
Conclusions 307
Abbreviations 307
Acknowledgments 307
References 308

ABSTRACT

The endogenous neurotransmitter noradrenaline (NA) exerts antiinflammatory, neurotrophic, neurogenic, and neuroprotective effects. Although binding to and activation of adrenergic receptors (ARs) on neurons accounts for many of these actions, ARs on glial cells also provide a target for its numerous and diverse effects. Under disease or damaging conditions, the levels of NA or of ARs can be reduced or functionally deficient, leading to dysregulation of homeostasis. In this

✉Correspondence address
E-mail: dlfeins@uic.edu

Noradrenergic Signaling and Astroglia. DOI: http://dx.doi.org/10.1016/B978-0-12-805088-0.00016-5
© 2017 Elsevier Inc. All rights reserved.

chapter, we review these systems and describe recent pharmacological approaches to directly modulate them in neurological diseases and conditions.

Keywords: Astrocytes; microglia; locus coeruleus; amyloid; inflammation

INTRODUCTION: THE LOCUS COERULEUS AND NORADRENALINE FUNCTION IN ALZHEIMER'S DISEASE

The locus coeruleus (LC) represents the major source of cerebral noradrenaline (NA) and projects to virtually all areas of the brain and spinal cord. The majority of axons arising from the LC terminate in the neocortex and hippocampus.[1–3] Half of all LC terminals end by contacting neurons, whereas the remaining half ends at microglial and astroglial cells forming "nonsynaptic" contacts. Activation of adrenergic receptors (ARs) on glial cells can elicit potent antiinflammatory actions, which help limit neuroinflammatory events throughout the central nervous system (CNS). LC-derived NA and β-ARs have also been implicated in the physiological modulation of memory formation and retrieval.[2,4–6]

During normal aging, LC noradrenergic neurons undergo damage, and it is estimated that up to 25% of those neurons and 50% of NA levels are being lost in elderly ($>$90 years age) adults.[7] In Alzheimer's disease (AD), however, the LC degenerates at very early stages of the disease, which results in a progressive loss of NA innervation of the neocortex and hippocampus. Loss of NA supply to these brain regions increases neuronal degeneration and aggravates learning and memory deficits. Principally, NA may influence neuronal networks by acting directly on neuronal β-ARs and thus modulating synaptic plasticity and ultimately learning and memory formation. Alternatively, NA may modulate glia activation states and the release of immune mediators. NA suppresses the glial release of mediators, which have been shown to suppress synaptic plasticity including nitric oxide (iNOS), interleukin-1β (IL-1β) and tumor necrosis factor-α (TNFα).[8–12] Furthermore, under physiological conditions, glial cells contribute to neuronal network integrity and functioning through synaptic scaling and pruning along with the release of trophic factors. NA may therefore modulate neuronal function indirectly through acting on ARs of glial cells.

NEUROINFLAMMATION IN ALZHEIMER'S DISEASE

The deposition of amyloid-β (Aβ), which derives from the sequential amyloidogenic processing of the amyloid precursor protein, is clearly linked to the risk of developing AD. However, in sporadic as well as familial AD, deposition of Aβ precedes cognitive decline and brain atrophy presumably by decades, raising the question, which mechanisms are driving neuronal dysfunction and demise, ultimately causing the development of memory failure and cognitive decline. Since Aβ aggregates act as a danger associated molecular pattern (DAMP) and activate various pattern recognition receptors on the surface of microglia and astrocytes, it can be hypothesized that formation of Aβ oligomers and deposits causes an activation of these cells that belong to the innate immune defense of the brain.

Experimentally, the accumulation of Aβ in murine models of AD, or the exposure of primary cultures of glial cells to Aβ, has been well documented to induce an array of inflammatory responses. These include the activation of inflammatory transcription factors such as NFκB, induction of proinflammatory cytokine expression, release of reactive oxygen and nitrogen species, and activation of microglial phagocytotic activities, collectively described as neuroinflammation. These neuroinflammatory responses are normally restricted in magnitude or duration by compensatory actions, for example increased production of antiinflammatory molecules. Interestingly, neuroinflammatory processes are also held in check by the actions of endogenous neuropeptides and neurotransmitters. Amongst those, NA represents a critical determinant since its extracellular levels are kept high, and its targets ARs are ubiquitously expressed on glial cells though the CNS including cortical areas subject to high amyloid burden.

ANTIINFLAMMATORY ACTIONS OF NORADRENALINE

In vitro evidence has convincingly demonstrated that NA exerts antiinflammatory actions on glial cells associated with reduced neuronal damage. The inflammatory status of both astrocytes and microglial cells is significantly reduced by exposure to NA, or to selective agonists of β-AR. Early studies demonstrated these actions following stimulation with bacterial endotoxin lipopolysaccharides (LPS) or with proinflammatory cytokines. Subsequently, antiinflammatory actions of β-AR activation were demonstrated following activation of glial cells with Aβ. As example, Aβ can increase the expression of various cytokines from microglia cells, including that of TNFα, monocyte chemoattractant protein-1 (MCP1), iNOS, and cyclooxygenase-2 (COX2); and those increases were all dramatically reduced by coincubation the cells with NA.[13] Similarly, in astrocytes, incubation with Aβ (either aggregated forms, or smaller oligomeric species) leads to induction of inflammatory gene expression and cytokine release; which are prevented by cotreatment with NA or selective β-AR agonists.

Similarly, treatments with NA, or drugs that elevate NA levels in vivo can reduce inflammatory events occurring in the CNS due to administration of various toxins including cytokines,[14] endotoxin,[15] or Aβ[13,16] (Fig. 16.1). NA levels can be increased by treatment with α_2-AR antagonists (e.g., 5-fluoro-methoxyidazoxan), which prevent activation of

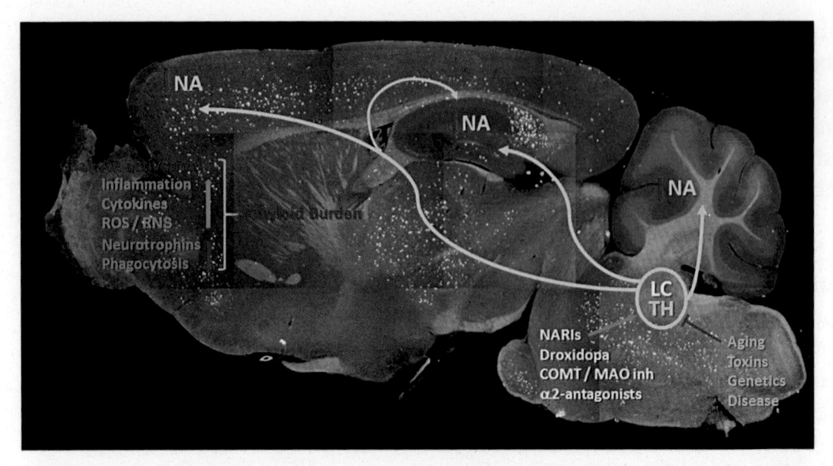

FIGURE 16.1 A schematic overlaid on a montage of sagittal sections through the brain of 5xFAD mouse, a commonly used model for Alzheimer's disease (AD), stained for amyloid plaques using thioflavin S (white). The cartoon illustrates potential sources of damage to locus coeruleus (LC) noradrenergic neuronal health and function. This includes aging, environment toxins (e.g., heavy metals), genetic background (e.g., variants in dopamine beta--hydroxylase gene), and disease (e.g., AD and multiple sclerosis). As a consequence, LC tyrosine hydroxylase (TH) activity is reduced, causing reductions in noradrenaline (NA) levels throughout the brain. Lower levels of NA, which normally binds to and activates adrenergic receptors (AR) on glial cells, allows for increased glial cell inflammatory responses, including production of proinflammatory cytokines, and of reactive oxygen (ROS) and nitrogen (RNS) species. Loss of adrenergic signaling can also lead to decreased production of neurotrophins, and reduced phagocytosis. Together these factors lead to increases in amyloid burden throughout the CNS. Several agents discussed herein can minimize the consequences of LC damage, including NA reuptake inhibitors (NARIs), the NA precursor droxidopa, inhibitors of NA metabolizing enzymes catechol-*O*-methyltransferase (COMT) and monoamine oxidase (MAO inh), and antagonists of the α_2-AR.

inhibitory pre-synaptic α_2-autoreceptors, which normally reduce NA release. Treatment of adult rats with 5-fluoro-methoxyidazoxan reduced inflammatory responses as well as neurotoxicity following intraparenchymal injection of aggregated Aβ. NA can also be increased by using NA reuptake inhibitors (NARIs) such as desipramine; however, many of those drugs also block serotonin reuptake and therefore whether their actions are due to increased extracellular NA cannot be discerned. Several nontricyclic antidepressants, such as atomoxetine and reboxetine, have greater selective actions on NA reuptake and have also been shown to exert antiinflammatory and neuroprotective actions in vivo.[17–19] A third way to increase NA is to reduce its metabolism, which is normally carried out by monoamine oxidases (MAOs) and catechol-*O*-methyltransferase (COMT).[20] It is known that MAO mRNA levels are higher in AD patients than in controls,[21] and that genetic variants in the MAO genes are risk factor to develop AD.[22] COMT inhibitors have been used for many years to reduce levodopa breakdown during treatment of Parkinson's disease[23], and recently, we showed that 2DNC, a blood–brain barrier (BBB)-permeable COMT inhibitor provided benefit in the experimental autoimmune encephalomyelitis (EAE) mouse model of multiple sclerosis (MS).[24]

Although promising, NA reuptake inhibitors, α_2-AR antagonists, and metabolic inhibitors may not be effective if endogenous NA levels are low, for example due to loss or damage to LC noradrenergic neurons. Alternatively, NA can be increased in a manner similar to that of dopamine, e.g., by treatment with an NA precursor such as droxidopa.[25] Droxidopa, currently approved to treat hypotension, is directly converted to NA by the enzyme dopamine decarboxylase, located throughout the body including in glial cells. Several in vivo studies have shown that droxidopa can reduce clinical symptoms in mouse models of MS[26] and transgenic mouse models of AD.[13,27]

LOCUS COERULEUS DAMAGE IN ALZHEIMER'S DISEASE

Observations that treatments to raise NA levels can provide benefit suggested an intrinsic deficit in noradrenergic function; and in fact loss and damage to noradrenergic neurons in the LC was observed over 50 years ago.[28] Catecholamines, including NA, are subject to metabolic breakdown which leads to formation of highly colored products, termed neuromelanin, that are visible by light microscopy. This allowed early investigators to easily quantify the number of LC neurons during aging,[29–31] and in specimens from patients with neurological disease including AD. This led to conclusions that up to 60% of the noradrenergic neurons in the LC are lost in AD patients versus controls.[32–35] Although concerns were raised regarding the methods of counting, as well as the criteria for patient selections, subsequent studies using immunohistochemical staining for markers of NA synthesis [tyrosine hydroxylase (TH) and dopamine beta-hydroxylase (DBH)] allowed confirmation of noradrenergic neuronal loss occurring in AD,[1,36–38] as well as demonstration that neuronal loss is significantly greater in LC than in nucleus basalis and substantia nigra in AD,[39] and occurs at early stages of disease.[40,41]

LOCUS COERULEUS DAMAGE IN MOUSE MODELS OF ALZHEIMER'S DISEASE

Knowledge that the LC is damaged in AD provided the rationale for testing the consequences of experimental lesion of the LC in mouse models of AD. Some earlier studies used an electrolytic approach to lesion the LC, however results from those studies are limited. Several groups made use of the neurotoxin 6-OHDA (6-hydroxy dopamine) to lesion the LC,[42,43] but again findings are limited. In contrast, a serendipitous discovery led to identification of DSP4 (N-(2-chloroethyl)-N-ethyl-2-bromobenzylamine), an alkylating agent that selectively reduces NA uptake.[44] DSP4 is converted to a more reactive molecule (an aziridinum) that is selectively transported through NA transporters into LC neurons, leading to cell death.[45,46] Initial lesion studies using DSP4 were carried out by our groups, and showed that following LC lesion (which was confirmed by immunohistochemical staining for TH), there was a significant increase in glial inflammatory responses to intraparenchymal injections of cytokines or to Aβ.[16,47,48] In DSP4 lesioned animals, the increased glial responses could be reduced by cotreatment with NA or with β-AR agonists, confirming that the loss of NA was responsible for observed effects.[16] Subsequent studies have documented that not only does LC lesion increase inflammatory responses, but also increases other indices of neuropathology, including neuronal death, behavioral deficits, and importantly amyloid burden.[9,13,49–54] The increase in amyloid levels has been postulated to be due to several causes, including an increased inflammatory milieu which can increase proamyloidogenic processing,[55] a reduction in microglial Aβ phagocytosis,[50] and reductions in Aβ degradation by proteinase such as neprilysin known to be regulated by NA.[50]

Additional evidence that LC noradrenergic damage and loss can contribute to exacerbation of AD type pathology comes from genetic studies in which, the rate limiting step in NA synthesis, was deleted from a transgenic mouse model of AD.[56] In the DBH knockout mice, there were deficits in hippocampal long-term potentiation (LTP), and in spatial memory; however surprisingly in these mice increases in amyloid burden were not observed. In similar studies, the same mice (PS1:APP mice) were generated lacking the transcription factor Ear2 that is needed for proper LC development.[57] Those mice showed increased inflammation, as well as spatial memory and LTP deficits; however as in the DBH null studies, amyloid burden was not significantly increased. A role for NA in mediating the observed behavioral effects was confirmed since treatment with NA precursor droxidopa ameliorated the increased responses.

CLINICAL TRIALS TO MODULATE NORADRENALINE LEVELS IN ALZHEIMER'S DISEASE PATIENTS

Despite increasing evidence that noradrenergic dysregulation contributes to neuropathology and behavioral defects in AD, and that treatments which raise NA (or reduce LC neuronal damage) can provide benefit in mouse models of AD, only limited clinical testing has been done. A meta review of selegiline, an MAO-B inhibitor suggested that this drug did elicit benefit on memory and cognition times, but only in a small number of trials[58], whereas a small trial of atomoxetine showed only modest effect in AD patients.[59] Testing of the NA precursor droxidopa has not yet been carried out in AD patients.

CONCLUSIONS

Although deficits in acetylcholine due to the degeneration of the nucleus basalis Meynert is currently the only accepted clinically relevant, drugable neurotransmitter deficit in AD, the degeneration of the LC and the subsequent loss of NA supply to its projection areas, in particular, the hippocampus and neocortex represents an additional contributing factor to neurodegenerative processes that can also be targeted with pharmacological interventions. Numerous studies demonstrate that NA loss leads to an increase in the production of inflammatory mediators, reduced clearance of amyloid burden, and increased neuronal death. The development of methods to restore depleted pools of NA, to reduce LC damage and loss, or to increase AR signaling therefore seems warranted and eventually clinical trials to test these possibilities should be considered.

ABBREVIATIONS

6-OHDA 6-hydroxy dopamine
AD Alzheimer's disease
AR adrenergic receptor
Aβ amyloid-β
CNS central nervous system
COMT catechol-*O*-methyltransferase
COX2 cyclooxygenase-2
DAMP danger associated molecular pattern
DBH dopamine beta-hydroxylase
DSP4 *N*-(2-chloroethyl)-*N*-ethyl-2-bromobenzylamine
EAE experimental autoimmune encephalomyelitis
IL-1β interleukin-1β
iNOS nitric oxide
LC locus coeruleus
LPS lipopolysaccharides
LTP long-term potentiation
MAO monoamine oxidase
MCP1 monocyte chemoattractant protein-1
MS multiple sclerosis
NARI noradrenaline reuptake inhibitor
NA noradrenaline or norepinephrine
TH tyrosine hydroxylase
TNFα tumor necrosis factor-α

ACKNOWLEDGMENTS

DLF is a recipient of a Research Career Scientist award from the Department of Veterans Affairs.

REFERENCES

1. Trillo L, Das D, Hsieh W, et al. Ascending monoaminergic systems alterations in Alzheimer's disease translating basic science into clinical care. *Neurosci Biobehav Rev.* 2013;37(8):1363–1379.
2. O'Dell TJ, Connor SA, Gelinas JN, Nguyen PV. Viagra for your synapses: enhancement of hippocampal long-term potentiation by activation of beta-adrenergic receptors. *Cell Signal.* 2010;22(5):728–736.
3. Shepard KN, Liles LC, Weinshenker D, Liu RC. Norepinephrine is necessary for experience-dependent plasticity in the developing mouse auditory cortex. *J Neurosci.* 2015;35(6):2432–2437.
4. Murchison CF, Schutsky K, Jin SH, Thomas SA. Norepinephrine and ss(1)-adrenergic signaling facilitate activation of hippocampal CA1 pyramidal neurons during contextual memory retrieval. *Neuroscience.* 2011;181:109–116.
5. Ramos BP, Colgan L, Nou E, Ovadia S, Wilson SR, Arnsten AF. The beta-1 adrenergic antagonist, betaxolol, improves working memory performance in rats and monkeys. *Biol Psychiatry.* 2005;58(11):894–900.
6. Zheng J, Luo F, Guo NN, Cheng ZY, Li BM. beta1-and beta2-adrenoceptors in hippocampal CA3 region are required for long-term memory consolidation in rats. *Brain Res.* 2015;1627:109–118.
7. Marien MR, Colpaert FC, Rosenquist AC. Noradrenergic mechanisms in neurodegenerative diseases: a theory. *Brain Res Brain Res Rev.* 2004;45(1):38–78.
8. Feinstein DL, Heneka MT, Gavrilyuk V, Dello RC, Weinberg G, Galea E. Noradrenergic regulation of inflammatory gene expression in brain. *Neurochem Int.* 2002;41(5):357–365.
9. Jardanhazi-Kurutz D, Kummer MP, Terwel D, Vogel K, Thiele A, Heneka MT. Distinct adrenergic system changes and neuroinflammation in response to induced locus ceruleus degeneration in APP/PS1 transgenic mice. *Neuroscience.* 2011;176:396–407.
10. O'Donnell J, Zeppenfeld D, McConnell E, Pena S, Nedergaard M. Norepinephrine: a neuromodulator that boosts the function of multiple cell types to optimize CNS performance. *Neurochem Res.* 2012;37(11):2496–2512.
11. Gyoneva S, Traynelis SF. Norepinephrine modulates the motility of resting and activated microglia via different adrenergic receptors. *J Biol Chem.* 2013;288(21):15291–15302.
12. Mori K, Ozaki E, Zhang B, et al. Effects of norepinephrine on rat cultured microglial cells that express alpha1, alpha2, beta1 and beta2 adrenergic receptors. *Neuropharmacology.* 2002;43(6):1026–1034.
13. Heneka MT, Nadrigny F, Regen T, et al. Locus ceruleus controls Alzheimer's disease pathology by modulating microglial functions through norepinephrine. *Proc Natl Acad Sci USA.* 2010;107(13):6058–6063.
14. Szabo C, Hasko G, Zingarelli B, et al. Isoproterenol regulates tumour necrosis factor, interleukin-10, interleukin-6 and nitric oxide production and protects against the development of vascular hyporeactivity in endotoxaemia. *Immunology.* 1997;90(1):95–100.
15. Feinstein DL, Galea E, Reis DJ. Norepinephrine suppresses inducible nitric oxide synthase activity in rat astroglial cultures. *J Neurochem.* 1993;60(5):1945–1948.
16. Heneka MT, Galea E, Gavriluyk V, et al. Noradrenergic depletion potentiates beta-amyloid-induced cortical inflammation: implications for Alzheimer's disease. *J Neurosci.* 2002;22(7):2434–2442.
17. Hashioka S, Klegeris A, Monji A, et al. Antidepressants inhibit interferon-gamma-induced microglial production of IL-6 and nitric oxide. *Exp Neurol.* 2007;206(1):33–42.

18. O'Sullivan JB, Ryan KM, Curtin NM, Harkin A, Connor TJ. Noradrenaline reuptake inhibitors limit neuroinflammation in rat cortex following a systemic inflammatory challenge: implications for depression and neurodegeneration. *Int J Neuropsychopharmacol.* 2009;12(5):687–699.
19. McNamee EN, Ryan KM, Griffin EW, et al. Noradrenaline acting at central beta-adrenoceptors induces interleukin-10 and suppressor of cytokine signaling-3 expression in rat brain: implications for neurodegeneration. *Brain Behav Immun.* 2010;24(4):660–671.
20. Eisenhofer G, Finberg JP. Different metabolism of norepinephrine and epinephrine by catechol-*O*-methyltransferase and monoamine oxidase in rats. *J Pharmacol Exp Ther.* 1994;268(3):1242–1251.
21. Emilsson L, Saetre P, Balciuniene J, Castensson A, Cairns N, Jazin EE. Increased monoamine oxidase messenger RNA expression levels in frontal cortex of Alzheimer's disease patients. *Neurosci Lett.* 2002;326(1):56–60.
22. Takehashi M, Tanaka S, Masliah E, Ueda K. Association of monoamine oxidase A gene polymorphism with Alzheimer's disease and Lewy body variant. *Neurosci Lett.* 2002;327(2):79–82.
23. Muller T, Kuhn W, Przuntek H. Therapy with central active catechol-*O*-methyltransferase (COMT)-inhibitors: is addition of monoamine oxidase (MAO)-inhibitors necessary to slow progress of neurodegenerative disorders?. *J Neural Transm Gen Sect.* 1993;92 (2–3):187–195.
24. Polak PE, Lin SX, Pelligrino D, Feinstein DL. The blood–brain barrier-permeable catechol-*O*-methyltransferase inhibitor dinitrocatechol suppresses experimental autoimmune encephalomyelitis. *J Neuroimmunol.* 2014;276(1–2):135–141.
25. Thomas SA, Marck BT, Palmiter RD, Matsumoto AM. Restoration of norepinephrine and reversal of phenotypes in mice lacking dopamine beta-hydroxylase. *J Neurochem.* 1998;70(6):2468–2476.
26. Simonini MV, Polak PE, Sharp A, McGuire S, Galea E, Feinstein DL. Increasing CNS noradrenaline reduces EAE severity. *J Neuroimmune Pharmacol.* 2010;5(2):252–259.
27. Kalinin S, Polak PE, Lin SX, Sakharkar AJ, Pandey SC, Feinstein DL. The noradrenaline precursor L-DOPS reduces pathology in a mouse model of Alzheimer's disease. *Neurobiol Aging.* 2012;33(8):1651–1663.
28. Forno LS. Pathology of Parkinsonism: a preliminary report of 24 cases. *J Neurosurg.* 1966;266–271:Supplement, Part II
29. Vijayashankar N, Brody H. A quantitative study of the pigmented neurons in the nuclei locus coeruleus and subcoeruleus in man as related to aging. *J Neuropathol Exp Neurol.* 1979;38(5):490–497.
30. Mann DM. The locus coeruleus and its possible role in ageing and degenerative disease of the human central nervous system. *Mech Ageing Dev.* 1983;23(1):73–94.
31. Chan-Palay V, Asan E. Quantitation of catecholamine neurons in the locus coeruleus in human brains of normal young and older adults and in depression. *J Comp Neurol.* 1989;287(3):357–372.
32. Tomlinson BE, Irving D, Blessed G. Cell loss in the locus coeruleus in senile dementia of Alzheimer type. *J Neurol Sci.* 1981;49(3):419–428.
33. Perry EK, Tomlinson BE, Blessed G, Perry RH, Cross AJ, Crow TJ. Neuropathological and biochemical observations on the noradrenergic system in Alzheimer's disease. *J Neurol Sci.* 1981;51(2):279–287.
34. Bondareff W, Mountjoy CQ, Roth M. Selective loss of neurones of origin of adrenergic projection to cerebral cortex (nucleus locus coeruleus) in senile dementia. *Lancet (London, England).* 1981;1(8223):783–784.

35. Mann DM, Yates PO, Hawkes J. The noradrenergic system in Alzheimer and multi-infarct dementias. *J Neurol Neurosurg Psychiatry.* 1982;45(2):113–119.
36. Szot P. Common factors among Alzheimer's disease, Parkinson's disease, and epilepsy: possible role of the noradrenergic nervous system. *Epilepsia.* 2012;53(Suppl 1):61–66.
37. Gannon M, Che P, Chen Y, Jiao K, Roberson ED, Wang Q. Noradrenergic dysfunction in Alzheimer's disease. *Front Neurosci.* 2015;9:220. Available from: http://dx.doi.org/10.3389/fnins.2015.00220.
38. Weinshenker D. Functional consequences of locus coeruleus degeneration in Alzheimer's disease. *Curr Alzheimer Res.* 2008;5(3):342–345.
39. Zarow C, Lyness SA, Mortimer JA, Chui HC. Neuronal loss is greater in the locus coeruleus than nucleus basalis and substantia nigra in Alzheimer and Parkinson diseases. *Arch Neurol.* 2003;60(3):337–341.
40. Grudzien A, Shaw P, Weintraub S, Bigio E, Mash DC, Mesulam MM. Locus coeruleus neurofibrillary degeneration in aging, mild cognitive impairment and early Alzheimer's disease. *Neurobiol Aging.* 2007;28(3):327–335.
41. Braak H, Thal DR, Ghebremedhin E, Del TK. Stages of the pathologic process in Alzheimer disease: age categories from 1 to 100 years. *J Neuropathol Exp Neurol.* 2011;70(11):960–969.
42. Shin E, Rogers JT, Devoto P, Bjorklund A, Carta M. Noradrenaline neuron degeneration contributes to motor impairments and development of L-DOPA-induced dyskinesia in a rat model of Parkinson's disease. *Exp Neurol.* 2014;257:25–38.
43. Srinivasan J, Schmidt WJ. Behavioral and neurochemical effects of noradrenergic depletions with *N*-(2-chloroethyl)-*N*-ethyl-2-bromobenzylamine in 6-hydroxydopamine-induced rat model of Parkinson's disease. *Behav Brain Res.* 2004;151(1–2):191–199.
44. Ross SB. Long-term effects of *N*-2-chlorethyl-*N*-ethyl-2-bromobenzylamine hydrochloride on noradrenergic neurones in the rat brain and heart. *Br J Pharmacol.* 1976;58(4):521–527.
45. Ross SB, Stenfors C. DSP4, a selective neurotoxin for the locus coeruleus noradrenergic system. A review of its mode of action. *Neurotox Res.* 2015;27(1):15–30.
46. Fritschy JM, Grzanna R. Selective effects of DSP-4 on locus coeruleus axons: are there pharmacologically different types of noradrenergic axons in the central nervous system?. *Prog Brain Res.* 1991;88:257–268.
47. Heneka MT, Gavrilyuk V, Landreth GE, O'Banion MK, Weinberg G, Feinstein DL. Noradrenergic depletion increases inflammatory responses in brain: effects on IkappaB and HSP70 expression. *J Neurochem.* 2003;85(2):387–398.
48. Kalinin S, Polak PE, Madrigal JL, et al. Beta-amyloid-dependent expression of NOS2 in neurons: prevention by an alpha2-adrenergic antagonist. *Antioxid Redox Signal.* 2006; 8(5–6):873–883.
49. Heneka MT, Ramanathan M, Jacobs AH, et al. Locus ceruleus degeneration promotes Alzheimer pathogenesis in amyloid precursor protein 23 transgenic mice. *J Neurosci.* 2006;26(5):1343–1354.
50. Kalinin S, Gavrilyuk V, Polak PE, et al. Noradrenaline deficiency in brain increases beta-amyloid plaque burden in an animal model of Alzheimer's disease. *Neurobiol Aging.* 2007;28(8):1206–1214.
51. Pugh PL, Vidgeon-Hart MP, Ashmeade T, et al. Repeated administration of the noradrenergic neurotoxin *N*-(2-chloroethyl)-*N*-ethyl-2-bromobenzylamine (DSP-4) modulates neuroinflammation and amyloid plaque load in mice bearing amyloid precursor protein and presenilin-1 mutant transgenes. *J Neuroinflammation.* 2007;4:8.

52. Rey NL, Jardanhazi-Kurutz D, Terwel D, et al. Locus coeruleus degeneration exacerbates olfactory deficits in APP/PS1 transgenic mice. *Neurobiol Aging*. 2012;33(2):426e1–e11.
53. Jardanhazi-Kurutz D, Kummer MP, Terwel D, et al. Induced LC degeneration in APP/PS1 transgenic mice accelerates early cerebral amyloidosis and cognitive deficits. *Neurochem Int*. 2010;57(4):375–382.
54. Hurko O, Boudonck K, Gonzales C, et al. Ablation of the locus coeruleus increases oxidative stress in tg-2576 transgenic but not wild-type mice. *Int J Alzheimer's Dis*. 2010;2010:864625.
55. Rossner S, Sastre M, Bourne K, Lichtenthaler SF. Transcriptional and translational regulation of BACE1 expression—implications for Alzheimer's disease. *Prog Neurobiol*. 2006;79(2):95–111.
56. Hammerschmidt T, Kummer MP, Terwel D, et al. Selective loss of noradrenaline exacerbates early cognitive dysfunction and synaptic deficits in APP/PS1 mice. *Biol Psychiatry*. 2013;73(5):454–463.
57. Kummer MP, Hammerschmidt T, Martinez A, et al. Ear2 deletion causes early memory and learning deficits in APP/PS1 mice. *J Neurosci*. 2014;34(26):8845–8854.
58. Birks J, Flicker L. Selegiline for Alzheimer's disease. *Cochrane Database Syst Rev*. 2003;(1):Cd000442.
59. Mohs RC, Shiovitz TM, Tariot PN, Porsteinsson AP, Baker KD, Feldman PD. Atomoxetine augmentation of cholinesterase inhibitor therapy in patients with Alzheimer disease: 6-month, randomized, double-blind, placebo-controlled, parallel-trial study. *Am J Geriatr Psychiatry*. 2009;17(9):752–759.

Index

Note: Page numbers followed by "*f*" and "*t*" refer to figures and tables, respectively.

A

A61603, 84–86
Acetylation, 128
Acetylcholine, 222–223
Adenosine triphosphate (ATP) production, 177, 222–223, 226–227
Adrenergic activation
  characteristics of, Ca^{2+} signaling in astrocytes *in situ* and *in vivo*, 111–112
  indirect effects on coupling within the astrocyte network, 136–139
    calcium signaling, 136–137
    diffusion of cAMP, 137–138
    diffusion of metabolites, 138–139, 139*f*
  triggers phasic Ca^{2+} and tonic cAMP/PKA responses in, 108–110
Adrenergic modulation of cytosolic Ca^{2+} and cAMP excitability, 107–112, 136
Adrenergic receptors, 186*t*
  direct effects on gap junction, 130–136
Aerobic glycolysis, 5, 146–148
  adrenergic signaling and, 158–159
  in developing brain, 146–148
  lactate release *vs* lactate shuttling-oxidation, 148–149
  metabolic pathways involved in, 147*f*
  preferential upregulation of nonoxidative metabolism of glucose, 146–148
  rate of oxygen consumption (CMR_{O2}), 146–148, 147*f*
  selectivity of gap junctional trafficking of molecules involved in, 154
AG1478, 265
Age-related alterations in gliotransmission, 89–91
α_1-adrenergic receptors, 84–86, 85*f*, 186
  for glial modulation of excitatory and inhibitory synapses, 99
  mediated facilitation of LTP, 95, 97
  temporal differences in, 85*f*
α_{2A}-adrenergic receptor (α_{2A}-AR), 226
α_2-adrenergic receptors, 186
α-adrenergic signaling, 31–34
Alzheimer's disease (AD), 5–6, 65, 202–203, 243, 275, 304*f*
  accumulation of hyperphosphorylated tau, 277
  animal models of, 11
  clinical trials to modulate noradrenaline levels, 306
  locus coeruleus (LC) function in, 302
    damage in, 305–306
  NA deficiency and, 215
  neuroinflammation in, 302–303
  noradrenaline (NA) function in, 302
  pathological developments associated with, 11
  preclinical stages of, 10–11
  progression, 11
AMES test, 291
AMPA receptor-mediated signaling, 93
Amyloid beta, 276–277
Amyloid precursor protein (APP), 11–12
Amyotrophic lateral sclerosis (ALS), 275, 277
  neuronal degeneration in, 277
A6 noradrenergic cell group, 168
Antidopamine-beta-hydroxylase-saporin (anti-DBH-saporin), 202–203
Antigen-presenting cells (APCs), 14–15
Antisense oligonucleotides, 192

Apolipoprotein E (ApoE), 11
as a cholesterol-carrying protein, 12
mediated effects of, 11–12
polymorphic alleles of, 11
Aquaporin 4 (AQP4), 243–244
Astrocyte–neuron lactate (ANL) shuttle, 148–149
Astrocyte-neuron lactate shuttle (ANLS), 190–191
Astrocyte noradrenergic signaling system, 106–107
Astrocytes, 6–7, 30*f*, 46*f*, 65–66, 242–243, 259
acidic cargo in 3xTg-AD, 13, 14*f*
activation in neurodegenerative diseases, 276–277
in AD, 13–15
adrenergic activation triggers phasic Ca^{2+} and tonic cAMP/PKA responses in, 108–110
adrenergic excitability, effects of, 117*f*
morphologic plasticity, 115–120
prevention of CNS cellular edema, 119–120
adrenergic modulation of cytosolic Ca^{2+} and cAMP excitability in, 107–112
adrenergic stimulation during culturing of, 44–46
β-adrenergic activation and stellation of, 116–119
cAMP-mediated reduction in swelling of, 120
Ca^{2+} signaling in brain, 84–87
development of brain edema, 258–259
diffusion of glucose within, 9–10
dye transfer among, 152–154
dysregulation of astrocytic vesicle dynamics in neurodegeneration, 12–15
effects of NA on, 168–169
expression of connexin (Cx) family, 128
glycogen-derived cytosolic glucose in, 112–114
G-protein coupled receptors (GPCRs), 106
G_q protein-coupled neurotransmitter receptors, 84
hypertrophic, 10
inflammatory response of, 274–275
influence of noradrenaline on, 161–162
inositol uptake in, 32
interaction between α- and β-AR pathways, 110–111
involvement of PKC in Cx43 degradation in, 134–135
lactate dispersal among, 160–161
L-lactate release by, 169–172, 174*f*
concentrations for activation of HCA1, 176
modulation of NMDA receptor function, 176–177
production of pyruvate, 176–177
transfer to neurons, 177
lactate uptake and shuttling, 155–156, 156*f*
metabolic effects of α- and β-adrenergic stimulation of, 34–43
glucose metabolism, 34–38, 36*t*, 37*f*
glucose turnover, 38–43, 42*f*
glucose uptake, 34
morphology and distribution in neuropil, 244
Na,K-ATPase activity in, 261
neuron interactions during memory processes, 8
noradrenaline activation, 114*f*
induced astrocyte stellation, 116–118
intracellular cAMP and glucose concentration, 114*f*
role of morphological plasticity, 119
noradrenergic system and, 243–245
in situ effects, 245–247, 246*f*
in vitro effects, 245
P2X receptors activation by, 93
processes, 244–245
βAR-induced expansion of, 247–248
GFAP-negative portions of, 245
in neuropil, 244–245
reactive tissue, 115–116
release of gliotransmitters, 87–89
retractions or expansions of, 7–8
role in coupling local blood flow to neuronal activity, 72–73
sleep-related memory mechanisms, role in, 193–194
striatal, 131
structural plasticity of, 8
synaptic membranes and, 8
as targets in noradrenaline action on synaptic plasticity, 189
treatment with forskolin (FSK), 116–118
white matter networks and, 69*f*
Astrocyte-to-neuron glucose shuttling, 157
Astrocytic α-adrenergic receptor subtypes, signaling pathways for, 31–34

Astrocytic atrophy, 10
Astrocytic Ca^{2+} signaling, 84–87
Astrocytic expression of mRNA, 26–28
Astrocytic glycogenolysis, 38, 295
Astroglia, 6–10
role in brain signaling and metaplasticity, 83–84
Astroglial adrenergic signaling, 89–91
putative mechanisms underlying, 98*f*
on synaptic plasticity in the neocortex, 96*f*
Astroglial α_1-adrenoceptors, 91–98
Astroglial atrophy, 10–11
Astroglial β-ARs regulate cell morphology, 7
Astroglial Ca^{2+}-signaling (A–C), 92*f*
aging-related decline in, 97
Astroglial chemokines, 280–282
Astroglial hypertrophy, 10–11
Astroglial morphological plasticity, 5
Astrogliosis, 275–276
noradrenaline regulation of, 277–280
Astrogliosis, presymptomatic stage of neurodegeneration and, 10–12
Atomoxetine, 303–306
Autism, 65
Autoimmune diseases, 275
Autoimmune encephalomyelitis, 71–72
Axonal degeneration, 294–295
apoptosis of oligodendrocytes and, 295
Axon–glial–vascular network, 65–66

B

5-BDBD, 93–95
β_1-adrenergic receptor, 259–262
antagonist on brain edema during ischemia/reperfusion, 264–265, 265*t*
ERK phosphorylation, 260*f*
β_2-adrenergic receptors
astrocytic
axonal degeneration, role of, 294–295
consequence of deficiency, 295
downregulation of, 291–292
focal inflammatory lesions, 294
underlying mechanism in PML and multiple sclerosis lesions, 293–294, 293*f*
progressive multifocal leukoencephalopathy (PML), 292–293
β_2-adrenergic vagus nerve signaling, 159
β-adrenergic signaling pathways, 28–31, 30*f*
of astrocytic Na^+,K^+-ATPases, 44, 45*f*
during culturing of astrocytes, 44–46
temporal differences, 109*f*
β-AR/cAMP signaling pathway, 113
Betaxolol, 192, 258–259, 261–262
Brain activation
adrenergic signaling, 158–159
diffusion of lactate, 150
excitatory and inhibitory effects of lactate and influence on brain noradrenaline release, 159–161
glucose utilization (CMR_{glc}) during, 148–149
impact of lactate spreading and release on functional imaging of, 151–152
lactate fluxes during, 149–152
lactate release *vs* lactate shuttling-oxidation, 148–149
perivascular routes for metabolite discharge from, 158
roles for gap junctional communication during, 157*f*
trapping of labeled metabolites of glucose, 151
upregulation of glycolysis during, 149–150
Brain edema, 258
cell damage, 263
effect of β_1-adrenergic receptor antagonist on, 264–265
Brain extracellular space, 247–249
Brain noradrenaline release, excitatory and inhibitory effects of lactate and influence on, 159–161
Brain noradrenergic system
pathways and receptors, 184–187
Brain vascularization, 5–6
Brain-derived growth factor (BDNF), 295
8-Bromo-cAMP, 278–279

C

Cajal-Retzius cells, 2–5
Calcium communication, adrenergic signaling and, 136–137
cAMP/PKA responses in astrocytes, 110–111
Carbachol, 132–133
Carbamazepine, 32
Carbenoxolone, 138
Cardiomyocytes, 132
MAPKs/PKA activation, 135
Catechol-*O*-methyltransferase (COMT), 303–305

CCL2, 280–282
[^{14}C]DG method, 150–152, 154
diffusion of lactate from a microinfusion probe, 152
efflux pathways, analysis of, 158
labeling patterns in cerebral cortex, 151
Central nervous system (CNS), 2, 222
inflammatory reactions in, 290
Cerebral gymnastics, 7
[6-^{14}C]glucose, 150–151
Channelrhodopsin-2, 172–173
CHO-K1-based HCA1 expression system, 175–176
Cholesterol synthesis, 12
Clonidine, 32–34, 193, 258–259
Conditioned place preference (CPP) test, 192
Connexin (Cx) family, 128
Cx26, 138
Cx43
ablation in EGCs, 231–232
effect of β-AR stimulation via PKA activation on, 132
effects of α_1-AR stimulation on, 135
half-life of, 133–134
lipophilic domains, 133–134
modulated by PKs (PKC and PKA), 132
mRNA and protein levels, 135
permeable to cAMP, 138
phosphorylation, 131, 133–134
PKC activation accelerated degradation of, 133–135
PKC-dependent phosphorylation/dephosphorylation of, 135–136
transcription factors controlling, 135
four membrane-spanning domains, 128
functional, in astrocytes and oligodendrocytes, 130
hemichannels, 137–138
hexamer of, 128
modifiability of, 133–134
oligomerization of, 133–134
phosphorylation, 130–131
Cx43, 133–134
phosphorylation sites, 128–130
PKA activation on Cx43 gap junctions, 132
Constitutive (cNOS), 278
Convection-based signaling, 5
Corpus callosum, 64
CX3CL1, 281–282
Cyclic adenosine monophosphate (cAMP), 7, 242–243, 277–279, 294
ARs and, 186–187
delayed action, 187
diffusion of, 137–138
effects of NA on, 169
forskolin-stimulated, 175–176
generation by agents, 137
mediated signaling pathway in LC neurons, 173, 174f
permeability of, 137
phospholipase (PL) C activation and accumulation of, 131
receptor-mediated excitatory effects of lactate on, 159
sea urchin sperm (SpIH) as reporter of, 137
Cyclic guanosine monophosphate (cGMP), 132
Cytosolic excitability, 106
Cytotoxic edema, 258

D

Declarative (explicit) memory, 7
Desipramine, 303–305
Dexmedetomidine, 32–34, 258–259
Diacylglycerol (DAG), 31–32, 242–243
1,4-Dideoxy-1,4-imino-D-arabinitol (DAB), 113, 191–192
Diffusion-mediated signal propagation, 5
Distal astrocytic processes, 8, 244
Dominant-negative SNARE domain (dn-SNARE), 87
Dopamine beta-hydroxylase (DBH), 202–203, 305
Droxidopa, 305
DSP4 (N-(2-chloroethyl)-N-ethyl-2-bromobenzylamine), 305–306
Dysbiosis, 232

E

Endothelial (eNOS), 278
Endothelin-1 (ET-1), 132–133
Enriched environment (EE), 82
Enteric glia, 224–226
Ca^{2+} excitability, 227–228, 229f
innervated by sympathetic nervous system, 225f
physiological roles of, 232
role in GI disorders/diseases, 231–232
selected roles of sympathetic innervation and, 228–231

Enteric glial cells (EGCs), 224
response to direct sympathetic input, 226–227
sympathetic (co)transmitter receptors expressed on, 227*t*
Enteric nervous system (ENS), 222, 223*f*
extracellular regulation of nucleotide/ nucleoside levels, 227
Epidermal growth factor receptor (EGFR), 259
stimulation, 259
Extracellular K^+ concentration, during ischemia, 262–263
Extracellular space (ECS), 249–250
macroscopic parameters of, 248
quantitative analysis on parameters, 250*f*
simulations of diffusion, 251*f*
Extra-synaptic P2X receptors, 93

F

F-actin, 244
[^{18}F]FDG, 150
Fluorescence-activated cell sorting (FACS), 26
5-Fluoro-methoxyidazoxan, 303–305
Fluoxetine, 32
Focal inflammatory lesions, 294
Follicle-stimulating hormone (FSH), 137
Fornix, 64
Forskolin, 278–279
Free cytosolic calcium concentration, 33*f*

G

Gap junction
astrocytic lactate trafficking via, 152–157
dye coupling, 152–154, 153*f*
carboxyl-terminus of, 135–136
channels, 135–136
clathrin recruitment to, 135–136
direct effects of adrenergic receptors on, 130–136
formation and degradation, 133–136, 134*f*
intracellular acidification, 161
mediated diffusion, 138
selectivity of gap junctional trafficking of molecules involved in glycolysis, 154
subtypes in glia, 128–130
phosphorylation and dephosphorylation of Cx43, 128–130
Gastrointestinal (GI) motility, 222, 231
Gastrointestinal (GI) system, 222
enteric glia, 224–226
enteric glia and sympathetic innervation, role in disorders/diseases, 231–232
enteric nervous system (ENS), 222, 223*f*
innervation of gut wall, 222–223, 224*t*
nonsphincter regions of smooth muscles, 222–223
Genetic diseases, 275
GenWay Biotechnology, 292–293
GF109203X, 32–34
GLAST (glutamate aspartate transporter), 243–244
Glia-derived ATP, 83–84
Glial connexins, 128, 129*f*
Glial fibrillary acidic protein (GFAP), 10–11, 224, 243, 292–293
GLT-1 (glutamate transporter-1), 243–244
Glucose homeostasis, 188
Glucose-6-phosphate, 138, 149–150
Glucose shuttling, 157
Glutamate, 7–8
Glutamate-evoked stimulation of CMR_{glc}, 148–149
Glutamatergic vesicle motility, 13
Glutamine uptake and metabolism, 161–162
Glyceraldehyde-3-P, 154
Glycogen
astrocytic L-lactate derived from, 191–192
central role of, 191–192
Glycogenolysis, 9–10, 44, 113, 161–162, 171, 194
α_2-AR-induced increase in, 113
induced by β-AR, 191
Glycogen synthase (GlyS), 187, 191
Glycolysis, 161–162, 190–191
Glycosylation, 128
α-glycyrrhetinic acid, 152
GM6001, 265
GPR81, 174–175
GPR109a, 174–175
GPR109b, 174–175
G-protein coupled receptors (GPCRs), 29, 174
astrocytes, 106
Gray matter (GM), 64–65

H

H-89, 265, 278–279
[^{3}H]dihydroalprenolol, 291–292
Hebbian memory formation, 8
HeLa cells, 137
Heparin-binding epidermal growth factor (HB-EGF), 258–259

Heterosynaptic metaplasticity, 83
Hippocampal CA1 pyramidal neurons, 193
Hippocampal noradrenergic depletions, study of
behavioral tests, 204–205
lesion and transplantation surgery, 204
microscopic analysis and quantitative evaluation, 207–208
morphological analyses
anti-DBH-saporin lesion, effects of, 214–215
estimates of dopamine-ß-hydroxylase-immunoreactive neurons and fiber density, 213*t*
lesion and of transplants, effects of, 212–216
neuronal loss, 212–214
Morris water maze (MWM) test, 205–206, 209–211, 209*f*, 210*f*
postmortem analyses, 206–207
results
behavioral analyses, 209–212
general observation, 208
motor performance, 208*t*
radial arm water maze test, 211–212, 211*f*
reference memory performance, 210*f*
working memory performance, 211*f*
subjects and experimental design, 203–204
Hippocampal noradrenergic neurotransmission, 203
H8 (PKA inhibitor), 135
6-Hydroxydopamine, 226
Hypertonicity, 265
Hypoglycemia, 113–114

I

ICI 118,551, 192
Idiopathic PD, 6
IL-1 receptor (IL-1R), 231
Inducible (iNOS), 278, 294–295, 303
Inflammatory bowel disease (IBD), 231–232
Infused inferior colliculus, 158
Inositol 1,4,5-trisphosphate (IP_3), 31–32, 242–243
Insulin degrading enzyme (IDE), 13–14
Interleukin (IL)-1β, 279–280, 282
Internal capsule, 64
Intestinal astrocytes, 224
Ischemia, 275
effect of β_1-adrenergic receptor antagonist on, 264–265
extracellular K^+ concentration during, 262–263
MAPK/$ERK_{1/2}$ signaling pathway during, 263–264, 264*f*
Isofagomine, 191–192
Isoproterenol (ISO), 28–29, 44, 108–109, 131, 136–137, 243–244, 259, 261–262
stimulated glycogenolysis, 44

J

John Cunningham (JC) virus, 292

K

K_{ATP} channels, 177

L

L-Lactate, 190–191
as a gliotransmitter feed forward signal, 172–174
potential signaling roles of, 174–177
release by astrocytes, 169–172, 174*f*
Lactate fluxes during brain activation, 149–152
influence of noradrenaline on, 158–162
L-Lactate production, 5
Lactate release *vs* lactate shuttling-oxidation, 148–149
excitatory and inhibitory effects and influence on brain noradrenaline release, 159–161
suppressive effects on neuronal signaling, 160–161
Lactate trafficking within brain, 151
Leukodystrophies, 65
Levodopa, 303–305
Lipopolysaccharide (LPS), 278–279
Locus coeruleus (LC), 106–107, 168, 242
in Alzheimer's disease (AD), 302, 305–306
anatomy and pathophysiology, 2–6
efferents, 2–5
fundamental LC-mediated functions, 2
impairments, 12
microinjection of L-lactate into, 173
neurons of, 2
noradrenergic damage and loss, 306
nucleus, 2–5, 4*f*, 202–203
deficit in, 6

vascularization of, 5–6
in Parkinson's disease, 275
synchronous activation of LC projections, effect of, 2
Lucifer yellow labeling, 153f, 131, 152–154
Lund concept, 258–259

M

Major histocompatibility complex class II (MHCII), 277–278
MAPK/ERK$_{1/2}$ signaling pathway, 263–264, 264f
MAPK signal transduction pathway, 259
MCT1 or MCT2 L-lactate transporters, 192
Medial cerebral artery occlusion (MCAO), 258–259, 262, 265t, 266t
EGFR phosphorylation, 263
MEK1 inhibitor, 135
Memory
additional source of energy by neurons (ANLS) during, 194
brain energy metabolism and, 190–191
noradrenaline (NA) and, 188–190
sleep and, 193
Metabolite trafficking, 161
Metaplasticity, 82–83
Methoxamine, 132–133
2-Methythiol-ATP, 136
Metoprolol, 158–159
MHC-II molecules, 14–15
Microglia, 65–66, 274, 278
anti-inflammatory effects on, 71–72
Miniature excitatory and inhibitory synaptic currents (mEPSCs and mIPSCs), 93–95
MK801, 176–177
Monoamine oxidases (MAOs), 303–305
Monocarboxylase transporters (MCTs), 170
MCT1–MCT4 transport L-lactate, 170–171
Monocyte chemoattractant protein-1 (MCP1), 303
mRNA expression, 26–28, 27t
Multiple sclerosis (MS), 65, 119, 290, 303–305
β_2-adrenergic receptors in
downregulation of astrocytic, 291–292
progressive multifocal leukoencephalopathy (PML), 292–293
progressive, 290–291
reactive astrocytes in, 295

N

Na,K-ATPase, $Na^+/K^+/Cl^-$cotransporter (NKCC1), 258–259, 261–262, 265–267
activity in astrocytes, 261–262
effects of β_1- and β_2-ARs on, 262
phosphoryation, 261–262
transmembrane domains, 261–262
Na^+,K^+-ATPases stimulation, 44, 45f
NA reuptake inhibitors (NARIs), 303–305
Natalizumab, 292
Ndufs4s gene, 11–12
Neurodegeneration, 6–10
LC-dependent deficit, 7
Neurodegenerative diseases, 275
Neuroinflammation, 106–107, 274–275
Neuromelanin, 2
Neuronal mRNA expression, 26–28
Neuronal synaptic signaling, 65
Neuropeptide Y (NPY), 226
Neuropsychiatric diseases, 65
Neurotrauma, 275
Nitric oxide (NO), 278
2-[*N*-(7-Nitrobenz-2-oxa-1,3-diazol-4-yl) amino]-2-deoxyglucose (2-NBDG), 138
Nitrosylation, 128
Nodal synapse, 66–68
Nodes of Ranvier, 66–68, 67f
Nondeclarative (implicit) memory, 7
Nonneuronal cells outnumber neurons, 106
Noradrenaline-evoked purinergic modulation, 94f
of synaptic currents, 93–95, 94f
Noradrenaline (NA), 2, 184, 226, 242–243, 275
action on glycogen metabolism, 187–188
action on synaptic plasticity
astrocytes as targets, 189
neurons as targets, 188–189
in Alzheimer's disease (AD), 302
antiinflammatory actions of, 277–278, 303–305
brain energy metabolism and memory, 190–191
in the context of astrocyte-neuron interactions, 188f
depletion in neurodegenerative diseases, 275–276
effects of, 7
excitation-energy coupling, role of, 9–10
mediated metabolism, 10
memory and, 188–190
regulation of astroglial chemokines, 280–282, 281f

Noradrenaline (NA) (*Continued*)
release, 83
Noradrenaline-to-astrocyte signaling axis, 168–169
Noradrenergic receptors, 242–243
Noradrenergic-rich LC transplants, 212–216
Noradrenergic signaling, 242
Noradrenergic stimulation of astrocytes, 161–162
Noradrenergic system, effects of
astrocytes, 243–245
in situ effects, 245–247, 246*f*
in vitro effects, 245
on extracellular space structure, 249–250
quantitative analysis on parameters, 250*f*
simulations of diffusion, 251*f*
Nuclear factor kappa B, 279
Nucleoside triphosphate diphosphohydrolase (NTPDase), 226–227

O

Oleamide, 152
Oligodendrocytes, 66–72, 224
Cx expression, 128
Oligodendrogliopathy, 295
Optic tracts, 64
OR51E2 orthologs, 176
Oxygen-glucose index (OGI), 150

P

P38 inhibitor, 135
Palmitoylation, 128
Panx1, 138
Parkinson's disease (PD), 5–6, 277
locus coeruleus (LC) in, 275
Pavlovian threat, 8
Pentose-phosphate shunt pathway (PPP) fluxes, 146–148
Peptidergic vesicle trafficking, 13
Perisynaptic astrocytic processes, 115
Perivascular-lymphatic drainage system, 158
Phenylephrine, 243–244
Phosphatase, and tensin homolog (PTEN), 263–264
Phosphatase-1 glycogen, 187
Phosphate-buffered saline (PBS), 204
6-Phosphofructo-2-kinase/fructose-2,6-biphosphatase 3, 171–172
Phosphorylation, 128
PI3K/AKT pathway, 263
Pigmented neurons, 2
Posterior funiculus (dorsal columns) of the spinal cord, 64
Postsynaptic P2X receptors, 93
Posttraumatic stress disorder (PTSD), 276
PPADS, 93–95
Prefrontal cortex (PFC), 189–190
Presynaptic sympathetic neurons, 222–223
Progressive multifocal leukoencephalopathy (PML), 292–293
hypertrophic astrocytes, 293
Propranolol, 135, 161–162, 192
Protein kinase A signaling, 159

R

Ramon-y-Cajal, Santiago, 7
Rapid eye movement (REM), 193
Ras/Raf/mitogen-activated protein kinase (MAPK)/$ERK_{1/2}$, 259
Reactive oxygen species (ROS), 259
Real-time iontophoretic (RTI) method, 248
Reboxetine, 303–305
Reil, Johann Christian, 2
Relapsing–remitting disease, 290–291
Reperfusion
effect of β_1-adrenergic receptor antagonist on, 264–265
extracellular K^+ concentration during, 262–263
MAPK/$ERK_{1/2}$ signaling pathway during, 263–264, 264*f*
Riboxetine, 193

S

S100β, 224
Saponaria officinalis, 202–203
Shuttling of lactate, 148–149, 155–156, 156*f*
Sleep-related memory mechanisms, 193
noradrenaline and, 193
participation of astrocytes, 193–194
Sleep–wake cycle, 193–194, 249
Slow wave sleep (SWS), 193
SOD1 gene, 277
Somatostatin (SST), 226
Spatial buffering, 243–244
Stellation of cultured astrocytes, 116–119, 247
NA-induced, 119
Store-operated channels (SOCEs), 32
Sudan black B, 292–293
SUMOylation, 128

Sympathetic ganglionic neurons, 222–223
Sympathetic innervation of gut wall, 222–223, 224*t*

T

Tetradecanoylphorbol acetate (TPA), 132
Tetrodotoxin (TTX), 93–95
Toll-like receptors (TLRs), 230
Tricarboxylic acid (TCA) cycle-derived amino acids, 149–150
Tripartite synapse, 8, 9*f*, 66–68, 115
Tumor necrosis factor alpha (TNFα), 230–231, 279–280
Tyrosine hydroxylase (TH), 305

U

U0126, 265
U-73122, 136–137
Ubiquination, 128
UK5099, 176–177
Uridine 5′ triphosphate (UTP), 227

V

Valproic acid, 32
Vasopressin, 259
Vicq-d'Azyr, Félix, 2
Vimentin, 243

W

Warburg effect, 5
Wenzel brothers, 2
White matter (WM), 64–65, 64*f*
  adrenergic mechanisms in, 68–70
  adrenergic signaling in, 70–71
    in pathology, 71–72
    physiological functions of, 70–71
    regulation of blood flow, 72–73
  astroglial expression of α_1-AR in, 68–70
  cerebellar and brainstem, 64
  glia and neuronal signaling, 65–66
  hypoperfusion, 72–73
  myelinated axons and glia, 65*f*
  neuroglial communication in, 66–68
  neurotransmitter signaling in, 65
  NG2-glia, 66–68
  oligodendrocytes, 65–66
  rapid signal transmission in, 65

Z

Zinterol, 191

Printed in the United States
By Bookmasters